Development Beyond Neoliberalism?

Development's current focus – poverty reduction and good governance – signals a turn away from the older neoliberal preoccupation with structural adjustment, privatization, and downsizing the state. For some, the new emphases on empowering and securing the poor through basic service delivery, local partnership, decentralization and institution building constitute a decisive break with the past, and a whole set of new Development possibilities beyond neoliberalism.

Taking a wider historical perspective, this book charts the emergence of poverty reduction and governance at the centre of Development. It shows that the Poverty Reduction paradigm does indeed mark a shift in the wider liberal project that has underpinned Development: precisely what is new and what this means for the ways the poorer parts of the planet are governed are here described in detail.

This book provides a compelling history of Development doctrine and practice, and in particular offers the first comprehensive account of the last 20 years, and Development's shift towards institution building, decentralized governance and local partnerships. The story is illustrated with extensive case studies from first-hand experience in Vietnam, Uganda, Pakistan and New Zealand.

David Craig teaches Sociology at the University of Auckland. **Doug Porter** is with the Asian Development Bank.

Development Beyond Neoliberalism?

Governance, poverty reduction and political economy

David Craig and Doug Porter

Routledge
Taylor & Francis Group

LONDON AND NEW YORK

First published 2006
by Routledge
2 Park Square, Milton Park, Abingdon, Oxon OX14 4RN

Simultaneously published in the USA and Canada
by Routledge
711 Third Avenue, New York, NY 10017

Routledge is an imprint of the Taylor & Francis Group

Typeset in Times by
Florence Production Ltd, Stoodleigh, Devon

British Library Cataloguing in Publication Data
A catalogue record for this book is available from the British Library

Library of Congress Cataloging in Publication Data
Craig, David, 1961–
    Development beyond neoliberalism?: governance, poverty
    reduction, and political economy/David Craig and Doug Porter.
        p. cm.
    Includes bibliographical references and index.
    1. Economic development – International cooperation.
    2. Economic assistance – Developing countries.   3. Poverty –
    Government policy.   4. Neoliberalism.   I. Porter, Doug.   II. Title.
    HD82.C65 2006
    338.9′1′091724 – dc22   2005016680

ISBN10: 0–415–31959–5 (hbk)
ISBN10: 0–415–31960–9 (pbk)

ISBN13: 978–0–415–31959–1 (hbk)
ISBN13: 978–0–415–31960–7 (pbk)

Contents

Illustrations

Figures

Tables

Acknowledgements

Existence is never a truly linear progression: all overlaps, merges, partly conceals, as the scales of a serpent. So every moment is at once a beginning and an end, as well as continuity, an assertion and a giving up, a revelation and disguise, a knapsack of the past trekked into new country.

Owen Marshall, *A Many Coated Man,* 1995

For us, travelling and Development have always gone together. They produce that press of half-formed ideas, questions and answers that pop at the corner of your mind as you whiz from one situation to another. It's a formidable way to accumulate unprocessed experiences, stories, cullings from project documents, papers, ideas for a book, a seminar, an exposé. Mentally pushing your case shut always leaves your thumbs white and bloodless, while the taxi presses the traffic on the way to the airport.

This book, like Development, is a tightly packed suitcase of ideas that we wrote over eight years in more than a dozen places. We worked separately and only for short periods together, in touch by email daily and more, and meeting every few months when, over several days, the whole book would go up on a white board or sheets of butcher paper or café napkins and notebooks, and all the arguments and cases would be reworked, more often than not from first principles. We wrote the book around these meetings: in Manila, Brisbane, Cairns, Thursday Island, Islamabad and Peshawar, Kampala, Singapore and Hong Kong, Auckland, Mullumbimby, Napier, Raglan, Phnom Penh and numerous points between. However others judge the result, it was for both of us a great and polychrome intellectual adventure, as well as a more personal attempt to make critical sense of our different working experiences.

Ideas, stories, cases, all piled up and sedimented over time, often they fell apart, and finally some of them coalesced. Much was simply sacrificed, in the end, to the inevitable constraints of space, and to efforts to muster as much clarity as we could. The original idea for the book included a chapter set in remote Aboriginal Australia. Another chapter could have included our experiences in Cambodia: instead, there'll be a book. The challenge was to weave swathes of ethnographic accounts together with historical narrative and at the same time to relate some fairly 'high theory' we found useful to convey the whole story as a piece rather than a series of 'case examples'.

The result is not a simple book. Obviously we think the issues are both enormously important and complex. To every complex question, someone wise said, there's a simple answer. And, they and we agree, it's wrong. Complex answers, of course, can be wrong too; especially since so often they are contingent on particular situations, times and places. For this reason, our aim has been to 'show' as much as 'tell' the argument.

It should be clear, given the long labour, that we've ramped up some huge debts to the people we've worked with and, not least, to those who have struggled to guide us through many earlier drafts. Our first obligations are to our partners, to Rae and Liz who did far more than simply put up with it all, the travel, the late nights, the frustrations and intellectual luxury of it all. Without their love and support the book would have remained just a collection of ideas scrapped about in emails and field notes. Rae buoyed the process all the way through, with love, life and common passion for Development's puzzles, joys and anger in Vietnam, various places in Africa, Cambodia, the Philippines and Pakistan. David is indebted to Liz, Nick and Jack for suffering and supporting through the serial distraction, travel, fatigue, overdoing it, all while themselves turning into, as Bill Murray had it, the most interesting people he's ever known.

In Vietnam, we learnt much from the smart and keen researchers at the former Community Health Research Unit at the Hanoi Medical School, the non-government organization (NGO) support training programme, the Danish Red Cross, from the Centre for Rural Planning and Development (CERPAD), and from all the NGO participants in Participatory Rapid Appraisal (PRA) and qualitative research methods training events, whence came some of the ethnographic material from Chapter 5. Phung Ha, Nguyen Bao, Ba Ho, and Nguyen Khoi deserve special thanks from Doug, for the struggles to find common ground in the late 1980s and early 1990s. Our debt to friends and workmates in Uganda will be obvious: Martin Onyach-Olaa for his skill and outrageously good humour, Jimmy Lwamafa for the start of it all, Francis Lubanga for pushing hard to test it all out. Workmates Per Tidemand, Laura Kullenberg, Roger Shotton, Paul Grosen, Leonardo Romeo struggle constantly with the kinds of issues this book tries to tame. As can be seen from the Pakistan chapter, so much of this book's argument found its most engaging overreach in Pakistan, and not surprisingly the list of people who became deeply involved in its story is long. Thanks to Hamid Sharif, Naved and Naseem Hamid who introduced Doug to the joys of the country, to Musharraf Cyan and Raza Ahmad, Javed Sadik Malik, Justice (rtd) Shafi ur Rehman and Shoaib Suddle who guided, cajoled and patiently explained the chicaneries of Pakistan's 150 years of governance accommodations. Nick Manning, Jackie Charlton, Zahid Hasnain, workmates in Asian Development Bank's (ADB) governance unit,

Iffat Idris, Shaista Hussain, Peter Robertson and special thanks to Ayesha Salman for editorial help. To the many colleagues in ADB, the World Bank, Department for International Development (DFID) and a slew of NGOs and local governments, the programmes supporting devolution and access to justice, a big thanks.

Some of our debts are to the academic institutions we've worked with, including colleagues and students at the Indigenous Health Programme at the University of Queensland (and especially its one-time Thursday Island Campus, in the Torres Strait), and Auckland University's Department of Sociology and the Public Policy Discussion Group (special thanks to the university's generous Staff Research Fund). And thanks are also due to Megan Courtney, Mark Allen, Ann Magee and Alan Rodgers-Smith at the Waitakere City Council, the Seila Task Force Secretariat (especially Scott Leiper) and the Cambodian Development Resource Institute for their support to write, workshop, debate, engage . . .

Germinating ideas and material were presented in conferences and seminars in Hong Kong, Paris, Oviedo, Brisbane, Lahore, Manila, Auckland, Wellington, Phnom Penh and Bath. Encouragement and/or engagement and debate about the ideas came from Ian Carter, Stuart Corbridge, Gerry Cotterell, Tobias Haque, Wendy Larner, Rianne Mahon, Gerda Roelvink, Mohammad Shahabuddin, several anonymous journal reviewers, and several years of social theory students at Auckland University.

The really difficult task, given our penchant for convoluted argument, was reading and directing our efforts and for this we are indebted to Musharraf Cyan, Robin Biddulph and Clay Wescott. But especially Ole Therkildsen provided wonderfully insightful comments at a crucial stage in the book, and two anonymous reviewers read and commented wisely and attentively on the manuscript. Zoe Kruze of Routledge worked hard to reach the deadlines and maintain production quality. For the final readings, when deadlines seemed impossibly pressing, we are hugely indebted to Peter and Rae for keeping up the pressure, good humour and coffee, and for reminding us that even this effort was not good enough.

Phnom Penh and Islamabad
May 2005

Disclaimers

1 Governing poverty
Development beyond neoliberalism?

> People rise from poverty when countries act on two pillars of development: building a good investment climate in which private entrepreneurs will invest, generate jobs, and produce efficiently, and empowering poor people and investing in them so that they can participate in economic growth. What's a good investment climate? Start with sound macroeconomic management and trade and investment policies that promote openness and raise productivity and growth. Add the elements of good governance, such as regulation of industry, promotion of competition, and prevention of corruption. Then set all that on a foundation of basic infrastructure and effective basic services, such as health and education.
>
> World Bank, *States and Markets*, 2002[1]

DEVELOPMENT REINVENTS ITSELF

'The Bank', the saying goes, 'always knows best'. But, says the rejoinder, 'what the Bank knows, changes'. Certainly the agendas set by the world's leading Development institutions, the World Bank included, have varied markedly over time. In this they have both led and, more often, reacted to wider political and economic changes, and especially to crises. Yet the late 1990s shift from frank neoliberal Structural Adjustment to a softer, more 'inclusive' Poverty Reduction and Good Governance agenda seemed particularly dramatic and rapid. And successful, at least in repositioning Development's lead institutions, and in creating a wider consensus around market-led growth and poor countries' integration into global capitalism. Even more prodigious, however, was Development's elaboration of a set of potentially harmonized, shared technical instruments with which to pursue Poverty Reduction and institution building goals.

By the early twenty-first century, the breadth of both the consensus and its technical elaboration were unprecedented. Within Poverty Reduction Strategy's (PRS) multifaceted frames, the free global flows of capital, poor people's participation, and competitively provided health and other services can be combined to promise opportunity, security and empowerment to the most peripheral places. Here, everyone can, indeed must be included: Washington's financial institutions can partner with sub-Saharan NGOs, with global accounting and audit franchises, Pakistani provincial governments, and Vietnamese commune authorities. Development's fiscal and security-oriented conservatives, market neoliberals, communitarian social developers, and governance technocrats all have roles in deepening this consensus and rolling out its practice across global and local institutions. More importantly, now, we are told, Development can finally work. A fresh commitment to harmonize around this consensus and apply its governing techniques, and a much greater investment of funds, will yield a substantial dividend for the poor.[2]

This is a notable turnaround: through much of the 1990s, Development was on the ropes. Post-developmentalists had dismissed Development as a pernicious discourse, a grand modernizing and colonial narrative reflecting and serving Eurocentric interests.[3] Gung-ho globalizers declared that the nation state was no longer a significant entity for governing territories, and that radical market openness and integration was the only plausible development strategy. When, mid-decade, private investment transfers began to eclipse official aid, liberals and conservatives alike saw (and applauded) international private capital taking Development's place.[4] Yet by decade's end, the triumphant Millenarian vision of global market integration was under siege. In some of Development's most conspicuous success stories, the 1997–1998 Asian crisis showed how a flighty drove of fast moving capital could undo the progress of decades in a matter of weeks. And Development's new atlas was showing the vastly uneven results produced by the first round of neoliberal reform – the 'balance the books, open up the economy, and hope' approach of 1980s Structural Adjustment. Two decades of development failure and zero net growth on whole continents had produced alarming peripheries of insecurity, disaffection and risk. Neoliberalism's trust in free markets and self-regulation were brought back into critical review. As the millennium loomed, critics of the Washington Consensus were raging in the streets of Seattle and Genoa to press home questions about the basic legitimacy of International Financial Institutions (IFI) – the World Bank, the regional development banks and the International Monetary Fund (IMF). Prominent figures within IFIs were publicly conceding that the structural adjustments of radical neoliberal reform had often delivered more shock than therapy: it was all too narrow, too IFI-led, too banker driven.[5] Once considered comprehensive, the neoliberal reforms driven home through Structural Adjustment conditionalities now seemed limited in scope and imagination. Too often, loan financed privatizations had perverse outcomes; too often, radical downsizing of the state had failed to produce a more efficient or effective set of institutions that could support market growth.

Poverty reduction: a shift beyond neoliberalism?

So Development's most recent change, like others before it, was partly impelled by crisis, and reactions to both insecurity and to IFI's apparent inability to contain it. In some political economy circles, this reaction to free marketism and its IFI avant-garde was plausibly described in terms of Karl Polanyi's 1944 thesis that a self-regulating market 'could not exist for any length of time without annihilating the human and natural substance of [its] society'.[6] Through the 1990s, and via failed programmes of Structural Adjustment, it was arguable that unregulated global markets were causing enormous disruption to rich and poor societies alike. The High Street protests, public debunking of narrow neoliberal orthodoxies and the calls for re-regulation, strengthening governance and social protection might be examples of what Polanyi termed an 'enlightened' reaction, a plural, broad based social reaction[7] against the insecurities unfettered markets routinely generated. For Polanyi, this reaction was a part of a much bigger, more powerful pattern he called the 'double movement'. In his conception, markets and societies always existed locked in a lurching relationship and struggle, which progressed unevenly (or could even be destroyed) as, in a two stage or 'double' movement, markets disembedded themselves from social constraint, and were then re-embedded and thereby secured and sustained, as well as constrained in the reactionary phase. In Polanyi's historical account of events from the industrial revolution to the mid-twentieth century, the first part of the 'double movement' had seen markets breaking out from – and enormously disrupting – social and territorial constraints of a pre-industrial age: the somewhat more bounded territories of rural, local, national production and consumption. In the second phase of the double movement, multiple 'enlightened reactionaries' within societies had acted to re-embed markets within social, governmental, security and regulatory contexts, forming unions, supporting labour laws and early sanitation and town planning, extending the franchise, buying British (or German or French) and pursuing rival empires with privileged internal trade. Ultimately, this second phase proved perversely potent: the alignment of reactionary, nationalist, socialist, local and imperialist movements seeking to co-opt the market had led to the collapse of liberalism in the Great Depression and the Second World War. By then, it seemed to Polanyi as to Keynes that the unregulated market might safely be consigned to history.

As the 1980s and 1990s had demonstrated, this was simply not the case: here again, another double movement round of liberalization saw markets tearing away at social, regulatory and territorial constraints, disembedding themselves in revolutionary projects recognized as neoliberalism and globalization. They also restructured and disaggregated the state through privatization and new contractual and managerialist arrangements in ways that would mimic markets. Through the 1990s, markets seemed to be triumphant:

but following the Asian/IFI crisis, and especially after 11 September 2001, it seemed that the security-reaction ambit might again return, and perhaps Polanyi would get the last word. But even if the reaction were truly Polanyian, how profound and powerful was it, and how might market forces (and especially market promoting international financial agencies) in turn recover? This time around, the enlightened reactionaries might succeed not in rolling back resurgent global liberalism, but in the equally Polanyian end of enabling its socially embedded rollout.[8]

Or just Development's new clothes?

In fact, as this book will show, before the 1997–1998 Asian crisis Development was already well on the way to rebuilding consensus. And, through the 1990s, it had been assembling an astonishing array of instruments to put it into practice. The 1997 World Bank Development Report (WDR), *The State in a Changing World,* was a thoroughgoing treatise on the ways in which a 'capable state' was needed to support markets. But more on this later. More evident at the time was that crisis and violent protest needed immediate response. Development's lead institutions had to be recast as more 'inclusive', more responsive and 'participatory', and thereby, somehow, more legitimate. Hence IFIs' quickly rebranded their products, for example, in 1999, relabelling overnight the Enhanced Structural Adjustment Facility (ESAF) with the nomenclature of Poverty Reduction and Growth Facility[9] (PRGF). And they redoubled efforts to show how recipient countries, 'development partners' and even the poor themselves endorsed what the IFIs were doing. Now, collaboration around IFI-led Development would be pinned to national Poverty Reduction Strategy Papers (PRSP) 'owned by' recipient governments themselves.

Poverty Reduction Development would from 2000 roll out on a broad, three-legged agenda of promoting economic *opportunity* through global market integration, and enhanced social and economic *security* and *empowerment* through innovative governance arrangements for local delivery of health, education and other poverty-reducing services. None of these alone would reduce poverty: but together, the consensus concurred, they should. Indeed, progress would be monitored against a range of poverty related targets and goals. Poverty Reduction's Millennium Development Goals (MDGs) were birthed within the United Nations Development Programme (UNDP), adopted by 189 nations in the Millennium Declaration in September 2000, and then reaffirmed by all United Nations (UN) members in the Monterrey Consensus and in the Johannesburg Plan of Implementation in 2002. With these targets came technical and institutional alignments, generating a breath-taking round of High Level 'harmonization' Forums, largely sponsored by the 'like minded' group of bilateral donors, the British, the Scandinavian countries, in Rome (February 2003), Marrakech (February 2004) and Paris (March 2005). The 2004 WDR, *Making*

Services Work for Poor People, made the urgent need for a new consensus even more explicit: neoliberal, market integration would not by itself produce the kinds of economic growth needed to lift the world's poor out of poverty. MDG targets would require focused moral commitment by rich and poor alike. Where poor countries and their citizens were able to demonstrate this, social services could be delivered in ways that supported Poverty Reduction's 'empowerment' and 'security' at the same time.

The strength of the new consensus was evident in the fact that these high level forums barely discussed the efficacy of its underpinning neoliberal market orthodoxy. Rather, the focus was from 2000 largely on its organizing rubrics and technical means. By 2000, for over 80 of the world's poorest countries, PRSPs would be presented as the primary strategic and implementation vehicle to reach the MDGs. They would allow global commitments to be communicated through national and local level arrangements that tied all Poverty Reduction spending to highly visible technical instruments that controlled budget-making and the transfer of fiscal resources, and could be used to deliver sanctions where necessary. In this way, the global, national and local levels could be 'joined-up' and clearly accountable to the poor. By this time, from Uganda to Uzbekistan, NGOs across the world had contributed to a variety of techniques that delivered 'Voices of the Poor' to policy-makers, and made it possible for the whole enterprise to appear to be legitimated by the poor themselves.

Not everyone was convinced. For all the consensus and partnership, PRSP's doubters have found numerous points of issue.[10] What is this, they ask, beyond structural adjustment in pro-poor, 'inclusive' neoliberal drag, or a mere technical elaboration of the notorious neoliberal formulism of the Washington Consensus?[11] Certainly, despite endless reference to local ownership and professed rejection of 'one size fits all', the three-legged PRSP formula of 'Opportunity, Empowerment and Security' varied little across countries.[12] Indeed, because they were so slippery they could easily be adopted by the World Bank's landmark 2000 WDR, *Attacking Poverty*; Cambodia's PRSP ('promoting opportunities, creating security, strengthening capabilities and generating empowerment . . .'); New Zealand's Third Way, social democratic Prime Minister Helen Clark, ('fairness, opportunity, security'); and, at least on 8 September 2001, even in George Bush Jnr's 'Priorities for Fall: Education, economy, opportunity, security'.[13]

Beyond rhetoric? The rise of governance

As this book will make clear, this was certainly more Liberalism, but it was also more than a new rhetorical garb for neoliberal Development. How much more, this book will debate: but in general, the major substance came under the apparently benign label of 'good governance'. At times, good governance seemed to be a fourth leg in the Poverty Reduction rubric, at other times, merely an elaboration of the 'empowerment' dimension. This is evident in Pakistan's 2003 PRSP. It faithfully renders the Opportunity, Empowerment and Security troika, but in four goals, weaving the three legs through with the various inflections of 'good governance'; stability, service delivery efficiency, devolved power, justice, and reducing regional vulnerabilities and inequalities:

- Achieving high and broad-based economic growth focusing particularly on the rural economy, while maintaining macroeconomic stability.
- Improving governance and consolidating devolution, both as a means of delivering better development results and ensuring social and economic justice.
- Investing in human capital with a renewed emphasis on effective delivery of basic social services.
- Bringing the poor and vulnerable and backward regions into the mainstream of development, and to make marked progress in reducing existing inequalities.[14]

In fact, the appearance and disappearance of governance as a fourth leg of Poverty Reduction is a little misleading. Rather, the most important innovations of the Poverty Reduction paradigm were eventually put together under the rubric of governance, and wider conceptions of 'institution building'. In the early to mid-1990s, governance reform tended to be restricted to protecting and building confidence around market and capital security: privatization together with anti-corruption measures would remove obstacles in the way of market forces, and promote a secure, rule of law environment for investments. But capable, corruption free governance, it soon became apparent, also held keys to putting a 'human face' on macroeconomic structural adjustment, enhancing investments in human capital, health and education, working with civil society for better delivery of social services. This promised healthier, more educated and engaging citizens able to participate in new market Opportunities. But its more expansive claim was that good governance would also create Security and Empowerment via a new, citizen responsive, capable state.

Crucially for this book's story, it was *decentralized* aspects of governance that offered some of the most enticing promises. The community, the locality, the territories of sub-national authorities became the domains where good governance seemed to offer most. Here, good governance rubrics promised a less corrupt institutional environment for local business, and better access to decentralized service delivery, by responsive agencies held accountable by informed clients. In slightly more technical terms, explained later, new fiscal arrangements for decentralized governance, acting together with multiple market actors and revitalized local juridical (or law based) mechanisms, could lower transaction costs and raise allocative efficiencies for businesses and services alike, creating multidimensional accountabilities for service providers, and better outcomes for poor people.

Thus, by the early 2000s, good governance had emerged from the brown manila folder of 'public administration and anti-corruption' and achieved a spectacular primacy in Development's headline strategies, including 2004 WDR, *Making Services Work for Poor People*, the UN's 2005 Millennium Project Report[15] and the Blair Commission's 2005, *Our Common Interest*, which framed entrenched poverty in Africa ('the greatest tragedy of our time') for substantive G8 attention. In this, we will argue, it aligned itself a little more with a somewhat Polanyian shift in international political economy. And at the same time, recast itself once again in the image of a much more durable political and economic project: a project with a history of promising relief from poverty to those who respected, above all, the rule of law, and the property rights of the powerful. The project, that is, of wider historical Liberalism.

Governance and the poor

This book tells the story of how Liberal conceptions of good governance and the need for stronger institutions came to dominate Development and Poverty Reduction programmes. Before we lay out the book's argument, we need to understand why this happened, and some general parameters of the recent relationship between governance and poverty.

As often, crisis has fuelled much of the resurgence of interest in governance. Strong impetus came with the early 1990s collapse of Soviet Union and Eastern bloc economies, attributed to a lack of institutional and regulatory frameworks to make new markets work. The same period saw the emergence of the criminal mafias and ethnic warriors in 'failed' states – the former Yugoslavia, Central Asia, Somalia. Poverty, it became clear, was not just a matter of humanitarian disasters or Structural Adjustment failures, but was linked to the breakdown of civic order into civil war, and the predatory rule of warlords and cronies running 'out of control'.[16]

At the same time, especially across Africa but also in parts of Asia and Latin America, other forces were undermining government. The pace of urbanization was outstripping the capacity of poor governments to deliver basic services, or to police and provide security in emerging urban settings.[17] A new phenomenon emerged, urban 'involution',[18] in which against all predictions of theory, urbanization was disconnected from economic growth. Urbanization into slums and resettlement zones was escalating at the same time as Structural Adjustment Programmes (SAPs) were prompting the collapse of import substitution industries, contractions in public sector employment, and the weakening effects of debt. The growth of informal sector economies in burgeoning urban slums meant that, for much of the 1990s, an increasing share of employment in poor countries was generated 'outside the rule of law', in zones of 'illegality'. Thus, the poor lived increasingly in ungoverned domains: they occupied land illegally or with uncertain ownership, and became dependent on unprotected subsistence economies and insecure, unregulated enterprises. Here they faced marginal profitability, and became enmeshed in criminality, racketeering, trafficking and desperate, illicit migration.

All this might be seen in hard-nosed political economy terms: struggles for scarce resources among growing populations, within markets that scarcely valued their labour. And then tearing away at whatever could be pulled from the public domain, destroying it in the process. Yet through the 1990s, a consensus emerged that much of this resulted from 'bad governance': governments not completing reforms, poor Public Expenditure Management (PEM), failing to produce good legal and other environments for growth. Bad policemen and greedy officials, abusing power to oppress the poor.

Certainly governance in many poor countries was appalling: but its failure needs to be seen not just in moral or technical terms. Here, it might be argued, was an under-resourced state: a leviathan in crisis consuming itself and its citizens. Under Structural Adjustment, even as the private sector sputtered, state sector cuts and the privatization of state business and assets meant less money for salaries and fewer opportunities to extract rent. Impoverished local tax bases and leaky public fiscal controls meant the salaries of frontline and administrative workers became unliveable. 'We pretend to work, and they pretend to pay us', the Vietnamese saying goes. With pay packets as low as $10 per month, officials had to get a second job, almost always more rewarding than their official one. 'On $10 a month, you're not corrupt, you don't eat', in the words of a Phnom Penh ex-policeman.

Increasingly, corruption focused on social sector budgets and projects that had been padded out with quick dispersing global aid transfers made to ameliorate or 'put a human face' on the fallout from adjustment. In some sectors, public 'services' in fact expanded, but in perverse ways, in areas where rent could be extracted (busy intersections, airports and customs checkpoints, lists of real and imagined soldiers and health personnel to be paid). Under such circumstances, anything or anyone that moves, especially across boundaries, becomes a target. Whether extracted from exporters or aid-funded road budgets, a share of the spoils would be passed up, the rest remaining for local patrimonial discretion. Rent is sought up and down government levels too. A province department, a local commune or a police station might become a patrimonial domain, sold or rented to a patron who would turn it into an extraction machine.

Thus starved of basic resources, in these ways the state literally eats into its basic legitimacy. The starving state becomes grotesque, its belly distended, its arms and legs feeble. Its territorial domains become fields of plunder, its markets are choked by rent seekers at every point of passage. Territorial patrimony, not market exchange, becomes the basis of wealth and poverty. Great inequalities are promoted in these situations. The rich typically are closely interwoven with patrimonial governance, and can leverage territorial and economic assets and opportunities: land, water, cheap labour, trading positions and natural resources. Their connections secure government contracts at high prices, and the protection of the bendable rule of law. Politics becomes a club of elite, landed interests, funding elections, cherry picking official positions, installing cronies and agents.

The poor under such circumstances lack secure access to resources, justice or political power. Their best hope (short of migration) involves not fighting patrimony, but aligning themselves to a local patron, and latching onto whatever opportunity and security that offers. Politically, this rural or involuted urban support base is readily co-optable, via 'treating' with minor infrastructure, enticing with opportunities for lower-level public employment, or straight vote buying. Thus even the poorest people perversely support patrimonial alliances that offer immediate reward and longer term stability. Crucially for this book, it also means that strong patrimonial governments have incentives to decentralize their governance, establishing mechanisms to directly channel largesse into local political domains. There, funds for local infrastructure can reinforce the position of local patrons. Patron-client relationships with contractors recycle funds back into higher patrons' pockets and consolidate political power.

It should be clear that under such circumstances, fighting corruption by punishing officials taking bribes or scrutineering budgets won't fix the problem. The grotesquely governed territories of patrimonial rule seem to require more radical surgery. And this, this book will show, is exactly what the current governance agenda offers. The remedy involves the breaking up of patrimonial territories and using markets to replace and reconstruct the institutions of governance. Markets here offer precisely what patrimonial governance does not. Services can be provided on the basis of what people choose, rather than what patrons want to bribe with. Competition for the delivery for services within legally enforceable contracts will lower costs. The free movement of goods, resources and people, unhindered by gatekeepers at territorial boundaries will promote the wider economic wellbeing.

Markets, in other words, could do just what Polanyi said they would: disembed social relations from their existing conditions, and turn things previously regarded as social goods into commodities. But for markets themselves to survive, as Polanyi also emphasized, they need to be embedded in regulatory and constitutional frameworks. In Development's new institutional programme, market-society relations could to be deeply transformed. They could be disembedded from corrupt, territorial patrimonialism, and secured through Liberal institutional frameworks, such as the universal rule of law (which governs everyone alike, and provides security for the poor). This security could be supported by other Liberal governance measures, such as freedom of information, participation in politics, and basic education and health services. It could also be reinforced by the social capital of 'softer' institutions: community participation and civil society partnerships, creating trust and providing institutional support for the market.

All this, as this book will describe in detail, would become core to Development's move beyond the raw neoliberalism of Structural Adjustment. But in moving beyond, Development was expanding the reach of markets well beyond the policy ambit of SAPs, and into areas that had not yet been neoliberalized. Good Governance, in other words, might mean the neoliberalization of social services, a marketization that in Polanyian terms might ultimately be more disembedding than re-embedding. On the other hand, talk of civil society, local partnerships, participation and community seemed fairly frail kinds of embedding. So what was going on? And would the result be disembedding and neoliberalization, or re-embedding community control and security for the poor? Or some messy hybrid of the two – an under-resourced, weakly Liberal governance, creating the form without the content of a Liberal order, yet giving cover to patrimony's preying on the poor?

THIS BOOK'S FOCUS: (NEO)LIBERALISM, POVERTY REDUCTION AND GOVERNANCE

This book recounts how, in response to 1980s failures and 1990s crises, IFIs began to work on a series of steps that would ultimately reinvent their work, their image and Development as a whole. But in telling the story of this reinvention, and the role governance played in the construction of the Poverty Reduction paradigm, it also reaches both to wider historical political economy (Part I, Chapters 2–4), especially the political economy of what we call Liberal Development, and its relation to other modes of governing poor and peripheral places. Our story then shifts its focus to specific cases of Liberal, decentralized governance in Vietnam, Uganda, Pakistan and New Zealand (Part II, Chapters 5–8). It concludes, in Chapter 9, with a critical summation of the current situation, two scenarios for Development and poor countries' futures, and a discussion of what alternatives need considering.

Part I: Liberal Development and governance from free trading to 'neoliberal institutionalism'

In Part I, we begin in Chapter 2 with a selective history of Liberalism in relation to Development's several phases to date. We sketch history from the East India Company forward through British Liberal colonialism and pre-Second World War imperial Development. We consider in turn the security-embedded Development of much of the Cold War era 1945–*c.*1980 (including an earlier 'Poverty Alleviation' phase), and the neoliberal, structural adjustment era that emerged as capital again broke free of its territorial

Definition Box 1.1: Liberalism

A political ideology and form of governance that has hybridized over time, but generally emphasizes the benefits of markets, the rule of universal law, the need for individual human and especially property rights. In its approach to poverty, it eschews major redistribution, and emphasizes moral discipline and (again) markets.

constraints (1980–1990). While scarcely comprehensive, this story shows Liberal developmentalism changing over time, lurching in Polanyi-style 'double movements' between security-embedding and market-disembedding approaches, all of which approached governance distinctively, but within hybrid Liberal frames.

The history of Liberalism's consistent prescriptions and various hybrids are, we think, the key to understanding much of past and current Development. As we'll see, in the wider historical Liberal orientation of 'liberty within the law', poor people and their countries' primary obligation has been to conform to standard 'universal' laws and rules governing economic processes. Special attention has typically been given to property rights and security of capital, and to ensuring governance processes help not hinder wider market processes. This has occurred while in general keeping the existing distributions of property, capital and corporate market power firmly off the reform agenda, and keeping the territorial powers of the state (social redistribution or coercion, economic nationalism, protection of nascent industries) firmly in check, or firmly aligned to core security interests. In return for such discipline, Liberal developmentalism has offered economic and social assistance, seldom very generous and often self serving, but never without crucial policy strings and surveillance attached. In poor countries, Liberal development has usually been aligned with security concerns, and can, we argue, be seen as in part a means of containing the poor politically. This we show was apparent in the processes of 'Indirect Rule', in nation states aligned to security blocs, in local communities serviced by some NGO, or in the localized, 'quasi-territories' popular in decentralized governance today. By the end of this book, it should be clear that Poverty Reduction and Good Governance deviate very little from this historical Liberal ambit. 'What the Bank knows, changes': true, but within crucial Liberal parameters.

Chapters 3 and 4 tell the story of the rise of Poverty Reduction and Good Governance in the 15 years 1990–2005. Here, we show that there has been a shift away from a 'conservative neoliberalism' – a 'negative' Liberalism concerned to get the state out of markets, deregulate and privatize, reduce social and bureaucratic spending. What has emerged,

---

Definition Box 1.2: 'Inclusive' neoliberalism and 'positive liberalism'

While retaining core conservative neoliberal macroeconomic and pro-market policy settings, 'inclusive' neoliberalism adds 'positive liberal' approaches emphasizing 'empowerment' to enable participation (and ensure 'inclusion') of countries and people in global and local markets. These include: institution building and an enabling state ensuring global market integration; building human capital via services (health, education); empowering and protecting the rights of the vulnerable through participatory voice and legal access; engendering moral obligations to community and work.

---

we argue, is a new Liberal hybrid, an 'inclusive' neoliberalism, market-oriented, but also involving many aspects of a 'positive' Liberalism of 'empowerment' and market enablement.[19]

This shift might in fact be seen as a part of wider Polanyian political economy shift, the early rumblings perhaps of a re-embedding phase following on from the (ongoing) disembedding of markets via neoliberal reforms. But if so, we think this phase shift is still in progress. It has involved elaboration of a variety of mostly enabling regulatory frames for markets, alongside bolstering legitimacy by claiming (and morally prescribing) social purpose for reforms (for example, social inclusion, gender equity, Poverty Reduction, environmental sustainability). But a darker Polanyian reactionary side is also visible in, among other things, rising nationalism, security obsessed neo-conservatism, military expansionism and rivalry, persistent protection in some crucial production sectors, anti-immigration backlash, and selective bilateral free trade agreements.

In Development since 1990, however, this phase has increasingly focused on 'institutions', including the laws, policies and rules that govern market and public sector activities for service delivery and 'participatory' engagement. In the words of Washington Consensus author John Williamson, the 1980s were a time when 'economists became convinced that the key to rapid economic development lay not in a country's natural resources, or even in its physical or human capital, but rather in the set of economic policies that it pursued'.[20]

Subsequently, the bare bones of frank neoliberal policy were seen as in need of a stronger institutional basis – the right 'institutional settings' at all levels, national through local. In its early reflection on events in Eastern Europe, 1991 WDR, *The Challenge of Development*, explained how 'institutions' could be the platonic guardians of market-led reforms and secure them against plunder and instability.[21] What was needed, it was increasingly clear, was not shaky and corrupt government further undermined by anti-statist reforms, but an *enabling* state that could support the crucial 'institutions' of law, financial and policy transparency and market information that the New Institutional Economics (NIE) held were basic to the emergence of efficient and competitive markets.[22] By 1995, a number of voices was cogently outlining a set of second-generation reforms focused on governance.[23] In 1997, the Bank's headline annual WDR, *The State in a Changing World* could point to the need for a strong, *capable* state that 'focussed on the fundamentals' of the 'good governance' agenda. From this deep institutional intervention, World Bank president James Wolfensohn and his chief economist Joseph Stiglitz explicitly imagined long-term economic and social change could be expected.[24] This was a crucial elaboration in what for critics such as Ha-Joon Chang was already Development's 'Age of Institutional Reform' (1980–2000), which he sees as extending into the current Poverty Reduction era, and which he among others has shown has delivered relatively little to many of the poorest countries.[25]

In this book, from Chapters 3 and 4 onwards, we improvise on Ha-Joon Chang in characterizing the last 25 years of Liberal Development as *neoliberal institutionalism*. We describe neoliberal institutionalism as a historical high point of Liberal hegemony in Development. Here formal, normative elements of Liberal policy and governance have 'crowded out' Development's engagement with sectoral, political, productive and other political-economic realities. At the same time, they have displaced substantive questions about the resourcing and returns from these forms of governance, such as 'Does enough money come out the end of such expensive systems to make any difference?'. Addressing these kinds of questions has been made more difficult by the rising complexity resulting from many such approaches, with different agencies and arrangements for accountability overlapping and often competing with each other. These together, via often paltry funds, pilot projects and resource-consuming partnerships with other fragmented agencies, often involve high transaction and opportunity costs. All this, we conclude, has little wider accountability to substantive questions of Poverty Reduction.

In projecting the possibility of comprehensive Liberal institutional reform, neoliberal institutionalist Development has, we argue, overreached itself, and not been able to apply institutional discipline effectively either to its own processes, or to those of many governments it engages. Yet the frailty of its disciplinary leverage is gradually becoming apparent, as is the amount of real resources needed to instigate such reforms. It has struggled (though perhaps not hard enough) to grasp how politically and economically embedded and pervasive are existing, more territorial modes of government (based on patrimony, rent seeking, but also sometimes on progressive social agendas).

Inclusive neoliberal Development has also been somewhat unexpectedly undermined by the many ways that ensconced territorial interests are able to 'game' and use new institutional arrangements to their own ends. In all this, there seems to have been over expectation about the kinds of efficiency and accountability outcomes that are achievable via its two principal points of focus, that is, from over-arching juridical and state sector reform (PEM, access to justice programmes, civil service reform, devolution and privatization of services), and from very locally scaled interventions involving civil society (community planning, co-production of local services). Similarly, there has, we think, been an overestimation of the leverage that a market or liberal democratic orientation to government can have on substantive 'pro-poor' outcomes. It has left itself vulnerable to obvious critiques: if this framing is all so crucial, how come China and Vietnam have been such successful Poverty Reducers?

All this, we argue, is mildly Polanyian, both in its somewhat shallow re-embedding of markets in institutional contexts, and its embrace of the 'soft' institutionalism of community participation and NGO partnerships. This scarcely constitutes a profound Polanyian 'double movement' reaction. Rather, to date, it needs to be understood primarily as a 'top-down' response, led by IFI and the Group of 7 (major industrial economies) (G7) interests urgently re-asserting their legitimacy in a time of crisis, and seeking to expand their own institutions' intellectual hegemony as leaders of Development consensus. Top-down, the actors have genuinely sought to provide markets (and wider neoliberal reform) with the minimal regulatory and institutional support and social legitimacy needed for market-oriented reforms to do their poverty-reducing magic. And to support these markets and their participants from the 'supply side', through providing services that will in turn offer the market healthy, educated workers. Finally, they have in fact used all of this to extend the reach of markets into poor societies. This means that 'Development beyond neoliberalism' will certainly involve further development of neoliberalism.

Hence the preponderance of technical institutional interventions, driven by consultants and contractors able to be deployed as technocratic experts. Hence also the lack of political coherence, and the easy alignment with otherwise illiberal authoritarian governments. As will become clear, Poverty Reduction strategies and decentralized governance proceeded powerfully

---

Definition Box 1.3: Neoliberal institutionalism

Current Development's priority emphasis on getting institutional dimensions right: policy, legal frameworks, governance, market mechanisms and participatory democracy. This is often at the neglect or expense of substantive sectoral and directly productive development and investments.

among such states, which ironically saw these Liberal institutional developments as a means to secure support, legitimacy and patronage, both internationally and in local contexts. What's not visible, then, is any phase shift in which markets are being subordinated to social issues of the kind Polanyi's wider social vision evokes. Rather, what we have is the establishment of what Bob Jessop has called a necessary 'flanking compensatory mechanism for the inadequacies of the market mechanism',[26] enabling markets themselves to preserve their priority, and a series of core neoliberal reforms to survive. In this, perhaps, a wider Polanyian task has been achieved, pre-empting the more powerful Polanyian reaction that might have followed further rounds of Structural Adjustment.

This is the general argument of Part I, where we show, step by step, how these core institutional and good governance elements emerged from a longer history of Liberal developmentalism, and evolved to take their current consensual and harmonizable form.

Part II: Cases from Vietnam, Uganda, Pakistan and New Zealand

Part II tells this story from a country case perspective. Here we will see how embryonic forms of the neoliberal institutionalist Development headlined by the 1997 WDR, *The State in a Changing World* were already emerging in practice in the early 1990s. In Vietnam by 1992–1993, the communist, totalitarian state was not so much fading and liberalizing as a result of its particular version of 1980s structural adjustment – known as *Doi Moi*. As far as many everyday Vietnamese were concerned, it was just becoming fractious and ineffective. Inequality was apparently increasing dramatically, public protest was rising about this, and about the collapse of public services and unbridled official corruption. Donors and the state concurred: new means were needed through which the economic transformation might be managed in an ordered, socially legitimate way. In this context, limited experimentation began with new approaches, including one whose story we sketch in Chapter 5.

In Vietnam and elsewhere, these approaches included PRA and Local Development Funds (LDF). PRAs offered a new, apparently participatory way of framing the poor in their places: hearing the voices of the most marginal, representing poor communities in globally legible ways, and in so doing, presenting poverty as something 'inclusive' neoliberalism could fix locally. The LDF, on the other hand, offered a direct means for donors and central government to channel to localities, resources for palliative services and basic infrastructure. And, at the same time, a way to get local buy-in for their efforts to deal with socially disembedding consequences of market-led transformation. When set against later, more sophisticated arrangements that developed in Uganda (Chapter 6 – mid-1990s to 2000) and Pakistan (Chapter 7 – 1999 to 2005), these instruments seem quite limited in ambition. And, not least importantly, quite frail when set against the longer political history of governance in Vietnam.

While being played out in Vietnam, PRAs and LDFs were actually part of a much wider shift in Development, into which they would later be more closely tied. But as we'll see in Vietnam, early 1990s attention to participatory planning, voices of the poor, engaging NGOs in service delivery, all in large measure occurred 'outside of government', in the realm of 'community' that was being framed up via PRA and LDF approaches. By cobbling together an accommodating donor-Communist Party litany of 'growth with stability', 'market socialism' and 'community participation', the intention was to create 'policy windows' that justified bypassing intervening levels of government, which were often seen as an opportunistic corrupter of efforts to respond to local needs. In fact, these approaches tended to be seen by donors as a way of achieving democratization by stealth: achieving change through administrative, *executive* or technical means what had proven difficult through *political* channels. Adherents of local PRAs hoped these instruments would clasp together community and government and foster a new accountability to Poverty Reduction that reflected a consensus about what the poor needed. But they were also frequently aware these innovations needed to be bound into more thorough-going restructuring of local to national institutional and fiscal arrangements, for any substantive link to be made with Poverty Reduction. As it was, PRA- and LDF-style democratizing efforts in authoritarian states were like water off a duck's back, even as they were co-opted as signs of regime openness.

But quite quickly, from the mid-1990s, a major step-up in sophistication in techniques for governing poverty was made to occur. By the mid-1990s PRAs were being elaborated not just in community-specific, but in 'whole of country' ways. The upscaling of PRAs into national Participatory Poverty Assessments (PPAs) would, by 1997, make it plausible that PRSPs would transform discredited instruments like National Five Year Development Plans into something that could deliver relief and political representation to the poor. Alongside, the limited LDFs were also being replicated and upscaled in wider systems of decentralization, sector-wide reform instruments, and techniques of managing global, national and local fiscal relations. This we show in Uganda, in Chapter 6. But as we will see, it was never a matter of this governance and Poverty Reduction doctrine arriving fully blown. Indeed, it was not until 2000 that Uganda's government was able to broker a deal with the World Bank that would overcome long-standing Bank scepticism about decentralization. Overcoming this trepidation, in Uganda as elsewhere, would require both the refinement of a host of techniques for reforming intergovernmental relations, and robust political accommodations between global promoters of Poverty Reduction and national governing regimes.

The difficulty for IFIs, as we explain in Chapter 4, was that while by the mid-1990s they well understood the need for a 'capable state', it was not yet clear how the policy appeal of NIE doctrines could be operationally applied. NIE sounded constructive ('building institutions'), but it was also premised on a mentality of governance that required the prior and effective *disaggregation* of government. 'Government', especially in its patrimonial, territorial forms was for NIE an 'obstacle' to the free flow of information and adequate competition. NIE promoted a disaggregation of forms of government that from the 1980s had popularly been seen as fetters on development, gave them to the market, and made would-be providers compete to deliver them. Earlier moves towards privatization may be seen as 'horizontally' disaggregating government: taking government functions at say local (or other) level, and contracting them out. In Development, NIE approaches argued for wholesale engagement of NGOs and private sector in service delivery. The result was an efflorescence of private providers, acting as NGOs and masquerading as 'civil society'. Public agencies too joined in acting as private contractors, competing to provide the state's regulatory, policy, enforcement and service delivery functions.

Hence one source of fragmentation. But NIE reforms also implied a 'vertical' disaggregation, in which the political, administrative and fiscal functions of the central state were delegated or devolved up or down governance hierarchies. Delegating up meant separating rule making and policy framing from executive and implementation, by shifting the former into separate, often internationally sanctioned frameworks and domains (such as the World Trade Organization (WTO) and its global rules). Devolving 'down' meant shifting mandates and tasks (and hopefully funding) to multiple levels of local, regional, state/provincial authorities. If it could be done locally, the principle of subsidiarity urged, it should be. Locally, enhanced voice, participation and multiple points of client exit would ensure resources were allocated efficiently and make the whole operation for accountable. And thus the proliferation of local 'accountabilities' was seen as the prerequisite for achieving greater accountability of the whole.

Neoliberal authoritarian accommodations and frail institutional overreach

NIE doctrine fitted well with the mid-1990s 'radical conservative' governments in the United States of America (US), UK and some other Organisation for Economic Co-operation and Development (OECD) countries. But this was only part of the story. To become current in poor country Development practice, these modes of reform had to find an accommodation with a range of other concerns. Here, alignment came with political projects to break old corrupt bureaucratic and elite fiefdoms, deliver better services to localities, and new MDGs for Poverty Reduction. Achieving all this required accommodations to be reached with national governing regimes that had sufficient political traction. At the same time, it needed the application of new disciplinary techniques. Altogether, it provided a much stronger rationale for decentralization.

Getting this kind of national 'ownership' became the *sine qua non* of second-generation governance reforms. Until Uganda became Africa's shining star in the mid-1990s, there were few opportunities for achieving political traction, disciplinary reform, and pro-poor decentralization simultaneously. As we'll see in Chapter 6, the Ugandan government's accommodation with what became Poverty Reduction over the 1995 to 2000 period shows that governing innovations do not just travel into and get laid down in poor countries at IFI's suggestion. Rather, Museveni 'pulled' IFIs in to help achieve his clear need to consolidate territorial control over a country fractured by two decades of civil war. The first rounds of structural adjustment in the late 1980s had seen an immediate blow-out in national debt and a growing public perception that he had capitulated to the IFI's 'imperial hand'. This, and growing worries about poverty and inequality, saw his popularity under threat, with supporters questioning his commitment to the people's empowerment strategy of local Resistance Councils. By directing and encouraging a doughty crew of local and international governance and decentralization entrepreneurs, Museveni arranged for the LDF approach to be trialled in Uganda. At the same time, he wrapped it into an apparent 'whole-of-government' reform that sported medium-term budget and expenditure management techniques (such as the Medium Term Expenditure Framework [MTEF]), and Sector Wide Approaches (SWAps) to health, water, roads and other service delivery. All this was rapidly tied to a nationally owned Poverty Eradication Action Plan (PEAP). This precursor to PRSP thus reflected a political accommodation, which gave crucial plausibility (and fundability) to Ugandan decentralization. As crucially, it provided the wider donor compliance his governing regime needed to resource its 'no-party' democracy.

The immediate result of this credibility was a dramatic increase in funds for Poverty Reduction spending; which, as critics concede, is common where such Poverty Reduction accommodation has occurred. Once the necessary degree of national ownership and the top-level policy and technical devices for ensuring expenditure discipline were in place, it then became possible to unlock the tremendous fiscal transfers of the Heavily Indebted Poor Countries (HIPC) initiative. Uganda was the first recipient of HIPC in 1998. At one step, 20 per cent was wiped off the country's external debt, and when marshalled together with huge increases in donor aid, saw unprecedented increases in public spending within the disciplinary corral provided by the PRSP and MTEF, and, as we'll see, in local government settings. But perhaps the most enduring result, irrespective of its impact on poverty, was the consolidation by one-party government of control and influence over their local political power bases. And, in all this, the co-opting and compromising of Liberal governance prospects.

For, the durability of these global-national accommodations aside, what we see in Uganda and then on an even grander scale in Pakistan, is a story of institutions in overreach, sending Liberal modes of governance spiralling over distance into some very illiberal places. Simply, neoliberal institutionalist governance modes are routinely prey to delusions about the scope of transformation possible. At the same time, they are often impossibly rational and elaborate, a perfect and expensive system laid down on a very fraught local situation. As we'll show in Part II of this book, the real difficulties of 'joining up' these often frail and always complex reforms is exacerbated by Development's endless predilection for experimentation, a penchant pushed along by large amounts of discretion and available resources. The reasons for Development's continual reach beyond the empirical reality of what can reasonably be achieved are many. Often it's plainly human. Tracking through the decade up to 2005, we'll see entrepreneurial agencies, career jockeys and innovative consultants stretching to be one step ahead in the next big thing; in Uganda, for instance, we'll show institutions trading on apparent success somewhere, publicly overplaying its impact, building an overblown reputation for certain kinds of work, and getting resourced to upscale their efforts at the first blush of success. As in Pakistan, we'll see this overreach happening because large tranches of money had suddenly became available in a security-threatened part of the world. And this being multiplied, as plural agencies sponsor a vast range of vertical programmes and grant systems, creating enormous mess and fraught accountabilities on the ground, yet all individually acting out Development's travelling orthodoxies.

Crisis, harmonization, accountability?

It is thus understandable that from the late 1990s tremendous efforts were being made by the international community to 'harmonize' their support for Poverty Reduction's Development. Harmonization efforts, from rallying calls around Comprehensive Development Frameworks (CDFs), SWAps and the PRSP in the late 1990s, to the MDGs in 2002 and thereafter a spate of OECD High Level Forums through to 2005, are at one level entirely reasonable responses to the disabling, multi-actor complexity that characterizes Development now.

But at another level, harmonization signals the increasingly shrill moralism around Poverty Reduction, which judges that poor countries should enjoy no rights to global assistance unless they had willingly adopted its prescriptions. By 2002, as we explain in Chapter 4, harmonization signalled the firm resolve to apply a hard, disciplinary and security driven edge. Cheap, fungible loans and discretionary resources – the critical lifeline of poor countries – would through harmonization become more selectively available to Poverty Reduction's willing adopters: to countries, that is, that were able to demonstrate they were already 'better governed', according to their scoring against a set of policy, fiduciary risk and expenditure management standards.

The kinds of Poverty Reduction harmonization achieved by Uganda in the late 1990s, and by Pakistan in the fraught post-9/11 environment in the early 2000s, were as we'll see closely tied to fraught security and legitimacy situations, which created strong incentives for accommodations between IFIs and the national regimes. But even as security aspects were strengthening state-IFI accommodations and harmonization at the national level, NIE governance approaches were doing their disaggregating work at local scales: and all in the name of Poverty Reduction. By the time the 2004 WDR *Making Services Work for the Poor* was released, the kinds of local regimes of NIE accountability that were being upscaled in places like Uganda and Pakistan were able to be presented in highly formalistic terms. As we'll see in Chapters 4 and 7, Development NIE's own simple three-legged rubric 'Inform (consumers), Enforce (contracts and the law), Compete (make multiple agencies compete for contracts to deliver services)' was presented as an Accountability Triangle, wherein (in one corner) consumer's informed voice and choice, together with (in corner 2) policy-maker's contracts and compacts with service deliverers (corner 3), would deliver more accountability for service delivery.[27]

Again, as we will see in Pakistan, these precepts appealed to governing regimes that lacked a democratic accountability process, and sought rather to achieve better, durable relations with the local constituencies via an apparently non-political, executive short-cut route, again explained in Chapter 7. In a high security stakes context, the incoming government of General/President Pervaiz Musharraf (echoing Museveni) had duly announced that the 'crisis of governance' could only be overcome by eschewing old, discredited political systems dominated by unresponsive political parties. Musharraf's first three years of government from 1999 began with a popular appeal to the poor, 'common man' who without doubt wanted the efficient delivery of social services and access to local justice. Devolution in Pakistan, in an extraordinarily short period of time, sought to apply the NIE 'Accountability Triangle' in its full glory, but again, not by importing its WDR formalism, but by a reconfiguring of 150 years of territorial governance along time honoured Liberal lines. Here, the local separation of judicial, executive and legislative powers (the heart of Liberalism, and reformed in 2004 WDR's Accountability Triangle) would superintend a radical disaggregation of government. Now, authority for managing services and providing social regulation (of citizen rights to land, labour, natural resources, public safety and security) would be in true Liberal governance terms assigned to a host of locally mandated public, private and civil society bodies.

What this kind of Poverty Reduction-inspired disaggregation delivers in terms of governing poverty is of course central to this book. But Pakistan's story, like Uganda's, is so fresh that we can barely sustain speculation about the future. But in various ways the reforms experienced from Vietnam in the early 1990s, to Pakistan most recently, were all given a prior and unprecedented rein in New Zealand, where we arrive in Chapter 8. While this relatively rich country may seem an odd destination for a book on governing poverty, New Zealand was in many ways the test tube for the kinds of reforms we see being played out in developing countries. There, they were initiated in two phases: a radical decentralizing and marketizing phase (1987–1998), and a second, 'joined-up inclusive' approach to social governance since 1999. New Zealand was, by 1993, an exemplar of Washington Consensus-style structural adjustment, and for long its NIE and New Public Management (NPM) governance reforms (defined in this chapter) have been a rod for poor countries' backs. But at the same time as its crystal clear approach to accountabilities was being internationally trumpeted, a backlash was already underway, as the fragmenting effects of such reform were painfully reconsidered. New Zealand offers, then, an occasion to recapitulate in an actual case the whole book's argument, a place to revisit the central themes developed through the 1990–2005 transit from Vietnam, to Uganda and Pakistan. And to ask, now, after six years of Poverty Reduction-style reform, what is there to show in terms of process, service delivery and social regulation, and most importantly, accountability around pro-poor outcomes?

This, before we draw the book to a conclusion in Chapter 9, with a set of critical arguments and possible ways forward we won't anticipate here and now.

THIS BOOK'S ANALYSIS: POLITICAL ECONOMY, POLANYI, LIBERAL AND TERRITORIAL GOVERNANCE

The book's presenting concerns, then, are the emergence of the Poverty Reduction and good governance paradigm, and its outworking in peripheral places. But in both history and cases, this book also aims to show a much wider set of issues and practices at work, and a much longer historical pattern of their development. The wider issues and practices at work are those of economic and political Liberalism, more recently reconfigured as 'neoliberalism', which we consider itself as a moving target, subject to ongoing hybridizations, from the frank, conservative neoliberalism of Structural Adjustment to the 'inclusive' neoliberalism of Poverty Reduction. The much longer historical pattern involves the extension of Liberal economic and governance modes to the poorest peripheries of the planet – the (neo)-liberalization, perhaps, of peripheral governance.

But our study also involves a counter-perspective, equally important, that requires we show the fraught relationship between these travelling

Liberal modes of governance and the actual local realities of patrimonial and territorial power, and territorially ensconced poverty where they end up. In these terms, as noted above, we need to consider that it has been overwhelmingly in authoritarian, one-party states, and for political reasons, that 'poverty reduction'-related decentralized governance has rolled out. Here we show how the interests of IFIs and authoritarian government have aligned in headline countries, reinforcing core doctrines and techniques, enabling them to be projected onto less able and stable states. A little more abstractly, but as crucially for the book's argument, we need to see how Liberal modes of governance relate more generally to other, what we call 'territorial' and socially embedded modes of governing.

To elaborate this argument, we develop a set of new arguments broadly based in political economy[28] and 'governmentalities' theory,[29] and building especially on perspectives from Polanyi and current decentralization theory. Through empirical cases, we show how Liberal modes of governance travel to and transform peripheral societies, creating new relationships between multiple actors.

In this book, we will elaborate a distinction between Liberal and Territorial modes of governing poverty (Table 1.1), and use it to do a number of jobs. Generally, we will use this distinction to expand our analysis of how Development's history has been characterized by lurches not just between market and society, but between different *territorialized* and *deterritorialized* modes of governance. In Polanyian terms, when liberal markets break out of social regulatory constraints they also break out of territorial boundaries. Liberal governance, then, might be expected to favour *de*territorialized approaches to its governance, or at least approaches that are territorialized (perhaps *re*territorialized) at levels of scale which enable rather than constrain markets, or don't impose national or other territorial constraints and burdens.

Liberal-territorial lurches in Development history

We will also use the Liberal-territorial distinction to show how Development has acted in much more territorially orientated ways in the past by, for example, framing minimum basic standards, safety, resource and other entitlements. Then, noting the lack of such resources in particular contexts, these approaches relied on territorially powerful actors, especially at nation state level. They favoured equally powerful, nationally leveraged redistributive instruments, such as by taxing or staple food subsidy, or by intervening in land, capital or water distribution, or by stimulating productive activity, protecting farmers or industries. In Chapter 2, for example, we show how in the generation after Polanyi, local and international Development, production and consumption, geo-politics and security, standards and targets were all largely imagined within the territory of the nation state. Practically, this meant bridging the national investment gap, aligning

Table 1.1 Liberal and Territorial governance of poverty

	Liberal governance of poverty: the (universal) market	Territorial governance of poverty: the (particular) place or society
Basic focus	Market mechanisms and their operation: universality, efficiency and security of these, integration of all into these	Actual territorial resource, production, market and population processes and their outcomes: position of social/demographic group *viz.* these
Obligations of governance	Getting the rules right: universal, technical, transparent, market enabling, non-restrictive of movement: enhancing power of the market and active subject	Engaging outcomes through distribution of power, assets and resources on the ground: in specific local/market/geographic conditions
Relation to space and place	Universal, place in relation to local and global markets	Place specific, territorial, historical, population oriented
Preferred governing techniques	Compliance with universal laws and contracts; separation of powers; state out of markets, and marketized decentralization; authoritarian government or 'technopols',[30] some participatory democracy	Political, state or patrimonial power, accommodated with markets; distributive control over resources via taxes and transfers; autocracy or representative democracy; plans, strategies; devolved authority, funds
Opportunity	Comes from integration into markets	Comes from position in markets and in relation to place, assets, resources
Security, stability and boundaries	Enforcement of rules wherever capital is circulating; risks managed	Security, stability, wealth and poverty of (particular groups in) territorial domain
Security, stability and subjects	Security for market actors: capital, property and contract rights, asset ownership	Security of place (e.g. work) in market/other contexts. Ability to strengthen this *vis* others, bargaining, political power
Empowerment and subjects: wellbeing and capability	Enabling market actors, through services, voice, education to be individually capable in safe communities	Wellbeing via adequate resource/income base (e.g. water, land, capital) for that place, and relative access to it
Empowerment and subjects: rights and entitlements	Universal Rights: freedom to contract and circulate, freedom to own accumulated wealth	Territorial entitlements: citizen's share of territorial wealth, services
View of poverty	Due to market failure, lack of market integration, lack of market capability, vulnerability to shocks	Weak position *vis* international/local markets, and *vis* political and historical distribution of assets, resources

the factors of production to achieve Gross National Product (GNP) takeoff, fostering national self-sufficiency through import substitution and tariff protection for nascent industry. Nations were conceived as being constructed through geographically specific 'building block' projects, sector-wide programmes or integrated area development efforts of the kind promoted in the 1970s as 'defensive modernization'.[31]

National industries for a range of reasons failed to compete well in increasingly liberal markets. Nation states themselves failed territorially, when wider cold war security arrangements promoted division, or when, already mal-territorialized by colonial intervention, they failed to cohere, and lurched into conflict. This period of statist territorial development was of course what neoliberalism overthrew in the name of getting nations to sign up to the universal Liberal doctrines of structural adjustment, a radical *deterritorializing* move designed to open up local territories to the new domain of the global market. But by the mid-1990s, nation states and societies were experiencing the political and economic effects of this profound disembedding and deterritorialization. Questions about how social life was to be governed could no longer plausibly be mapped and resolved at the scale of the nation state.[32]

At the same time, it's possible to see neoliberal governance's radical decentralizing via privatization, NGOs, etc. as a *reterritorialization* that was being enacted in favour of both global and local markets.[33] This had important implications for other scales of social and territorial accountability and outcomes. At the beginning of reforms, marketizing was what mattered: territorial aspects were subordinated to letting agencies from anywhere compete to deliver narrowly specified services for clients from nearly anywhere. Outcomes were often perverse: individual needs and rights often fell through the cracks between multiply contracted agencies delivering narrow classes of services. As the New Zealand case story in Chapter 8 will show, the 'inform-enforce-compete' rubric of NIE governance needed sharp revision. Elsewhere, similar Polanyian stories appeared. Markets were tearing at wider, territorial governance accountabilities to basic responsibilities of social regulation; inequality, secure access to basic citizen entitlements like land, education, irrigated water, public safety tended to be neglected in the drive to efficient service delivery. This prompted in turn the new inflections of 'inclusive' neoliberalism (that we discuss in Chapter 3), advocating among other things the reterritorializing services at community level via 'local partnerships'.

PRSP processes can in these terms be seen as a mild return to aspects of territorial governance of poverty: voice for the local poor, local partnerships, and the reintroduction of nationally aligned processes. So can the MDGs, with their national and lower territorial scaled indices of measurement, around which donors harmonize their activities. Other mild reterritorializations under Poverty Reduction include new attention to aspects of national and local political ownership, national donor and budget coordination moves, and the reinvention of roles for the local state in local participatory planning processes, decentralized service delivery, coordination and monitoring. But at the same time, the fragmentation of Development appears ensconced. And, while the NIE ambit is still being powerfully advocated, further fragmentation seems just as possible as more cohesion.

In this book, we show that what are currently emerging from these ongoing processes of deterritorializing, rescaling, localized partnerships, place based strategies and more are not substantive reterritorializations of accountability. Nor, we might add with a nod to theory, does all this constitute a profound neoliberalization of peripheral spaces and places. Rather, what emerges from hybrid, NIE meets inclusive, joined-up or weak local harmonization approaches are a plurality of what we will call *quasi-territorializations*. These are vague and ineffectual operationalizations of territorial aspects of poverty that are perverse in both their plurality, and in their failure to enable substantive practical approaches to the basic factors of poverty. These, as we see below range from the vaguely 'local' communities of interest groups imagined by 'community' and even 'neighbourhood' developers, to the 'watersheds' currently in vogue among natural resource and environmental programmes, to the areas, communes, districts or even provinces approached by development agencies in stand alone, placed based development projects.

Finally, in labelling 'quasi-' and 'more substantive' territories, we should make it clear that we are not here making a blanket argument for stronger territorial forms over weaker forms, and much less for the territory of the nation state over deterritorialized markets or civil society. What we are interested in however, and what is crucial to our conclusions, are forms of local, regional, national and global governance that are substantive enough in whatever way is necessary to be able to deliver better outcomes for poor people. Here, being able to allocate whatever slim resources are available via state or donor or local revenue contexts in efficient, population-equitable and/or pro-poor ways is a crucial outcome. By this simple standard, it should

---

Definition Box 1.4: Quasi-territories of 'joined-up' NIE governance

These include community, districts and communes (with revenue sources not matching responsibilities, with insufficient or bounded executive authority), 'areas' (as in Area/Integrated rural development programmes), neighbourhoods, watersheds, 'local-local' dialogue, moving government 'closer to the people'; health action zones, regional (or very local) devolved funds, local strategic plans, local wellbeing strategies, 'community wellbeing', PRA, local partnerships.

be clear by book's end, that many of Development's current quasi- and other territorializations are simply not adequate, are too vaguely and narrowly conceived, and produce not just ineffective delivery and account-ability, but actually create and compound gross inefficiencies and risks. These measures, as we will argue, do indeed need a measure of smart reterritorialization if they are going to accountably deliver anything for the poor.

CONCLUSIONS: EXPLAINING OUR CRITICAL STANCE

From all this, it is clear we have a critical ambit: for a number of reasons endemic to Liberalism and specific to its current incarnation, the Poverty Reduction and good governance paradigm is we think weak in its concep-tion, and frail in its execution. Perversely, it has become at once under-resourced and too expensive for what it achieves, even when judged within what we see as too narrow Liberal policy and outcome parameters.

But alongside this critical ambit, we are also mindful that Poverty Reduction and Good Governance is, in terms of being able to form final judgements of its effectiveness, a reasonably recent reformulation. Correspondingly, there exist few clear success and failures that can yet be judged. What will also become clear is that our cases consist largely of the IFIs' own leading examples, places where this approach has most spec-tacularly been rolled out. Three of four (Vietnam, Uganda, New Zealand) are in recent years absolute or plausible economic success stories, and two of them (Vietnam, Uganda) have achieved spectacular reduction of poverty over the last decade. Variously (and sometimes unrelated to the Poverty Reduction model), they have seen a commodity and tourism boom, and accelerated consumer spending and (private debt fuelled) housing and infra-structure construction (Vietnam and New Zealand). Meanwhile corporate institutions have poured in private investment (Vietnam) and in Uganda's case unprecedented volumes of public aid have gone to services dubbed 'pro-poor'. In Pakistan's case, governmental restructuring and donor largesse and demand have been driven to exceptional heights by the War on Terror. But there, despite spectacular recent economic growth, poverty has yet to shift. Political factors – especially the preponderance of author-itarian governments enacting decentralization programmes – have also been exceptional, and cannot be extrapolated from.

It should be clear, then, that we are not about to argue from our cases (nor from their limited case histories we construct) that Poverty Reduction and good governance can't or won't work. We have picked the early imple-menters to examine, yet there and elsewhere, it is impossible to discern whether the results will come from good Liberal governance, or from a com-bination of a strong state, exceptional levels of international support, and gradual liberalization. Further, as this book's critics will perceive, what we don't address are other factors that in both Liberal and Territorial perspective drive real poverty reduction: basic territorial integrity and security; geographic and demographic realities; growing productivity and competitive/comparative advantage, all within advantageous commodity/surplus chains and trading/tariff relations in wider capitalist markets; and the composition of local and international labour markets, gender relations, and other social factors shaping income distributions.

Demonstrating ultimate success or failure of Poverty Reduction could never have been our main point. But what we *have* tried to make clear are both the universal and the particular: the sources of the universal doctrines, techniques and programmes within which Poverty Reduction and good governance are enacted, and the local contexts, with the complexities and unintended outcomes arising when this paradigm is rolled out. For this reason we have gone not to where this is thin on the ground, but to where Poverty Reduction, decentralized governance and community partnerships have made a splash, where we can see the whole 'inclusive' neoliberal apparatus of Poverty Reduction elaborated. But that said, even in the success stories, within the particular contexts we focus on, there's not always much joy. There is considerable cost, there are opportunities foregone, and perversely, a great deal of cost and effort born by the poor and their close allies. And here, it's worth again reflecting that if it is thus where enormous resources are available (Uganda, Pakistan, New Zealand), what is it like elsewhere?

What this book will show is that Development has always been about pushing hopes forward on a long string of (Liberal) assumptions. Now, Poverty Reduction and Good Governance are again conveying normative assumptions into peripheries, to be tossed about by potent peripheral political and economic realities. Poverty Reduction adds new dimensions to this process and its rising complexity by pulling a thin institutionalist veil over fundamental (often territorial) aspects of poverty, and making frail compromises with territorial governance around community, local partnership and some kinds of decentralization. Poverty Reduction's current promise is that by gathering all these instruments around *Making Services Work for Poor People* this service delivery route to empowerment will build the basis for a new system of governance, one responsive to the poor, delivering them services efficiently and equitably. In some cases this may happen. But just as primary education and health, clean water and paved roads have transformed many people's lives, there are just as many examples where healthy, educated and determined people have run smack into the wall of privilege and exclusion and the simple reality that the productive economy is beyond their reach or just not there at all. In any case, for all Poverty Reduction's impressive institutional scope, the poor and their governments are firmly made responsible for their poverty, and the poor themselves are destined to be governed in local, disciplinary spaces, supplementing what they may prise from global markets with whatever scant resources flow from the now extraordinarily elaborate (but by no means all joined-up) Institutional Development Machine.

Yet our point is not just critical. We hope, rather, that while our critical perspectives and descriptions of rising complexity and fraught accountabilities will resonate with those caught up in the Institutional Development Machine on all sides, it will also clear away a little of the obfuscation that machine bestows on its servants, subjects and scrutineers alike. For, we are sure, the linked perversities of Liberal and illiberal peripheral governance and poverty are not going away any time soon.

Part I

Liberal Development and governance from free trading to 'neoliberal institutionalism'

2 Historical hybrids of Liberal and other Development, *c.*1600–1990

Markets, territory and security in Development retrospect

From colonial rule to decentralization, agents of commerce, empire and now Development have sought to frame and open up peripheral territories for commercial and moral enterprise, property and rent seeking. Creating financial, physical and institutionally secure territories of commerce meant overwriting existing territorial boundaries: deterritorializing and opening up the old, then reterritorializing in new frames and alignments. This not just in the beginning, where tribal and other territories were reinscribed into trading relationships and empire, but throughout Development history, up to and including the inclusive and institutionalist neoliberal governing arrangements we will consider in Chapters 3 and 4. But the territories and economies they sought to open up and secure were never the smooth spaces or repositories of exploitable resources imagined on the map: and nor did existing political and economic territorializations simply disappear because a gunship or a loan conditionality had arrived on the local political and economic horizon. Rather, hybrid local accommodations would be reached. Whether this was achieved by direct or more Liberal indirect rule, reinvented tradition, integrated area development programmes, or tough loan conditionalities, the places they colonized and developed were only ever partially transformed into local versions of European kingdoms, developmental states or open economy free trading nations.

Everywhere, then, there would end up being compromises, accommodations and frank failures, Polanyian lurches between Liberal market openings and security authoritarianism. But everywhere, too, there would be ongoing overreach, unproven doctrine and recent retrospect thrown forward onto new, diverse situations. To describe this later phenomenon we evoke the image of the retrospectoscope, a mythical device whose use will nonetheless be all to familiar to those doing Development. This machine always sees the world backwards, distilling all too sharp and clear lessons and doctrine from recent and longer history elsewhere in Development (or wider economics and governance), and then projecting them into new, apparently analogous situations. From Lugard to Rostow to the Washington consensus and beyond, as we will see, retrospectoscopic perspective has focused and propelled the travel of a great deal of Development vision.

In this chapter we too reach back, though with a slightly wider frame of reference, to show Development's Liberal and territorial genealogy. But our aim is not to construct comprehensive historical narrative;[1] rather, to set up the discussion for later chapters, we have picked situations that resonate in contemporary Poverty Reduction. They too are hybrid accommodations of Liberal and territorial, travelling and patrimonial governance, configured within wider, variously Polanyian swings in political economy. Our selective history is also weighted towards Poverty Reduction Development's mostly 'Anglo' (British/American) antecedents that were variously visible in the piratical and increasingly security-fraught free trading of early Empire; in Adam Smith's faith in markets and fears over the revolutionary ferment of 'unsocial passions';[2] J.S. Mill's romantic and increasingly 'positive' Liberalism of capability; and, closer to the present, in the post-Second World War 'embedded Liberal' welfare regimes of the New Deal/Keynesian/non-communist welfare state. The first section of this chapter sketches British imperial enterprise and governance, and moves through Fredrick Lugard's *The Dual Mandate* which, from the 1920s, was the Empire's *post hoc* governance manual. In the second section we shift forward to post-Second World War American 'establishment' or 'embedded' Liberalism (the creation of Bretton Woods institutions, the Truman doctrine, the Marshall Plan) through Harrod-Domar financing gap models, into Rostow's *Stages of Economic Growth*, to Robert McNamara's 'Defensive Modernization and Poverty Alleviation' following the US war in Vietnam. In the third, the chapter ends around 1990, following the ascendancy of footloose capital and its minimalist governing frameworks: neoliberalism, marketized governance economics and the Washington Consensus.

LIBERALISM AND (BRITISH) IMPERIAL DEVELOPMENT IN THE COLONIAL RETROSPECTOSCOPE

> . . . famine has never arisen from any other cause but the violence of government attempting, by improper means, to remedy the inconvenience of dearth.
>
> Adam Smith, *The Wealth of Nations*, 1759 (1976)

As the Oxford English Dictionary reminds us, long before Liberalism's political maturity in the nineteenth century, liberality was a characteristic 'befitting free men, noble, generous': most of whom sociologically belonged to urban, emerging middle and property owning classes. Liberty's earliest champions, Locke, Montesquieu, Smith, Rousseau, Jefferson and Franklin, were a mix of moral philosophers and bourgeois radicals, who pitched their political and ideological struggles – and their rationalist, utilitarian, free trading and rights frameworks – against the historical, traditional interests of territorialized property.

Early Liberal pamphleteers, constitutionalists and political philosophers couched their arguments in universal humanist terms and legal constitutional frames that partly sanitized their revolutionary anti-royal fervour. This Liberalism was modernist and forward-looking, its activists were prepared to overwrite the traditional territorial dominance of aristocracy and squirearchy, and frame universally applicable laws, property and citizen rights. But this wasn't just a struggle over rational rights and rules: class and territorial struggles involve real resources. What was ultimately politically feasible were various accommodations with the territorial powers that, (at least in Europe) even after a couple of revolutions, wouldn't go away. In this accommodation, traditional property rights were usually saved, and the middle classes were safely and securely included in governance via a gradually expanded franchise. Then as today, Liberal governance didn't agonize over questions of how asset, privilege and opportunity had been acquired. Nor were Liberals radically redistributionist. Rather, reforms stabilized existing property dispositions by incising the status quo in property-for-market rules.

By a trick of history, Liberal law also ended up ascribing the same property rights to corporations as to individuals:[3] the uneven and enormous powers of corporations of all shapes and sizes thus became the major vehicle of market power and property ownership that was legitimated by Liberal doctrines of equality and universality. As far as most of the law was concerned, they were just one market player among many, whatever their scale and reach. As Marx knew, where equal rights exist, force decides; the same emancipatory and universal doctrine of equal rights became a bastion of ideological legitimacy for market interests operating in Empire's new territories. Thus Liberalism emerged as the remarkably resilient, ideologically effective counterpart to capitalist expansion. Freedom, democracy, respect for human rights, security, the rule of law, and even inclusion and social justice could be enfolded and hybridized under its wings. This, even while it was legitimating the virulent marketizing of social relations, unequal market exchange and contest between the poor and socially unaccountable corporates, and the accumulation of vastly unequal property on a global basis.

The British empire began a very long way from the kind of pomp-and-circumstance militant Toryism it became, once Kipling and the 'white man' jingoists had puffed it in the imperial imagination, long before dominions, nation states, colonies and protectorates had been lined up to colour the globe pink. Territorial empire began as a reaction, a game of catch-up governance that was often reluctantly rolled out in response to trade or finance opportunity and competition, national-cum-imperial prestige, and European security issues and humanitarian concerns. From India to New

Table 2.1 Generic Liberalism

Liberal politics: agendas and strategies	Liberal governance: techniques and tactics
Defending property rights of middle and merchant classes, international and local capital and corporations	Framing and enforcing universal laws of human and contractual rights, to which every individual and social entity everywhere (kings, countries, the poor, the property-less) must subscribe; enforcing security around these; keeping existing market/power/property/resource inequalities and arrangements (corporate power, competitive advantage, capital dominance) off political and international agendas.
Enabling capital access to (and from) markets	Free trade, economic integration, free movement of capital; promoting global rules, market 'institutions'; strategic use of security blocks and domains.
Legitimating/securing Liberal politics versus other ideologies/agendas	Representing statism as failed; defining patrimony as corruption; showing how Liberalism responds to and reduces poverty; stressing 'ownership' and 'commitment'; monitoring, services, minimal safety nets for poor; 'enabling' poor to participate in markets.
Defending Liberal arrangements from state/ territorial/political interests	Privatization, decentralization, fostering (local) competition and voice; demanding conformity with international (Liberal) norms and frameworks; using participatory and executive means to govern; favouring technical rather than political means; making loans and grants conditional on Liberal reforms/security alignments; managing social dislocation by efficient safety nets; demonstrating that asset redistribution is inefficient.

Zealand to Africa, traders, slavers and land jobbers raced well ahead of politicians and colonial office officials. In doing so, they presented officials with difficult obligations and overstated incentives to step in, formalize boundaries and rule, secure trade and mitigate the abuse of locals. What, then, to do with already present religions and systems of rule and ownership? Should they be indulged, or perhaps erased altogether so as to impose a fresh, clean Liberal order based on 'individual liberties, [and] respect for the rule of law'? Many options were debated, and tried. From early colonial experimentation to the present, we see Liberal principles reaching variously convenient accommodations with expediency. It was soon understood that the 'thin white line' of imperial power had to be made to rest not just on British force and fortitude, but more firmly in the crafting of a colonial culture of governance, as we will see, often from the vestiges of already existing pre-colonial governing orders.[4]

By the late-Victorian imperial heyday, the die was already cast. It had begun more than two centuries before, in the East India Company's trading outposts. Private trading interests, defended by private militias and armed company merchantmen, held official charters that gave them trade monopoly and rights to piracy and punishment of whoever interfered with trade.[5] It was raw market enterprise, owned and financed from the heart of the City of London, but it was always generating territorial conflict. And, getting caught up in European wars and imperial competition, which set British traders abroad against Portuguese, Spanish, Dutch, French and other commercial/national interests. At the East Indian edges of trading empire, early exchanges produced hybrid accommodations: forms of custom, dress and ways of doing business. By the mid-1800s, however, the piratical-cum-military power of these companies overwhelmed their Eastern imperial hosts. By the time Robert Clive had turned the East India Company's operations into a much more definite domain for plunder, and Warren Hastings had formalized consequent arrangements constitutionally, the broad pattern for Malaya, New Zealand and equatorial Africa had been established. Commercial overreach and social disruption would lead to cries for military, governmental and moral security, prompting the drawing of sharp lines on vague maps, sending a military emissary and backup, and inviting humanitarianism to tidy up edges and legitimate a morally shaky enterprise. What emerged were not Westphalian nation states, but various other arrangements: protected and 'free' trading entrepots, possessions, protectorates, colonies, all territorial domains not meant to stand alone, but to be subordinate sites of accommodation, access and security within wider imperial relations and rivalries. Their governance and territorial boundaries were formed, indeed malformed, to these ends.

Commercial, security, moral and fiscal concerns dominated Imperial discourse about colonial states. Through the latter nineteenth and early twentieth centuries, Liberal colonialists fought running battles with shifting, converging Great British Empire conservatives on the right, and 'radical' missionary/Fabian socialists on the left. Liberal and conservative colonial interests both wanted to make colonial governance small and cheap, or at least fiscally neutral. For the conservative, latterly staunch imperialist Disraeli in 1852, the colonies could be 'a millstone around our necks'.[6] Gladstone, Victorian Liberal potentate, saw colonial governance in tight 'moral pocketbook' terms, while Joseph Chamberlain moving to the Colonial Office in 1895 was soon 'cooled by a douche of cold water' from Treasury.[7] Eventually, however, the costs of governing rose anyway.

But even if imperialism didn't pay, it might be a moral imperative. Liberal and evangelical humanitarians about in the periphery saw demoralization and harm in expansion, but feared *laissez faire* would result in 'fatal impact'. Church Missionary Society envoy Samuel Marsden wrote from 1830s New Zealand of 'no laws, judges, nor magistrates; so that Satan maintains his dominion without molestation'.[8] Many in the Colonial

Office fought the Wakefields, as they sought to re-establish the English class system and capital-land-labour relations over the top of free-for-all quarry economies and tenacious Maori sovereignty in New Zealand. And before 1890, in what became Uganda, British and French missionaries, out ahead of traders, had created such conflict through competitive evangelization, that fear of instability propelled a reluctant British government to supersede the bankrupt British Imperial East Africa Company.[9] Yet Liberal and conservative imperial doctrine converged on the notion that most trade would bring moral and material benefits for both colonizer and colonized.[10]

The 'governance minimalism' of Liberal colonialism often got it into trouble in these peripheries. Its free marketism too was frequently calamitous: the exacerbation of catastrophes including the Irish potato famine and the great El Nino famine of 1877 in India can fairly be laid at its door.[11] The advent of the railway had facilitated at once the development of grain as an export cash crop, and its movement away from drought stricken areas to speculator's grain depots. In 1877, while the famine raged and Indian grain exports to Britain reached an all time high, soldiers separated the starving from dockside granaries. As in Ireland, Liberal Viceroy Lord Lytton considered any intervention to save starving peasants a sin against the invisible hand, and, mid-famine, infamously staged 'the longest and most colossal meal in world history', a week long feast for 68,000 officials, satraps and maharajahs celebrating Victoria's elevation to Empress of India.[12]

Then as now, the hypocrisy of Liberal trade could be obscene, as Britain's largest trading partner in the late 1800s discovered. India was where Millite Liberalism initially triumphed as the governing rationale, displacing conservative Orientalism (which favoured retaining more traditional governance). But what in Manchester was called free trade in Berar India meant the disembedding of cotton production from *Balutedari* reciprocal social orders and fixing them into extortive pro-Lancashire governance and tax arrangements and monopsony purchasing. Meanwhile, India's own development of cotton manufacturing was assailed by the insistence of a Manchester Liberal lobby on free import into India and tariffs at home on Indian cotton manufactures. As expansive railways brought floods of English cotton to India's interior, Indian cotton farmers' children went naked. Again, Lytton appears as mad Liberal rationalist, when in the El Nino famine of 1879, he 'overruled his entire council to accommodate Lancashire's lobby by removing all tariffs on British made cotton, despite India's desperate need for more revenue in a year of widespread famine and tragic loss of life throughout Maharashtra'.[13] In fact, as we will see in Pakistan, Liberal governance, hybrid indirect rule and military occupation would each be deployed in different times and places, as security and frontier issues jostled trading and humanitarian concern. But as Davis notes, the verdict on the entire Indian Liberal colonial experience can be read in its per capita growth legacy: none between 1757 and 1947.[14]

Governance was required, both to contain indigenous rights and to expand those of settlers. As in East Africa, 'closely connected with European colonization is the question of native rights': but the natives 'must not be allowed to straggle over huge areas'. Rather, said Sir Charles Eliot in his 1905 The East African Protectorate, they 'must be protected from aggression . . . but with this proviso . . . I think we should recognize that European interests are paramount'.[15] Perhaps the best that Liberal empire managed was a Treaty with New Zealand Maori, guaranteeing them a measure of undisturbed 'sovereignty' over land and resources. Even this became irrelevant for a century after 1863, when land-grabbing agricultural interests precipitated conflict and formed militias that soon received imperial military backing. Provoked and shamed, Liberals and conservatives alike in the late 1800s reluctantly accepted the cost of colonial governance. Many, like Chamberlain and even Disraeli, developed a powerful taste for empire.

All this colonial misadventure is nonetheless remembered as Liberalism's golden age, the Pax of Britannica, the stable convertibility of gold standard, Polanyi's 100 years of peace. The following period of nationalist embedded competition, including Liberalism's collapse, Chamberlain-esque investment in basic infrastructure and security, the rise of Tory imperial protectionism and Imperial trading preference, and the embedding in a post-war security order of newly independent nation states, was for the colonies a better time. This, even if English manufacturers actively opposed colonial industrial development, and existing unequal terms of exchange (raw colonial commodities for core manufactures) were frequently misrepresented as mutual, free, and complementary. By the 1920s, 'complementary' colonial Development's time had come.

'A rough idea of the provinces': Lugard, Indirect Rule, *The Dual Mandate*, and Trusteeship at the edges of Empire

We develop new territory as Trustees for civilization, for the Commerce of the World.

Joseph Chamberlain, from the dedication
verso of Lugard, *The Dual Mandate*, 1922

The British Empire does not stand for the assimilation of its peoples into a common type, it does not stand for standardization, but for the fullest, freest development of its peoples along their own specific lines.

General Jan Smuts, *Rhodes Memorial Lectures*, 1929

The 'fullest, freest development', by which Smuts meant separate but equal institutional development, is now remembered as the recipe for apartheid. When Smuts spoke, however, its most evident expression was in the British colonial practice of Indirect Rule, which we want to show is a paradigmatic antecedent of current Liberal development doctrine. As we will see in Lord Lugard's key text *The Dual Mandate*, Indirect Rule was an accommodation between Liberal governance and territorial power; it was a 'decentralized despotism'[16] that claimed *The Dual Mandate* of being good for both core and periphery whilst projecting the explicit interests of security and trade.[17] In this, it created semi-autonomous, quasi-territorial protectorate domains into which a European patrimonial, indeed sovereign, imagination was projected. In these domains, indigenous semi-subjects received less than full Liberal rights, but most importantly, were stabilized and secured in their local places.

As Mamdani notes, late colonialism brought a wealth of experience to Africa. 'By the time the scramble for Africa took place, the turn from a civilizing mission to a law and order administration, from progress to power, was complete'.[18] For the young military/missionary adventurer Frederick Lugard, however, learning to govern in Africa meant mixing Indian experience with governance initiative based on first principles. Drawn into Africa by moral abhorrence of the slave trade and a desire for destiny, he was quickly caught up in boundary definition projects: militarily pioneering the formation of Protectorates, waiting for the Colonial Office and its army to come in behind. Arriving from India at the peak of Uganda's religious conflict, he issued guns to the Protestants and precipitated the bloody slaughter of Catholics in the1892 Battle of Mengo. But how to rule and how to create order? Quickly, he learnt that colonial rule required turning existing chiefs into a convincing ruling class, 'entitled to hold sway over their subjects not only through force of arms or finance, but also through the prescriptive status bestowed by neo-tradition'.[19] Thus, in the newly defined domains of Uganda and then Nigeria, he was impelled into crafting systems of governance not from scratch, but rather, as had been learnt from the construction of the Raj during the dying years of Mughal India, from re-inventions of existing traditions of governance, through which indirect rule might operate. Famously, he reinvented not just traditional governance, but crafted a marvellously formal simplification that travelled back to impact on British governance too.[20] As we will see, Lugard's paradigmatic colonial policies became most clear and influential after *The Dual Mandate* was published in 1922, when he had had repose to reflect and objectify comprehensively. *The Dual Mandate* was in these terms a classic 'retrospectoscopic' view. It pared down, distilled and selectively recovered experience from one place and projected this forward as a formalist analogy relevant elsewhere in more universalist terms. In 1920s and 1930s British colonial Africa, it was referred to precisely as the practical technical manual of governance: a travelling rationality, deemed 'suitable for export' to the rest of the imperial administration. Indirect rule was appealing because it promised the difference, in practice, between the chronic expense of 'domination', and 'governing', by instrumentalizing and directing others' capacity for action. In this, it showed that British colonialism more than any other keenly understood the governing possibilities of 'culture'. For those so governed, the most lasting effects of indirect rule were psychological: for where direct foreign control tends to unite and forge identity through resistance, indirect rule encourages people to despise indigenous leadership and undermine its legitimacy.[21]

In 'Nigeria', so first named on Lugard's letter of appointment as governor on 1 January 1901, the rights to rule the country had to be bought from the private Royal Niger Company, for whom military conflict with the French had proved too much, despite a profitable open charter on settlement and trade. As Colonial Secretary Hicks Beach described to a House of Commons considering funding the new arrangements, 'the company had founded an empire', a barely mapable domain of a million square miles and 30 million people of starkly differing ethnic and religious affiliations.[22] Anticipating his governorship, among Lugard's first actions was the drawing of the outlines of his domain with 'coloured chalk and pencil' on a series of maps he annotated 'Northern Nigeria as we took it over 1 January 1900. The greater part quite unexplored. The rough idea of the provinces is indicated'.[23]

For Lugard writing in 1922, '[t]he principles which should guide the controlling powers in Africa' were recognizably three-legged. These were expressed in the Berlin Act of 1885, which Lugard approvingly quotes Keith describing as 'aimed at the extension of the benefits of civilization to the natives, the promotion of trade and navigation on the basis of perfect equality for all nations, and the preservation of the territories affected from the ravages of war':[24] inclusion in civilized society, open integrated global trade and security. On the ground, what Mamdani recognized as a much more substantive 'regime of compulsions' operated. Here, Africans had to be liberated from slavery and incorporated into a different, Livingstonian three-part ambit, Christianity, civilization and commodity, which led the way in the cotton and commodity famine aftermath of the American civil war.[25] Despite that 'Europe is in Africa for the mutual benefit of her own industrial classes', he believed that the 'genius of the English' was that 'the benefit can be made reciprocal, and that it is the aim and desire of civilized administration to fulfil this dual mandate'.[26]

As in today's Poverty Reduction, concerns about open markets, well-being, basic infrastructure, good governance and building human capital are to the fore:

> By railways and roads, by reclamation of swamps and irrigation of deserts, and by a system of fair trade and competition, we have added to the prosperity and wealth of these lands, and checked famine and disease ... We are endeavouring to teach the native races to conduct their own affairs with justice and humanity, and to educate them alike in letters and in industry.[27]

To be fair, while the high moral sentiment is there, for the most part, in page upon page, precept upon precept, *The Dual Mandate* firmly addresses the technical over the political.

Lugard's Africa was a fluid, uncertain world of threats and contests, over which white rule is urged to provide the obverse: discipline, stability, planning and truth. The dual mandate was to protect commercial interests, but the security threats posed by further rapid adjustment are always prominent among concerns: warnings about 'effects of sudden emancipation', 'avoidance of sudden change', that 'change must be gradual': or, economic transformation, political order, social stability.

Like his British Raj forebears, Lugard realized that to maintain order with a miniscule staff and budget he would have to rule 'indirectly', or through existing native institutions, most notably Fulani chiefs, and associated 'native courts'.[28] Here, Lugard's retrospective, Eurocentric knowledge of the stages of social evolution would point the way to good governance: some of 'the finer Negro races' had 'reached a degree of social organization which ... has attained to the kingdom stage under a despot with provincial chiefs of the feudal type'. Worse, Lugard's mistaken colonial recognition of local best practice would become the rubric for restructuring governance elsewhere. Lugard's disciple-biographer Marjorie Perham could claim that Indirect Rule 'demanded the utmost possible adaptation to the immense varieties of African society'.[29] Like today's advocates claiming a 'no blueprints approach' for Poverty Reduction while pressing its claims as a 'comprehensive development framework', Lugard stridently rejects one size fits all, even as he lays out Imperial precepts: 'Principles do not change, but their mode of application should vary with the customs, the traditions, and the prejudices of each [administrative] unit'.[30] In practice, Mamdani notes, the British worked with a single model of customary authority across Africa that mirrored images of traditional European monarchy and patriarchy. These artificial archaic reterritorializations 'presumed a king at the centre of every polity, a chief on every piece of administrative ground, and a patriarch in every homestead or kraal'.[31] Ironically, this model first ran into trouble close to home in 'hydra-headed' Yoruba and then dense forest 'acephalic' Ibo territory, both polities adapted to surviving the plunder and pillage of slave trading.

One of Lugard's enduring legacies (potent in 2004 as we will see in Pakistan) was a departure from the Liberal code of separation of political, executive and judicial powers. Elsewhere, in India, for 160 years, from 1786 to 1947 (and beyond) Liberal principles and the practical needs of administration made an uneasy concession to convenience.[32] In both Indirect Rule, and the core Native Courts system, regardless of local precedent, Lugard united the role of ('traditional') political leadership with executive and judicial power. It was, Lugard claimed, 'obviously unavoidable'; for the 'separation of these functions would seem unnatural to the primitive African, since they are combined in his own rulers, and a system which involved the delay caused by reference, even in minor cases, would be detested'.[33] By fusing legal, military and political power in the personages of chief and protector, revenue extraction and unequal exchange were legitimated. These themes and legacies we will see reaching right to the present day. Combining offices and functions was also a matter of governing by (travelled) analogy:

> The government is constituted on the analogy of the British Government in England. The Governor represents the King, but combines the functions of the Prime Minister as head of the executive. The councils bear a resemblance to the Home Cabinet and Parliament ... [Only at the local, depoliticized and technical level was any separation useful: there, the detailed work] ... is carried out by a staff which may be roughly divided into the administrative, the judicial and the departmental branches.[34]

Decentralization too was justified in *The Dual Mandate*, as it is in doctrine today by principles of subsidiarity and allocative efficiency: 'The Man who is charged with the accomplishment of any task, and has the ability and discrimination to select the most capable of those who are subordinate to him, and to trust them with ever increasing responsibility, up to the limits of their capacity, will be rewarded not only with confidence and loyalty, but he will get more work done, and better done [*sic*], than the man who tries to keep too much work on his own hands ...'[35] At the same time, 'In applying the principle of decentralization it is very essential to maintain a strong central coordinating authority, in order to avoid centrifugal tendencies, and the multiplication of units without a sufficiently cohesive bond'.[36]

The Dual Mandate found its way into 'every British African headquarters, central and provincial'.[37] As Chanock remarks 'through it events were understood and guided, and because of this certain things could happen and others could not'.[38] But time quickly got the better of Lugard's prescriptions and in twenty years, the politics of white rule would be fatally impaled on rising national political activism.[39] Typically, Lugard himself sought to contain such notions within an evolutionary scheme: 'If there is unrest, and a desire for independence, it is because we have taught the value of liberty and freedom, which for centuries these peoples had not known. Their very discontent is a measure of their progress'.[40]

Empire then advanced both by direct rule and integration of territories and subjects, and by more Liberal, less taxing accommodations, involving setting up quasi-autonomous territories bearing just enough resemblance to imperial structures to be seen as legitimate. At the same time Indirect Rule, 'rule through its own executive government', was from the outset linked to the reterritorialization, that is, the incorporation, of these regions into peripheral, 'traditional' and subsidiary zones within the global capitalist economy. Here, the boundaries of the colony or protectorate, the regions of 'tribal authority', the zones marked out for peasant cotton production or local rule, all formed 'quasi-territories' wherein the locals could be securely enough contained.

By the end of the nineteenth century, between Britannia's rule of the waves (and the money markets) and empire's 'dotted-lines-on-the-map' local reterritorializations, a regime of free trading developed that was plausibly global. But territory-market relations were even then deep in the throes of another twist. As Polanyi relates, imperial territories became increasingly embedded in imperial rivalries, exclusive imperial trading regimes and ultimately imperial wars. After the Liberal order fell to pieces, these same places would re-emerge as sites for new doctrines of national liberty, territorial independence and autonomous development, once again security-tied to trade. Meantime the development results, as ever, were uneven: well beyond the Second World War, former colonies from New Zealand to Africa remained narrow, commodity driven economies, some better off than others, some more independent than others, but all tied for economic and security reasons viscerally to European Empire. And all rocked horrendously by core economy shocks and deflations, in the 1890s and again in the 1930s.

THE QUEST FOR FREEDOM AND SECURITY: THE TRUMAN DOCTRINE, EMBEDDED LIBERALISM AND LIBERAL ESTABLISHMENT DEVELOPMENT

> Not only has the dominion, in common with the rest of the world, suffered such devastation of unparalleled depression, but economic changes of such magnitude have taken place so that the very psychological outlook of the people has changed also. Today the common conviction in New Zealand and in other countries too, is that economic forces cannot be allowed to operate without restraint or regulation. There is a determination that such forces must be rationally controlled as far as is humanly possible to control them and that the sole aim and object of such control should be the provision of the highest possible standard of living consistent with a nations natural resources and its ability to utilize them efficiently. I feel certain that the realization of this objective insofar as it can be realized within the bounds of individual national economies will be a big step towards a more ordered, just and peaceful economic system in the wider sphere of international relations.
>
> Hon. Walter Nash, *New Zealand*
> *Minister of Finance*, 21 July 1936

Karl Polanyi's retrospective, 1944 account of the collapse of the Liberal order described how its pillars, the balance of power, the gold standard, the free and self-regulating market, the Liberal state, and pax Britannica, were each assailed by reactionary forces of national protection, security paranoia, geopolitical and populist power.[41] His account of the 'double movement' described the limits of Liberal market economies and broad-based social backlash that acted both to re-embed and ultimately undo Liberal market arrangements nationally and internationally. This may have been a valid analysis of 1930s and 1940s Europe, but what sorts of embedding were possible for the post-Second World War territories of the colonies?

The colonial world features only on the periphery of Polanyi's analysis. Indeed, his Eurocentrism meant he could describe the period from 1815 to 1915 as 'a hundred years' peace'. In his analysis, the colonial world was a 'not yet developing' terrain in which 'imperialist rivalries' were the last 'disruptive strain' (after first world 'unemployment', 'tension of classes', and 'pressure on exchanges') that was contributing to the desta-bilization of core political economy, and generating 'double movement' reaction.[42] After the Second World War, the core powers lurched towards a Cold War standoff settlement, within which economies on both sides would prosper. But insecurity concerns, generally heightened and over-spilled everywhere. By the late 1940s in Iran, Greece and Western Europe, in the 1950s and 1960s in South East Asia, Latin America and Africa, the world would focus on the shaky margins and security borders.

Certainly the cogent lessons of insecurity in the great depression and the Second World War left their mark in primary concerns with multilateral governing of global finance markets, and of newly independent states. What emerged was a security order tied directly to both multilateral and nation state levels of scale, within which Liberal market and governance arrange-ments would be ensconced. As Ruggie's influential (1982) discussion sug-gests, post-war 'embedded' Liberalism was a 'multilateral nationalist' compromise. Under 'embedded' Liberalism, Liberal institutions (markets, trade, law, rights) were ensconced within two primary domains, both with potent territorial dimensions. One was the nation state, with its active national political and economic interventionism (national industries, pro-tection, Keynesian management, import restriction and substitution). The second was wider Cold War economic and political security arrangements: North Atlantic Treaty Organization, Bretton Woods, the UN, General Agreement on Trade and Tariffs (GATT), preferential arrangements for Japan, reconstructing Europe, and the British Commonwealth and wider security bloc rivalries.[43]

The peripheral nation state would emerge as a key territorial scale, a 'power container' aligned within wider security relations. In this container of economic territory, growth could be measured (GNP terms), investment requirements could be calculated (via Harrod-Domar investment gap models), investments could be made in national enterprises, and governing elites' ideological allegiance and economic dependence could be secured.

How these new 'states' would fare in this new context was not at all clear: but competing ideologies all promised modernity, prosperity, productivity, technology, education, national self-sufficiency, and union with other strategically aligned (or expressly non-aligned) 'democratic republics'.

In retrospect, the results were of course grotesquely uneven: for every nationalist success story, there were failed states where colonial mal-territorializations and pre-colonial ethnic territorialism caused seismic damage to the thin lines on the map. In the periphery, both 'freedom and democracy', and 'socialist self-sufficiency' turned out to be codes for brutal, self-serving governance protected by security bloc largesse. For many, the period was hardly Liberal at all, as Liberal concerns were not so much 'embedded' as submerged in mercantilist or plan-driven economic politics. Trade, capital flows, governance, citizen rights and what was then referred to as 'poverty alleviation' were subordinated to ideology and security relations in ways that concealed and abetted core countries' own economic nationalism and protectionism, unequal exchange, tied aid, and new global orders of dependency. In this context, Development emerged as a crude, unformed device, its ideological doctrines readily reducible to simple practice formulae and naive framings of territory, inputs and outputs, and expected benefits.

Bretton Woods, debtor adjustment and the Liberal establishment

> The period . . . 1941 through 1952 . . . was one of great obscurity to those who lived through it . . . The significance of events was shrouded in ambiguity. We groped after interpretations of them, sometimes reversed lines of action based on earlier views, and hesitated long before grasping what now seems obvious. The period was marked by the disappearance of world powers and empires, or their reduction to medium sized states, and from this wreckage emerged a multiplicity of states, most of them new, all of them largely undeveloped politi- cally and economically. Overshadowing all loomed two dangers to all, the Soviet Union's new-found power and expansive imperialism, and the development of nuclear weapons.
>
> Dean Acheson, *Present at the Creation*, 1987[44]

In the US, and only the US, by the end of 1944 the elements of the next thirty years of 'embedded' Liberal hegemony were all in place. Here was a Fordist mass-productivism, a powerful creditor nation with free enterprise (though not free trade) ambitions; the national social security economy of the New Deal; the potent, upscaled and marvellously rational experience of national wartime production dirigisme; the beginnings of Keynesian demand management; a powerful national industrial military complex closely linked to government; and a rising middle class enjoying a social wage. All this, yet no apparent imperial designs: rather, the talk was of a free alliance with other nation states, abetted by shared trade. But here, too was the lurching between hubris, overreach, and recoil, between cranky unilateralism and reluctant multilateralism. As Acheson remarks, the new order from the start was something groped after, developed in only partial awareness of its real, underlying conditions. In this the unseen, contradictory confidence of national productionism facing the uncertainty of international security would nonetheless result in a remarkably stable international order. At least, in the core of that order.

With British financial and trading power crippled by the war and overhung by Clause 7 of the Lend Lease arrangements, the Bretton Woods agreements were the result of an uneven Anglo-US contest of ideas and policy leverage. American political-economic mastery saw the establishment of two Washington-based and weighted institutions. Both were products of reaction to pre-war financial chaos and failure of multilateral imagination and will. Both ultimately up-scaled the relative successes of the past, the Commonwealth and Schachtian trading alliance convertibility; a rationalized and limited 'gold' currency standard; and cheap capital for infrastructure-led recovery. The International Bank for Reconstruction and Development (later the World Bank) ultimately emerged as a highly discretionary policy instrument for its backers, an institutionalized Marshall Plan 2 that bankrolled strategic, reconstruction and development initiatives of US-aligned nations. The IMF seemed more important because its functions were simpler and clearer. It held the restricted task of maintaining a stable, convertible currency regime based on both gold and the dollar. The goal, said US Treasury Secretary Morgenthau, was to 'drive the usurious moneychangers from the temple of international finance'.[45]

As a stabilization fund, rather than Keynes' hoped-for global clearing bank, the IMF allocated drawing rights and political power according to subscription. As Keynes among others emphasized, this shifted moral hazard and adjustment responsibilities firmly onto debtors, rather than creditors, thus safeguarding above all the US position as global creditor, and with this, bankers' and capital interests.[46] From this point on, nations in balance of payments deficit drawing down subscriptions or seeking support should expect conditionalities and (structural) adjustments in return. It set a pattern for the future: 1980s structural adjustment, for instance, would squarely lay responsibility for Development's problems at the feet of poor countries, just as Poverty Reduction in the late 1990s would firmly and with great moral purpose deposit poverty's problems at the doorstep of 'local community'. Countries that could maintain trade surpluses came under no similar countervailing pressure. We may never know what creditor adjustment leverage might have achieved. It would certainly have produced a political roughhouse. Economically powerful states would have been subject to both the combined multilateral pressure of their weaker debtors and dependant trading partners, and to the contradictory disaffection of politically powerful protectionists at home. As things turned out, slow progress with restoring convertibility to current and capital accounts, and related, initially modest development of international financial transfers meant that the harshest impact of debtor adjustment would be delayed thirty years. Nevertheless the Bretton Woods institutions would enjoy a period of considerable stability, which would assist their mission and credibility in maintaining financial stability, security and growth.[47]

As the 1940s moved on, uncertainty over Soviet spheres of influence quickly turned mutual paranoia into mutual expansion and containment policies, with the atomic bomb adding gravitas to fear. By 1946 core business in US and US aligned governance centred around security, around thoughts about what poverty and starvation in Western Europe or Japan might mean for the growth of communism. In both strong and wrecked core economies, the US and Europe, the working together of security, reconstruction and production concerns was remarkably successful. The rest of the world was always either peripheral or opposed to this bloc, and exercised what limited productive, financial or security leverage this position allowed.

This section tells the story of some of the development doctrine and practice which emerged in this period. We follow others in reflecting the privileged place of powerful, mostly male, American Liberal establishment figures: New Deal Liberals, the East Coast establishment, the Bretton Woods architects and author of *The Stages of Economic Growth: A Non-Communist Manifesto*, Walt Whitman Rostow, and Robert McNamara, who moved from being President of Ford Motor Company, to Kennedy and Johnson's Secretary of Defense, to lead the World Bank's first Poverty Alleviation phase.

The Truman doctrine

Britain, exhausted, crippled by war debts, facing an uncertain imperial future, signalled in 1946 it could no longer fund military and other aid to leftist insurgency-troubled Greece. The political machine around Harry Truman exploited growing paranoia over Russian expansionism to secure, via the 'Truman doctrine' announced in 1949, massive aid to Greece and Turkey, and, almost incidentally, a wider international development mandate.[48] Truman's famous speech, crafted by Dean Acheson's State Department to be 'clearer than truth', was designed to scare an isolationist Congress into expensive internationalism.[49] It is thus a founding document of Development in an age dominated by issues of containment, oil politics, and accusations of terrorism and imperialism:

> The very existence of the Greek state is today threatened by the terrorist activities of several thousand armed men, led by Communists, who defy the government's authority ... Since the war Turkey has

sought financial assistance from Great Britain and the United States for the purpose of effecting that modernization necessary for the maintenance of its national integrity. That integrity is essential to the preservation of order in the Middle East . . . At the present moment in world history nearly every nation must choose between alternative ways of life. The choice is too often not a free one. One way of life is based upon the will of the majority, and is distinguished by free institutions, representative government, free elections, guarantees of individual liberty, freedom of speech and religion, and freedom from political oppression. The second way of life is based upon the will of a minority forcibly imposed upon the majority. It relies upon terror and oppression, a controlled press and radio; fixed elections, and the suppression of personal freedoms. I believe that it must be the policy of the United States to support free peoples who are resisting attempted subjugation by armed minorities or by outside pressures. I believe that we must assist free peoples to work out their own destinies in their own way. I believe that our help should be primarily through economic and financial aid which is essential to economic stability and orderly political processes.[50]

In the Truman docrine, security fears powerfully linked 'independent' national development to active multilateralism, in ways that the US Congress would fund.[51] Its central contradiction, between national economic and political independence and peripheral subordination to the security and industrial needs of hegemonic power, looks back to *The Dual Mandate* and forward to PRSP, emerged in US Senate hearings. Acheson's ally Senator Connelly 'helpfully' prompted that: 'This is not a pattern out of a tailor's shop to fit everybody in the world and every nation in the world, because the conditions in no two nations are identical. Is that not true?' Mr Acheson: 'Yes sir, that is true . . . [requests] have to be judged according to the circumstances of each specific case.'[52]

The resulting Marshall Plan committed billions of dollars to the reconstruction of Europe. Expressly designed to rescue and wrest Western Europe from Communism and into American influence and markets, the Marshall Plan couched its offer as open to all Europe, indeed as something to be 'European owned'. As Marshall announced:

Our program is not directed against any country or doctrine, but against hunger, poverty, desperation and chaos . . . Before the US government can proceed much further there must be some agreement among the countries of Europe as to the requirements of the situation . . . The initiative, I think, must come from Europe.[53]

Soviet reaction followed the script: threatened by capitalist encirclement, Stalin prompted 40 years of mutual paranoia, reaction and overreaction. The International Bank for Reconstruction and Development cut its teeth on loans for rebuilding Western Europe. 'Wise Man' John McCloy, Bank President just eight months into its life, argued its investments in Europe would solve the linked problems of creating markets for US trade, curing the dollar surplus and stopping communism.[54] The World Bank's first loan, $250 million to France, was concluded in close collaboration with the State Department, coming, 'not coincidentally, only hours after the French government forced communists out of government'.[55] The US subsequent bolstering of Bank and IMF into premier development institutions arose in part from relatively weak US leverage over the UN organizations. The World Bank's ability to 'move the money' rapidly to build strategic alignment, backed and window dressed by 'technical assistance' attuned to Liberal agendas has, as we will show in chapters to follow, always been well masked by protestations and redefinitions of an 'apolitical' role.

Beyond security, the famous fourth point of Truman's 1949 inaugural address embodied US imagination about the Promethean power of technical knowledge and industrial production to carry the world beyond colonialism:

> Fourth, we must embark on a bold new program for making the benefits of our scientific advances and industrial progress available for the improvement and growth of underdeveloped areas. More than half of the people of the world are living in conditions approaching misery. Their food is inadequate, they are victims of disease. Their economic life is primitive and stagnant. Their poverty is a handicap and a threat both to them and more prosperous areas. For the first time in history, humanity possesses the knowledge and the skill to relieve the suffering of these people ... our imponderable resources in the technical knowledge are constantly growing and are inexhaustible ... The old imperialism – exploitation for foreign profit – has no place in our plans ... Greater production is the key to prosperity and peace. And the key to greater production is a wider and more vigorous application of modern scientific and technical knowledge.[56]

Henceforth, leaders of Western and newly emerging post-colonial states were assured that 'the technical and organizational imperatives of Western industrial development' would weaken and destroy any obstacles and impediments that may stand in the way of transformation.[57] This three-legged rubric, national economic productionism, progress through scientific technique, societal modernization for peace, conceals as much as it reveals, hiding market power and tied aid behind the universal rationale of production, the political behind the technical, and security compunction behind normative evolutionary social progress. As we will see repeatedly in Development history, these three legs would soon come together in another powerful simplification, this time perpetrated by an establishment Liberal at work deep in the US executive.

Rostow's retrospectoscope: lining up the national territories to secure freedom

Professor Walt Whitman Rostow had taught economic history at Columbia, Oxford, Cambridge and MIT before moving into government as special assistant to Presidents Kennedy and Johnson in National Security Affairs. His notorious tract, *The Stages of Economic Growth: A Non Communist Manifesto*, was by Rostow's own standards a grotesque pop simplification, like *The Dual Mandate*, a retrospective account designed to legitimate here-and-now development doctrine, defined as national productive moderniza-tion aligned to the US. Like later simplifications, its prominence has come as much from its recycling by critics as by actual hegemony in practice.

Rostow allowed his thin book to be 'a generalization from the whole span of modern history', which 'provides the significant links between economic and non-economic behaviour which Karl Marx failed to discern'.[58] He rep-resented the stages of capitalist growth as more scientifically attuned to the psychological realities of history and economics than Marxism. Like sub-sequent dot point summaries, it travelled exceedingly well, going through fifteen printings in five years. Disingenuously acknowledging the limita-tions of doctrinaire history, Rostow claimed that he aimed to both 'drama-tize not merely the uniformities in the sequence of modernization, but also, equally, the uniqueness of each nation's experience'.[59]

Colonialism's crisis of legitimacy provided a compelling rationale for Rostovian Development. His schema also provided a way to incorporate the newly ascendant political class of 'national' liberation movements into a new security and governance ambit whilst maintaining colonialism's old formal container boundaries. The nationalist political ambition of inde-pendence struggles could typically be split into capitalist and communist varieties, while favouring the strong, often the brutal, over the statesman. Nations were born with scant bureaucratic expertise, large independence armies, and cold war debts and allegiances. The left liberationist/nation-alist self-sufficiency leanings of many post-colonial regimes raised the stakes, but also the scope of national economic transformation. Rostow recognized the vital political power of new 'national' elites noting that 'the take off awaited . . . the emergence to political power of a group prepared to regard the modernization of the economy as a serious, high order polit-ical business'.[60] The new elites were vanguard national subjects, the bearers of a new governance rationale: 'The idea spreads not merely that economic progress is possible, but that it is a necessary condition for some other good purpose: be it national dignity, private profit, the general welfare, or a better life for the children'.[61]

This is the context within which 'colonial dependencies' emerged as 'developing countries'. Rostow's governance strategy relied on capturing these new elites' belief structures with his simple development schema. It wasn't just hearts and minds, however. 'Embedded' Liberal development would also reach a potent economic accommodation with tiny national elites, by sponsoring state-focused investment, building national symbols of infrastructure and industry, national plans and flag-carrying airlines, and generally by ensuring allegiance wavering elites were receiving clear signals about where their economic future lay. Following Rostow's doctrine, 'free' and independent nations built their own capital reserves and productive capacities as they lined up on 'Rostow's Runway', in an international development regatta that comparatively charted their progress against the territorial standard of the GNP. Again, this ably obscured the ways in which capital movements and location or production and trade were potently tied to wider, highly asymmetric security, financial and trade regimens.

Or rather, these were powerfully legitimated, as Rostow's doctrine formulized and transported what Easterly describes as one of Development's great 'stylized facts'.[62] The Harrod-Domar or 'financing gap' model was arguably the core travelling simplification of post-war development economics. Widely operationalized to legitimate loan after loan to debt-ridden countries, its core assertion was that countries' development demanded topping up of national investment capital deficits. Like many before and after him, Rostow picked it up (from Arthur Lewis[63]) and wielded it as his primary determinant for growth: to achieve the takeoff stage, it was necessary to increase investment from five to 10 per cent. Ideally this should be achieved through saving, though, as Easterly demonstrates, for development economists ever since any source would do: aid, loans, machinery, foreign or government investment.[64] Unfortunately, the whole model was a disastrous parade of mistaken assumptions. In making national boundaries equivalent to operational investment units, the IFI-based designers of take-off imagined that the financing gap model showed how national investment deficit financing could be formulaically converted to growth. At the very least, these simplifications helped justify 'moving the money', enabling otherwise unconscionable wishful thinking and crony industrialism, even simple plunder.

The colonial lines on vaguely defined maps, though shaken, proved both durable and unsatisfactory, corralling rival ethnic groups into the same power containers. They provided a bounded shape into which Rostow projected all the expectations and devices of embedded Liberal national statehood: Liberal democratic politics, nationally oriented productive industries, the flag, the five year development plan, the trusteeship and good governance of a professional, rationalized bureaucracy, concern for the welfare of the vulnerable and of course a security-cum-military force anxiously policing the borders. But post-colonial territorial states were also a boxing ring for security rivalries, a honey trap for IFI lenders and their 'national' counterparts, flypaper for dodgy national and area development projects, and a fiscal sump for bad loans. In fact, it wasn't long before civil conflict showed that post-colonial nation state governance was often a thin fabrication over very differently territorialized, visceral politics of tribal, regional and elite rivalries. Embedded Liberal arrangements did offer developing countries some leverage at an international level, though this had both positive and perverse effects. Genuine trade and industrial development opportunities were bound up with inclusion within a sphere of security influence. Meanwhile, local arrangements were distorted by these same links: loan arrangements and trade concessions created perverse incentives for 'political stability', the continuation in office of those with power to hang on. With the elaboration of post-colonial security and stability concerns, domestic military and patrimonial civil services created perverse incentives to request, deliver and soak up the spoils of loans, and create devastating balance of payments problems.[65]

This territorialized, security driven governance joined to Fordist delusion about nationalist production was ultimately a poor model for everybody. It meant insisting on industrial production within national boundaries, while negotiating in the realpolitik of international commodity chains and trade restrictions. A series of quasi-Fordisms emerged that were distorted by regimes of unequal exchange, disaggregated production processes and transfer pricing. Semi-peripheral countries (Latin America, New Zealand) got full or partial assembly lines for core industrial products that were strictly dependent on national markets. Few peripheral countries went beyond domestic commodity provision, controlled and owned by state capitalists and most saw the costs of underlying vulnerability mount as they faced unstable commodity prices, chronic and acute problems in the terms of trade in commodities, invisibles and imported high end products, debt and currency crises. There was no policy coherence from donor countries: rather, there was frank developmental hypocrisy. Aid, which might have transferred productive advantage to peripheries, operated against and as an excuse for preserving tariff protection, or was closely tied to core corporate expansion, the sale of products and productive machinery and technical expertise both at home and in developing countries.

The impending demise of Fordism, modernization and its territorial 'other', socialist self-determination, was not at all envisaged in Rostow's schema. For Rostow and the embedded, establishment Liberalism, poor countries future was either post-industrial nirvana, or communist engulfment. In Vietnam, where the overreach of establishment Liberalism's security hubris was played out, Rostow shared the Fordist state's wake (and the bombing) with both Truman's Wise Men, and the last great Fordist, Robert McNamara. Vietnam was part of the death throes of security-territorialized economics, throes that would ultimately bring down the gold standard, the Bretton Woods system, and the Soviet economic and security block. But here, the ultimately crucial security was to prove not military, or even productive, but financial; and soon finance's very different territorial predispositions became apparent.

Crisis and the legitimating turn: McNamara, defensive modernization and poverty alleviation in the 1970s

> ... the irreducible fact remains that our security is irreducibly related to the security of the newly developing world, and our role must be precisely this, to help provide security to those developing nations which genuinely need and request our help and which demonstrably are willing and able to help themselves. The rub is that we do not always grasp the meaning of security in this context. In a modernizing society security means development ... Security is development, and without development there can be no security.
>
> Robert McNamara, *The Essence of Security*, 1968

The US had emerged from the Second World War as the greatest power on earth, and was carried to Vietnam in the fullness of that power. But by the 1970s, the US and the 'embedded' Liberal order were on the back foot. A failing (and expensive) sphere-of-influence security war in Vietnam, declining corporate profits and fractious industrial relations, the emergence of Japan and West Germany as increasingly dynamic and competitive nation states; these factors were accompanied by a decentring of production along network lines into the beneficiaries of foreign direct investment (FDI), initially Latin America, subsequently South, East, and North Asia. Nixon's 1971 delinking of the dollar from gold, the oil shocks of 1973, stagflation and the crisis of Keynes-esque (not Keynesian) demand management, rising Soviet confidence leading to détente compromise and more pressure, all these secular trends and shockwaves would undo the Fordist convergence.

Defense Secretary Robert McNamara's Defensive Modernization paradigm was a fine rationale for his shift from Pentagon to World Bank, enacted as personal and other 'crisis piled on top of crisis'.[66] In *The Essence of Security* (1968), McNamara delineates the limitations of military defence of an order which failed to win 'hearts and minds', and reach an accommodation with fraught post-colonial politics. His *In Retrospect* (1995) closes with the 'lessons of Vietnam', that '[e]xternal military force cannot substitute for the political order and stability that must be forged by a people for themselves'.[67] Rather, security would emerge from a free partnership, wherein '[f]irst, we have to help protect those developing countries which genuinely need and request our help and which as an essential precondition are able to help themselves'.[68] Like all good Liberal subjects, these countries would first be seen to stand alone, and then by their own volition choose development and security in alignment to the wider order. Concern for 'political stability' meant uncertain continuation of pro-American regimes.[69] It was a card of uneven strategic value, but all sorts of nations played it, and the lion's share of the World Bank's International Development Association (IDA) money went to countries of core US strategic interest.[70] Thus geo-political contest drew both concessions and discipline.

McNamara, still characteristically naive, can-do pushy, uncontrollably rationalist, sought both humanitarian and technical redemption from Development's most powerful job. McNamara's comprehensive strategic leadership and hiring meant his 'Poverty Alleviation' agenda permeated many World Bank formations, at least until 1981. McNamara's leitmotif, the fascination with techniques, had been aided by teams drawn from RAND Corporation and The Brookings Institute. The infamous strategy of 'maximum pressure with minimum risk' was a direct military translation of the economist's cost-benefit ratio. Indeed, as one historian of the period remarks, 'The ability to control events precisely – rather than what effect those operations might have on the enemy – became a principal criterion for approving operations'.[71] This made it possible to give the 'move the money' targets an all too sharp 'evidence basis' (underwritten by apparently convincing social cost-benefit analysis) that were increasingly backed by the evident sophistication of planning tools (the Logical Framework, multiple objective planning, programme planning and budgeting systems).

America's geopolitical and fiscal bruising had by the beginning of the 1970s translated into a lowering of certain kinds of foreign policy and aid expectations. Soviet détente and dirty geopolitical struggles with backyard Latin American leftist regimes meant America needed clean hands to show the world. The crisis of confidence was palpable not just in post-Vietnam foreign policy, but in the ways the critique of trickle down and concepts like Redistribution with Growth, or Basic Needs worked their way through core Development institutions. Poverty Alleviation was the stance of a defensive, modern order still expressing its own Liberal virtue, desperate for geopolitical legitimacy. Sensitized by critique, but up to its eyeballs in lending to nasty political regimes, the World Bank could also rhetorically proclaim that a better, more humane order should still be pursued. It spectacularly increased lending in the names of the poor, wrestled with 'the social factor', and then fell back on the much more modest, immaculate legitimations provided by Basic Human Needs.

McNamara's drive to find Bankable loan targets with a primary poverty focus led the institution into uncharted territory: lending mounted for Integrated Rural Development (IRD), population control and urban site and services. What in retrospect were mal-territorialized, economic quagmires (both the IRD projects and the urban social infrastructure) were deployed into with all the World Bank's burgeoning investment and technical firepower. Rationality provided its own foundations, and the wings for travel. Ayres, in a contemporary review notes:

> The very thoroughness of the World Bank's approach often tends to lend a spurious solidity to the project as defined: from the language and content of the appraisal reports the projects appear overly definitive in areas of uncertainty, overly specific and detailed, overly assured in their assertion (seemingly normal to the World Bank) that something will happen as projected if the document says so.[72]

Ayres' account, and the remark of one caught up in it all, 'I don't know what the hell the goals are, but I'm moving ahead with projects',[73] captures the internal uncertainty about these overreachings, describing projects and their bureaucratic armatures as 'islands of rationality' and programmes 'pushing on a string'.[74] To the extent that they became briefly paradigmatic, the Basic Needs, or Basic Human Needs approaches brought the Bank and multilateral agencies into dialogue and some consensus with academic and NGO critics. The debates would easily be outflanked and swept aside by the powerfully emerging neoliberal agenda, which promised address to systemic causes of poverty, burgeoning debt, instability and governmental grotesque much more directly.

Poverty Alleviation was nonetheless both Liberal and territorially aware in ascribing the poor a residual position in a global order, marginal and vulnerable, lacking things that a benevolent international order should by rights be able to provide according to a global yardstick. Along with the UN, the non-aligned movement, the Club of Rome, the leaders of the North–South debate and calls for a New International Economic Order, it demonstrated, however, both the power and profound limitations of a Liberal rights framework. Law, justice and rights, the vulnerability and needs of the poor and the distinctive characteristics of the nation's interests, all in the 1970s signalled that the system was sick. If the many imagined apocalypses were to be avoided the system had to be called back to its Liberal justice intent and rhetoric. None, however, could have imagined the barbarism waiting at the gates.

FINANCIAL CRISIS AND THE RISE OF NEOLIBERALISM: STRUCTURAL ADJUSTMENT AND THE WASHINGTON CONSENSUS

By the early 1980s, speculative mobile capital came to drive the world financial order in ways not seen since the 1920s and 1930s and the death throes of a previous Liberal order. From the 1970s onwards the Bretton Woods institutions and their embedding of international finance were, like nation states, locked in powerful systemic contradictions and struggles with an increasingly muscular and mobile capital. Burgeoning Euro-dollar markets and other arrangements enabled capital to do an end run around territorial constraints. National governments found themselves unable to reasonably manage production, demand and employment only at the level of the territorialized nation state. As the 1980s progressed, national governments had to face up to the deterritorializing of their domains. Each had to open up and out to a global logic that required them to secure their interests on an expanding terrain in which the need to attract mobile capital in a globally networked system of production, and international market competitiveness was paramount.[75] Lurching on the advice of 'technopol revolutionaries' (politicians alert to new economic and technical doctrine)[76] or driven to adopt neoliberal market settings, they would become politically polarized. But at the same time, they would find themselves technically let down by a lack of prescience about the shocking structural fallout of these new settings, the slow recovery of the 1980s, the bad reform sequencing and exposed to the hubristic claims of transformation promised by NPM and NIE. As we will see, the molten deterritorializing caused by new engagements within the global economy was matched by local fragmentation, as the marketizing of governance by NPM, NIE and the like saw battles formerly waged in public institutions begin to be fought in the extra-institutional space of the social market.

As the global economy as a whole became less governable, local opening and integration meant exposure to unprecedented risk. In retrospect, as the potently retrospective Williamson suggests, it was necessary to find ways to 'crisis proof' national economies.[77] What emerged, painfully and inadequately, was a set of crisis alert, risk averse and IFI-endorsed national policies. These are remembered since 1989 as the Washington Consensus, and the chillingly technocratic enactments of Structural Adjustment. Unattractive in name, unloved and often regretted in practice, they would eventually be woven into the 'Golden Strait Jacket', the best ring-of-confidence defence against the brutal reactivity of the Electronic Herd.[78] But in the early 1980s, the lack of upfront analytic or political clarity about the destabilizing effects of financialized and footloose capital in practice was matched by the incapacity of the major players to analyse its systemic effects. Where regulatory moves were implemented they were reactionary, immediate and short-term. They were the products of sharpened pencils, and embodied all the systemic, long-term developmental and global trustee vision of New York finance markets and bankers. That all this referenced the ugly end of an era was however not widely apparent, except perhaps to the millenarian minds of Chicago School neoliberalism.

The most immediately apparent new turns and hybrids were political, both in government and core institutions. The return of conservative leadership in core countries (the Regan-Thatcher axis), joined at the hip to political techno-gnomes, provided the political occasion for a sharp and disciplined implementation of Liberal dogma. Within Bretton Woods, the rupture was marked by the departure of McNamara and his chief economist Mahbub ul Haq. They were replaced by eminent Wall Street banker Alden Clausen, with rising neoliberal economist Ann Kreuger as influential Vice President for Research. The scene was set for the radical reopening of nations to the perversities of capital flows and exchange rates, the roll back of regulation and state ownership, the apparent discrediting of national productive strategies, and the quick restructuring of industry. While Pinochet's authoritarian right Chilean government captured 1970s neo-liberal imaginations, it was a Labour government in semi-peripheral New Zealand that after 1984 provided neoliberal governance with its international apogee. What mattered in both cases was doctrinal and executive capture, and a crisis impelled policy and executive revolution; short sharp, bloody, messy, but still technocratic. New Zealand's experience is well remembered. There, a 'post-Rogernomics stagnation' (after the Minister of Finance Roger Douglas) occurred in the late 1980s. Quick, badly sequenced reform ran into a worldwide recession, and tight money and an inflamed exchange rate fed an industry and employment shakeout. This was however a minor rash compared to the systemic decline of peripheral countries that got thoroughly 'structurally adjusted'. For these countries, the more 'adjustment credits' (a form of policy based lending) they received, the more likely they were to be a basket case by the end of the 1990s.[79]

Despite the global mobility of capital, the focus on national economies persisted. But this too was gutted as these strategies moved into the guillotine framework of neo-classical New Political Economy (NPE), as we explain below. Salient indicators shifted from gauges of national production and wealth to policy settings which focused on the current account, the capital account and interest and currency rates, all assessed in terms of their scope for attracting and securing capital. In this new regatta the most spectacular reformers, the semi-peripheral countries like Chile, New Zealand, the Asian Tigers, had their moment in the international spotlight. Here, it was thought, Liberalized global integration had been empirically implemented. The Asian Tiger reality was quite different, but the extent of the deviance wasn't clear until it ran into frontal assault from larrikin capital and its neoliberal agents in the late 1990s.

Chile, however, does provide a telling example of reflexive crystallization and amplification of Liberal doctrine in the period, and the role of security arrangements in creating a context. Years later, Pinochet's Chile would be pulled out to remind folk that Liberal economics scarcely needed Liberal politics. In fact, then as now, capital reached some of its best accommodations with authoritarian governance. After the US-backed coup ousted leftist, World Bank abandoned Allende in 1973, the military government offered the Chicago Boys a disciplined, open research laboratory to elaborate its wider reform agenda.[80] The World Bank lent to Chile with a passion, over Mahbub ul Haq's objections. Here, as elsewhere, the 'freedoms' of Liberal political order were mashed between militaristic security and radically open markets.

From the beginnings of the 1980s debt crisis, hegemonic finance capital cast its rubric. The crisis was a familiar instance of a bubble-burst shock followed by myopia about the real systemic ills and underpinnings of instability. No-one realized the extent to which middle income Latin American economies driven by an accumulation of private debt would be caught by rippling monetary conservatism in US, or suffer the effects of bank panic.[81] At first, it was seen as short-term. Across the spectrum, from Harrod-Domar borrowers to US bankers myopically focused on 'getting my money out', all saw this as a recession to be managed rather than a systemic shift. In the obverse of their counterparts in the later 1997 Asian currency crisis, they sought to portray the crisis in more limited terms than it was: a liquidity issue, not a solvency one.

Ultimately, everyone knew Bretton Woods' solutions would become a recipe for lopsided, debtor adjustment, especially as debt-risk grew – in Latin America combined foreign debt blew from $2.3 billion in 1970, to $75 billion five years later, and by 1983 it stood at $340 billion.[82] But few could have anticipated what a potent yet blunt and perverse political instrument debt would become. Structural adjustment's lopsided first-generation solutions paraded in the technical clothes of policy based lending, and were implemented by technocrats apparently only concerned that policy should be visible and confidence engendering, technical and measurable.

The debt crisis was a result not of policy but of underpinning capital and human realities, precisely of the kind Keynes feared. It was a matter of lots of money being about (a bubble in finance available for developing countries from petrodollars recycled through Northern banks), 'inept borrowing country policies made worse by vested interests of governing elites'; and 'commercial bank cupidity and naivety'.[83] Greed, and bounded rationality, finance outstripping reason and regulation, pushed by brokers with little memory and few systemic responsibilities: this is the stuff all bubbles are made of. Nasty deflations are too. Across Latin America, all the deflationary ingredients of an 'IMF riot' rolled out: devaluations, demand compression, the removal of subsidies on staples, the related raising of commodity exports undercutting debtors and other agricultural producers' terms of trade. Reactions and roller coasters followed, as regional economies entered a semi-peripheral spiral that made post-war dependency relations look beneficent. Debts were socialized wholesale, deficits monetized, leading to more inflation. As in the 1990s, the semi-peripheral, unevenly integrated countries would be the high profile winners and losers from the shakeout. The truly peripheral would have debilitating experience of having different, but still second-hand, mis-scaled doctrines ineffectively applied to them.[84]

Through all this, neoliberal micro-doctrine travelled from Chicago to Chile, back via Chicago and Washington to the rest of Latin America, back to Washington, and out to the rest of the globe, typically via the persons of various multilateral missions.[85] Especially in their incarnation as Policy Based Lending and Structural Adjustment Credits, they had a particularly doctrinaire aspect. And it was in their most bare, retrospective, 10 dot point formalism that they ultimately attracted the most reaction. In sketching core elements of the Washington Consensus, John Williamson coined a term that would, as we will see, anchor and focus international opposition to neoliberalism, and invigorate the turn to Poverty Reduction. In an innocent 1989 seminar called 'What Washington Means by Policy Reform', Williamson set out to identify '10 policy instruments about whose proper deployment Washington can muster a reasonable degree of consensus'.[86] If it was a playful sketch, it represented a doctrinaire time: as Williamson later noted in a 1999 retrospect:

> One can view [the Washington Consensus] as an attempt to summarize the policies that were widely viewed as supportive of development at the end of two decades when economists became convinced that the key to rapid economic development lay not in a country's natural resources, or even in its physical or human capital, but rather in the set of economic policies that it pursued.[87]

Williamson's gambit here was in fact not the creation of a dot point orthodoxy: ironically, quite Lugardian, he intended to describe the consensus as a point of departure for a wider discussion about how individual country's application of the IFI policy formula saw some moving while others adopting only after inconsistent urging. Given the extent to which the Consensus adumbrated in the paper has been demonized as formulaic[88] (Williamson writing in 2002 describes some critics being unable to mention the Washington Consensus without foaming at the mouth),[89] the paper's tone is remarkably tentative and candid, reflecting what Williamson and his defenders would describe as the shifting, contested state of Washington orthodoxy.[90] In retrospect, Williamson's 10 points were either provisionally arrived at (and subsequently disputed or discarded), or were more confidently expressed, 'mom and apple pie truisms' about the virtues of a measure of fiscal discipline, and trade liberalization, and promotion of FDI.[91] In practice, Washington Consensus rationalities were scarcely applied with Williamson's speculative nuance. Ravi Kanbur offers a clear account of the missionary zeal and one-sided, limited exchange involved:

> There is no question in my mind that in the 1980s, and to a certain extent well into the 1990s, many saw the main task as being storming the citadel of statist development strategies. In this mindset, nuances were beside the point – intellectual curiosities which paled in comparison to the benefits of rapid and deep movements away from the former paradigm. And, moreover, Washington institutions were deeply suspicious of the real intentions of those they were dealing with. They suspected, perhaps rightly, that those on the other side were hell bent

on preserving the status quo. In this setting, a negotiating stance, rather than a dialogue based on mutual comprehension, was appropriate. So the negotiators from Washington always took a more purist stance, a more extreme stance than even their own intellectual framework permitted (they were all surely well schooled in the theory of the second best). 'Give them an inch of nuance, and they'll take a mile of status quo', seemed to be the mindset and the stance. 'If you want 28 enterprises privatized, start by asking for 56', seemed to be the opening gambit. Is it any wonder then, that those on the other side came away with the impression that those from Washington had a consensus and one which did not match Williamson's nuanced formulation?[92]

In the core countries, reform, instability and attendant social disruption was partially offset by spending in other areas (Reagan's defence blowout, rising unemployment under Thatcher), so that at no point did state expenditure decline, and across the 1980s and 1990s there was no decline in state spending as a proportion of Gross Domestic Product (GDP) in most OECD countries. Unemployment rose in all OECD countries, and inequality especially in Liberal welfare regimes like New Zealand, Australia and the US. Almost everywhere, the long surfeit of pain over gain was simply sickening, as programmes of structural adjustment cut swathes through productive and public sectors.[93]

Neoliberal institutionalism: the new political economy of the state, and the governmentalizing of civil society

Policy reform, it turned out, was the first phase in a longer period of abstract governance thinking on Development practice. Here, from 1981–2005, in governance reform programmes in OECD and Development contexts, neoliberal policy formulae would be rolled out and embedded not in Polanyian social contexts, but in increasingly elaborate convolutions of what Liberal economic historians called 'institutions'. Looking back, the whole period can be seen as an 'age' of institutional development,[94] characterized by what we might call 'neoliberal institutionalism'. In closing this chapter, we will look briefly at some of its core elements.

As command over resource accumulation and use shifted away from the cabinets of independent nations, and out into the trading and boardrooms of financial corporations, the new era of conservatism, the 'counter-revolution' in development was underway.[95] The 'hollowing out' of functions that had been scaled to the nation state illustrates to two features of the 1980s orthodoxy, both part of a determined displacement of state institutions from the roles earlier accredited to them as 'the engines of development'.[96] One, already apparent, was 'external delegation' of state functions to international regulatory bodies like the IMF. A second was 'internal delegation' to non-state organizations in the form of NGOs and

Table 2.2 Core elements of neoliberal institutionalism, 1981–2005

Policy	Getting core macroeconomic settings 'right', getting microeconomic prices (and related market mechanisms) 'right', privatizing state provision into markets, keeping state out of markets.
Economy	The right policy-institutional environment will enable market to integrate globally, driving growth.
Governance	Rise of NIE and NPM. Authority and service delivery shared across state, market and civil society (NGOs). Institutional strengthening of budget/expenditure, contracting and auditing capacities, information transparencies, plural accountabilities to clients, law, media.
Decentralization	Competitive or co-production of local services, supporting state, market and civil societies capacities at decentralized levels; 'civil society' and local partnership, participation and voice.
Society/poverty	Disaggregated into local target groups and communities, preferably to 'citizen choice', to be addressed via market rather than centralist/redistributive mechanisms.

private sector agencies.[97] Both are features of what we have described as neoliberal deterritorialization of governance. The rise of NGOs, which we briefly deal with here, is a fascinating example of Polyani's 'enlightened reactionaries' at work, pitching themselves as both effective representer of the 'grassroots' and civil society (and thus as a counterbalance to both market and state failures), and a market-reliable deliverer of services to the marginal. It is also a telling tale of the ways such agency can be captured, and ironically become part of the exacerbation of market-oriented disaggregation of governance and accountability. Overall, what emerges is a remarkable accommodation, wherein each side would take advantage of the opportunities offered by neoliberalism's promotion of privatization and deregulation.[98]

By 1981, statist development was on the ropes, with the World Bank ready to lay the blame for Africa's woes at the feet of the nation state.[99] This 'internalist' analysis of poverty was being urged to go further by a more far-reaching academic exposition in the form of the 'NPE'. The classic text, also published in 1981, was Robert Bates' *Markets and States in Tropical Africa*.[100] Here, the answer to Africa's problems lay not just in reconfiguring the state, but in dismantling state power so as to undermine the instruments used by elites to accumulate political and material wealth, and thereby free the peasantry to take advantage of the new market opportunities. As the views of US Liberal political science and economics about the proper functioning of the economy gained new currency, the view emerged that governance should be liberated from state control and privatized in the same fashion as the market. Just as the nation state fetters efficient resource allocation by the market – the decentralized decisions of individuals – so too must governance problems be made contestable, disaggregated and opened to 'civil society'. Now a plethora of guru-led theoretical perspectives advanced, each advocating a radical re-assignment of state tasks to a much wider array of agencies and marketeers: the NIE of Douglass North, Coase and Demsetz, Buchanan's Public Choice theory, Williamson's transaction cost analysis, all laced with doses of Agency Theory and the NPM.[101] Perhaps more immediately, resort to NGOs was precipitated on the ground as the social impact of austerity measures became apparent. In sub-Saharan Africa, per capita incomes declined in real terms by 30 per cent between 1980 and 1988.[102] With states less able to meet even their basic service delivery responsibilities, rising disaffection and a spate of high profile famines – Ethiopia 1984–1985 – prompted urgent attention to the need to push services and safety net measures into contested domains in ways that gave recipients sufficient resources and incentives to stay put.

In the longer term, and in terms of this book's wider logics, the rise of NIE in governance and the governmentalizing of NGOs and civil society would lead to both a disaggregation and fragmentation of governance; in function, mandate, and funding, and an undermining of territorialized accountabilities. In the mid-late 1980s, all this was still to come. Meantime, in initial phases of the relationship, both NGOs and their sponsors focused on the gains the new pluralization and relationships seemed to offer. Erstwhile official aid-chastising NGOs put themselves up to a ready audience, claiming that their aid was more likely to reach the poorest groups. Who, after all, had better, more authentic links with local people? Government bodies were forced to work through diplomatic and official channels; voluntary aid must be more cost-effective. NGOs could act in regions where, for political reasons, governments are not able to assist and besides, non-government agencies were more flexible, innovative and responded more quickly to need than official agencies.[103] However, the dream of cheaper, and more legitimate governance in the periphery continued to prove elusive. But NGOs fed the hope with ample evidence that they indeed had a 'comparative advantage' – as always, despite that these claims were seriously contested by empirical evidence.[104] NGOs profile grew phenomenally, with OECD registered numbers moving from 1,700 to 4,000 from 1981 to 1988.[105] This was repeated in developing countries. In Kenya for instance, women's groups alone numbered 26,000 in 1988, up from less than 5,000 in 1980. Voluntary contributions and official financing through NGOs also accelerated. By 1988, NGOs provided approximately US$5.5 billion in financing, against the World Bank's US$4 billion.[106] And, after a long courtship in the 1980s, World Bank financing of NGOs jumped by more than 300 per cent in 1989.[107] This was far more than just a matter of co-opting the sharpest critics into 'responsible' engagement – although judging by the proliferation of NGO-IFI committees, task forces, conferences and joint communiqués in the mid-1980s, this was surely important too. NGOs went into these relationships largely with their eyes open; internal debates worried about co-option, and the perils of becoming what Sen referred to as 'shadow state', but for the most part, they were persuaded by arguments that apparently placed them at Development's top table.[108]

The immediate effect of these developments was to circumscribe the practical autonomy and shake any remaining sense of critical sovereignty for the nation state. NGOs ironically became neoliberalism's shock troops in this process. Again, it's important to note where the essential frame-work elements arose: NGOs won approval as contractors and partners in terms of a core neoliberal governance frame and Liberal welfare regime context (in New Zealand, the UK). And only later, as we will see, did the unanticipated fragmenting effects that occurred as this orientation visited on poor, distant countries become apparent. Yet for all this hollowing out and shadow state formation, everywhere the state remained. As a shadow of its former self, to be sure, but also as a useful and sometimes critical framing device, into which all the new armatures of NIE could be plausibly pitched from a distance, owned, implemented and ultimately joined-up to reconfigure the state in the interests of the markets. And, as James Fergusson brilliantly depicted in *The Anti-politics Machine*, reframed and reimagined as the playing field for IFIs and their partners. Soon these developing country states would be rearmed with depoliticized, partially reterritorialized Good Governance ambits that would reconstitute localities, governance and poverty in ways that made it possible for Liberal developmentalism to once again plausibly position itself as a comprehensive, inclusive and joined-up address to Poverty Reduction.[109] This is the focus of Chapter 3.

3 The rise of governance since 1990

The capable state, poverty reduction and 'inclusive' neoliberalism

The triumph of Liberalism represented by the fall of the Berlin Wall and talk of the 'end of ideology' contained the usual ironies of hegemonic Liberal governance: ideology is never so ideological as when it is seen as natural, consensual and merely technical matter. As we will show in this chapter, the great achievement of the Liberal project over the 1990s was precisely the forging and embedding of an ideological, political and technical consensus both globally and with governing regimes in key developing countries.

In Development this did not simply involve, as might have been anticipated, a capitulation to the There Is No Alternative neoliberalism of the anti-statist 1980s. It was achieved, in the context of crisis and reaction, by political and institutional agents keen to broaden Liberal governance's ambits to be more 'inclusive', and more technically innovative, elaborate and, it was imagined, integrated and 'comprehensive'. The earlier orientation to liberal markets and global integration, and experimenting (and overreaching) further with disaggregation and decentralization was retained. Thus in a number of influential core OECD countries, and in Development, a more 'inclusive' neoliberalism would ultimately combine institutionalized openness to capital markets and sharp, decentralized and disaggregated governance around service delivery with grand, overreaching poverty and 'inclusion' strategy at international, national and community levels.

This chapter and the next, chart this progress over the period 1990 to around 2005 through an overview of selected landmark policy documents. In this chapter, the story will be told in terms of the politics, political economy and institutional and public policy domains affecting Development. In Chapter 4, the story will reach back again to the mid- and late 1990s, and move forward telling the more technically elaborate story of the development of governance programmes and techniques in areas of budgetary disciplines, intergovernmental relations and decentralization. Finally, how this actually looked on the ground over the last 15 years, will be told in the progression of chapter cases in Part II, beginning in Vietnam in the early 1990s, and ending in Pakistan and New Zealand in 2004–2005.

The first section of this chapter sketches the wider contexts that would drive the policy, political and technical changes described in these two chapters. Turning back to the early 1990s, we see the grand claims of global summits jostling with the security threats of 'failed', 'criminalized' and insecure states, to reinforce the paradox that global embedding of the market required a 'capable state' to deal with the ultimate threat of local social instability. In the second section, the Asian financial crisis provides the crucible for Poverty Reduction and clinched a consensus beyond Washington that hinged on deepening the institutional ambit to include aspects of political 'ownership' of adjustment reforms, and on technical innovations including PPAs that, among other things, enabled the poor to be framed in ways that would 'localize' and depoliticise poverty. Ironically here, as we will show, the poor could have a prominent role in legitimizing this new consensus by participating in IFI poverty assessments. Beyond this, we will outline the ideologically expansive consensus of Poverty Reduction's three legs – Opportunity, Security and Empowerment – that would become a global pre-qualification framework for development assistance that appealed to Liberal, conservative and social democrat alike. Then, in the third section, we consider wider shifts embodied in the 'positive' Liberalism and 'inclusive' neoliberalism of the Third Way, many of which spilt over into Development.

GOVERNING THE NEW WORLD (LIBERAL) ORDER AND ITS PERIPHERAL DISORDERS

Crisis and re-embedding: international politics, economy and development 1990–2005

The first public signs of a shift beyond the raw neoliberalism of Structural Adjustment came when the World Bank put *Poverty* smack in the middle of the 1990 WDR cover. But in 1990, the consensus in Washington was unprepared either for the decade of financial shocks that would follow, or to embed the kinds of social and institutional programmes that would be rolled out in reaction to them. Then, triumphal, lacking plausible alternatives, and with prospects of rapidly developing markets in the former Soviet Union, bare Liberal economic doctrine sat explosively alongside the remaining imperatives of political victory in the Cold War. Now, not least to prevent a return to socialism, 'shock therapy' privatizations and economic rationalism would expose the economies of the former Soviet Union to oligarchic capture and unregulated markets, shredding the remaining territorialized Eastern Bloc planned economy arrangements. As incomes and employment plummeted, life expectancy in Russia fell by an astonishing seven years in less than ten years.[1] Picking over the rubble, architects and critics of the reform would lament lack of foresight over the absence of market institutions in the former Soviet republic and, by 1995, begin to rethink and retool.

All too soon, however, the IFIs would be overwhelmed by the repercussions of another capital bubble in emerging Asian economies, and in the thick of this new crisis, conflate institutional reform needs with (short-term liquidity) crisis management. This would more than repeat the 1980s scenario, where financial crisis was ironically used to reinforce the overall 'open economy' orientation, further opening capital markets and facilitating the narrow bankerization (with its 'easy in, desperately out' penchants) of capital movements. Now 'securitized' capital, speculative and financialized as never before, would bring ratings agencies' sharp and narrow rationalities into the heart of political economy, and financial security fears would drive political commitment to economic conservatism with a vengeance. Meanwhile, short and long-term private capital flows into emerging economies would, like popularized technology and wider share markets,[2] boom and bust, and come to be seen as both an aid-dwarfing saviour and an aid-destroying devil. So for much of the 1990s capital seemed barely governable at all: by 2002, in places where the state had most sharply disciplined itself to remove sovereign financial risk (notably Argentina) the mechanisms that were meant to protect had become a part of the problem.

The 1990s was for all that a contradictory decade, as both marketization and reaction played out, sometimes within the same political and institutional settings. From the outset, variously enlightened reactions and moves were made to re-embed and re-regulate the neoliberal project in big picture, multilateral answers to emerging global issues. NGOs' commitments to participatory processes were beginning to be taken up by multilateral agencies and IFIs. At the same time, neoliberal reform was in many places only just hitting its straps. But as more privatizations rolled out, reactions too gathered pace. Moves by states to re-regulate the market gathered conviction and political support, especially in natural resource management, transport and utilities.[3] The re-emergence of a fiscally chastened centre left in Liberal, core economy politics produced some surprising outcomes. The most evident political harbinger of this was the election in 1992 of New Democrat Bill Clinton, whose policy ambit included both 'the economy, stupid' (ultimately involving a programme of deficit reduction that made fiscal hawks like Alan Greenspan proud), and an attempt at major health care expansion. Ambushed and railroaded by the Republican Revolution, the Clinton executive signed off on the decentralizing of wel-fare services down to sharply responsibilizing state agencies. Internationally, success was as mixed and contradictory.

Nasty 'failed state' security crises in Africa and Europe proved that multilateralism needed teeth, but initial commitments to the Doha round and Kyoto showed possibilities. Wider change was again promised and only partly delivered as what became known as the Third Way received a critical boost in the election of 'radical centre' governments in former bastions of conservative and radical neoliberalism, including the UK (1997) and New Zealand (1999). Here, commitments to both modern 'joined-up' governance, and wider partnerships across communities would see sharp targeting of benefits, and welfare regimens combining moral discipline with apparently comprehensive 'inclusion'. By 2005, Third Way Blair's moral 'Liberal interventionism' and upholding of a special security relationship with a now neo-conservative US would have committed his Labour government in Iraq. The Clinton era would by 2005 seem a high point of global Liberal multilateralism, its universal intent heavily qualified by security reaction and rising unilateralism, as a new administration pursued goals abroad (oil and terror security, and political-economic alignment in rising sphere-of-influence tension contexts with China, masked as freedom and democratization).

In retrospect, neoliberalism in the 1990s sat on a seesaw. Liberal projects both conservative and social democratic swayed between the deeper disaggregations of government wrought by privatization and NIE on the one hand and, on the other, shallow re-embeddings of market-based welfare regimes around community and individual responsibility. Often, these reforms were driven by the same political and institutional agencies, their broadly Liberal programme able to hybridize both inclusive/positive and market neoliberal positions. The result was unevenness, and often just plain dysfunctional governance: in the absence of a comprehensive and authoritative Liberal project, things took on what Ash Amin[4] has called a 'heterarchic and topological' aspect: multilevel governance, interagency arrangements, wildly differing in their composition from place to place.

So it was in Development, and especially in what then came to be called Good Governance. At the head of the 1990s, the determining factor behind Development's ups and downs was already regarded as 'the effectiveness of the state', although then, as now, as Joachim Ahrens remarks, 'there are still not clear or settled ideas about how effective governance should be suitably defined, let alone how key governance issues can be appropriately incorporated into externally-financed programs of policy reform'.[5] Somehow, however, political and technical boundaries would have to blur a little, and policy leverage would have to be strengthened, if CDFs and institutional frameworks were to be embedded. Ultimately, Good Governance would emerge from the crisis of 1997–1998 as paradigmatic, within the broadened, three-legged Liberal consensus in IFIs. Crisis would impel the consensus makers into a closer strategic harmonization and accommodations with national governments, and into broader institutional and local government/community engagements. This accelerated by promoting both right-looking disciplinary governance and by appropriating social democratic symbols of legitimacy – participation, empowerment, the community, the poor. As we will see in Chapter 4, the pre-crisis 1997 WDR, *The State in a Changing World* spelt much of this out in compelling detail.

By the 2000 WDR, *Attacking Poverty* it was core business. And, by the 2004 WDR, *Making Services Work for Poor People,* the state would have been reimagined in ways few in 1990 would have anticipated.

Global Governance summitry: imagining the globe as the new territory of Liberal governance

'New world order' imaginings had seen a flurry of multilateral networking involving private actors and regional bodies, NGOs and social movements working at a global scale. It was hoped that a new 'global governance' policy initiative, if played right, could harness a wider constituency and bring about a worldwide response around neoliberal economic reforms.[6] The new period of 'global summits' began in 1990 with the World Summit for Children, and road-showed from there: the UN Conference on Environment and Development, the Bruntland Report, in June 1992 that heralded a 'new era' of sustainable development, and soon after the World Conference on Human Rights in June 1993. Then, as whole regions of peripherality had their turn in the spotlight, came the Global Conference on Sustainable Development for Small Island Developing States in May 1994 and, harking back to the fears of the 1960s, the International Conference on Population and Development in September 1994. 1995 saw the Fourth World Conference on Women, the World Summit on Social Development. The round closed with a nod to the problems of burgeoning urban areas at the UN Habitat II in Istanbul in June 1996.

These events echoed the scrutiny the World Bank had to bear in the early 1970s, when the end of earlier era prompted a festival of summits and conferences on resource security, poverty and development.[7] At that time, the dethroning of the Fordist, Keynesian state ushered in the 1980s era of deterritorializing, formalist neoliberalism. Likewise, these new summits colluded to reinforce the paradox that 'just as the responsibilities have piled higher at the doors of the nation state, there has been a more pervading sense of the impotence of states as instruments of progressive social change'.[8] The summits were a response to the appearance of cracks and the dark side of the new world order. The threat posed by the collapse of the Soviet Union and the emergence of the criminal mafias and ethnic warriors prompted worries about 'failed states', 'criminalized states' and 'disrupted states' that were seen to be responsible for 'not only the intensification of civil wars, humanitarian disasters and displaced populations but the rise of shadow economies and increasingly powerful non-state actors such as organized crime networks and a thriving global black market'.[9] Anti-state forces had apparently reached the point of being 'out of control',[10] at the same time as economic orthodoxy and the fluid internationalization of markets, finance and communications was heralding the 'end of state sovereignty'. In a dystopic twist on Marx, globalization was seen as causing a 'withering away' of the state in favour not of a socialist or market utopia, but a wreck of rogue reterritorializations, including tribal domains, shanty-states, and nebulous and anarchic regionalisms.[11] By 2001, of course, this security angst would be blown up into the spectre of terror.

Would-be builders of a global Liberal governance order were faced with the prospect of doing more with less. Flows of development aid were falling noticeably.[12] Private capital flows were expected to fill the breech and indeed they did increase dramatically, rising between 1988 and 1997 from $36 billion to $252 billion.[13] But it became apparent that only 10 countries had received 78 per cent of FDI.[14] Elsewhere, starved of fresh FDI, strapped with debt and their capital accounts in flight, HIPC watched and waited for debt relief. By the time the 'capable state' was wheeled in by the WDR in 1997 to manage the politics (and screw down the discipline) around a new generation of HIPC structural adjustment, it was again carrying a colossal obligation. This revived state had to deal with the 'ultimate threat' of adjustment induced social instability, while, at the same time, creating 'ownership' for the widening remit given to, yes, structural adjustment. Ramped up before the adjusting state was the demand that it confront the factors that impeded adjustment's operationalization: 'weak institutions, lack of an adequate legal framework, weak financial accounting and auditing systems, damaging discretionary interventions, uncertain and variable policy frameworks and closed decision making, which increase risks of waste'.[15]

Disciplining the culprit: governance into the early 1990s

That the state was the principle culprit for the failures of Development had been established 15 years before 1997 WDR, notably in the case of Africa, in the 1981 Berg Report.[16] This posed political challenges that IFIs were then unprepared to deal with; nor were they geared up with the technical analysis or professional cadre needed to advance any reform beyond the iconoclastic, anti-statist tones of NPE. Adjustment lending in the 1980s now appears to have been 'based on nothing more precise and scientific than a preference for free markets and an instinct for the way that the local economy works'.[17] In retrospect, calling it an instinct was generous. Actual practice appeared to reflect the work of 'badly prepared protagonists of modest ability employing data of dubious quality and entering upon a series of battles over very complex policy questions'.[18] Of course, there was no contest; 'the blind side with the money normally won'.[19] Taking a dim view of the state, the structural adjustment reforms of the 1980s meant that to the extent that IFIs had a 'governance policy', it was in negative gear and could largely be read off the imperatives of fiscal stabilization. In other words, it meant PEM, (for which in the 1980s read spending cuts), civil service rationalization (read cost cutting contractions), and restructuring (privatization) to 'improve the competitiveness of the policy environment'. The partially realized, but grotesque and failed statism and territorialism of the Development Decades before the late 1970s was replaced by scorched earth rationalism: or would have been, had these reforms themselves not been partially realized, compromized by security accommodations, and hauled back through yet another round of partial reterritorializations.

In fact, it was not until 1989 that the 'crisis of governance' was first thoroughly set up in a controversial but humble World Bank report, *Sub-Saharan Africa: From Crisis to Sustainable Growth: A long-term perspective study*.[20] The study presented a circular paradox: effectively implementing reforms to create a capable state required that governments stay the course with necessary 'corrective measures', and yes, this required an already strong and capable state. As the 1991 WDR noted, there was no question that 'fundamental structural change is (still) needed to transform African economies' to save them from capricious states much remarked by 'lack of accountability, patronage and nepotism, and corruption ... bribery, nepotism, and venality ... can cripple administration and dilute equity from the provision of government services – and thus also undermine social cohesiveness'.[21] Failure was still not to be attributed to the model, but again to the state, which was failing to enable the embedding of market reforms. Market friendly reforms had thus to go 'deeper' and 'wider'. A 'second generation' of reforms (which we elaborate in Chapter 4) was needed to more fully 'embed' them in all aspects of social, political and economic affairs.[22] Achieving this required much more than the 1980s Structural Adjustment refrain of 'political will' by governments; it was, for the first time of wider policy significance, a problem of 'governance'. Here, crucially, *government* and *governance* were set apart; 'The absence of good governance has proved to be particularly damaging to the "corrective intervention" role of government'.[23] This statement proved signal in later elaboration of the PRS Initiative, as we will see in the second and third sections of this chapter. But first, it is necessary to look at the central pre-occupations of Good Governance policies as they were unfolding before this occurred, during the early 1990s in the Multilateral Development Banks (MDBs). Later, Chapter 4 will take these up to the present.

Finessing the MDB charters: not politics, but the rules of the game

The World Bank's *Long Term Perspective Study* published in 1990 defined governance in explicitly political terms: 'the exercise of political power to manage a nation's affairs'.[24] This political rendering of the 'governance' agenda was later finessed in favour of a more technical presentation. Though explicitly aware of the political nature of 'governance problems', the Bank's first treatise on 'the governance dimension', *Managing Development*, in 1991 set out the problem in the moral and technical terms of 'misgovernment'.

These reforms received a seminal formalization as 'second generation reforms' in a 1994 Moises Naim paper, 'Latin America: the Second Stage of Reform'[25] where the crucial orientation moved from 'changing rules' to 'changing institutions'.

In the meantime, it was corruption that set poor countries apart from the rest:

> It tends to thrive when resources are scarce, and governments rather than markets allocate them; when civil servants are underpaid; when rules are unreasonable or unclear; when controls are pervasive and regulations are excessive; and when disclosure and punishment are unlikely.[26]

Later, after some hesitation, the World Bank became the self-declared champion of the anti-corruption cause. In 1996 the World Bank's President pledged to commit to fighting 'the cancer of corruption'.[27] In 1999, James Wolfensohn took it further, claiming that '[a]s far as our institution is concerned there is nothing more important than the issue of corruption'.[28] But while early 1990s MDB definitions of governance signalled these concerns, their authors seem on a current reading to be peering out at the world from the strongly operational imperative of implementing the structural reform, macroeconomic and capital protection programmes needed to maintain the integrity of financial systems.

Technical and institutional overreach predominated, as proposed reforms were borrowed from theory or leading cases elsewhere, depoliticized and dehistoricized. When Liberal governance theory is harnessed to policy, analysis usually begins and ends with normative assumptions, so that what happens in between is anticipated as mere contingency. Then as now, structural, historical and political economic explanations of corruption were rarely entertained: corruption is a moral problem for now, not the product of, say, (post-)colonial experiences which centred productive resource and security-pushed loan transfers on the state, creating enormous incentives for rent seeking. So, good governance became synonymous with technocratic systematization: 'sound development management', and 'creating a sound investment environment'. Experience had been that projects may be technically sound 'but fail to deliver anticipated results for reasons connected to the quality of governance',[29] and since these projects were more often concerned with macroeconomic adjustment, good governance was seen as the 'essential complement to the sound economic policies' being carried forth by projects and adjustment lending.[30] This circular operational focus persists to this day, though as will be told in Chapter 4 it is now advanced through a far more embracing and robust set of frameworks. But while heavy with the language of accountability, the essential focus was and remains dealing with forms of corruption that might impact on what we will later describe as 'transaction costs'.

Two key documents defined what are still regarded in MDBs as 'core governance concerns'. In 1992, the World Bank published *Governance and Development,* and followed this up in 1994 with *Governance: The World Bank Experience.* Both documents were prompted by the need to steer a technical course through rapidly expanding and broadening operations that had been pressed forward by a sense of political crisis. This was forcefully presented in 1992 when the IMF's Managing Director, pushed no doubt by the rising social and political protest in Latin America,[31] stressed the need for 'democratizing social decisions' and a de-politicized strategy, 'for want of a better term, "good governance" [that is] accountable and active governments that enjoy the trust and support of their societies'.[32] The critical need was still to transform the state, but as we explain, a new approach was needed to 'nurture a political consensus in support of these reforms, (for which) governments require considerable skill'.[33]

Considerable skill was needed at the operational level too, where careful delineation of separate domains for the political and the technical might allow for the separation in government in ways that would let each get on with business. As *Governance and Development* made clear, reflecting a caution by its General Counsel, the agency's Articles of Agreement prohibit the World Bank from interfering in the political affairs of member states. Article IV, Section 10 of these Articles reads 'the Bank and its officers shall not interfere in the political affairs of any member; nor shall they be influenced in their decisions by the political character of the member or members concerned'. *Governance and Development* quotes Webster's *New Universal Unabridged Dictionary,* to find that governance has three distinct aspects: (a) the form of political regime; (b) the processes by which authority is exercised in the management of a country's economic and social resources; and (c) the capacity of governments to design, formulate, and implement policies, and, in general, to discharge government functions. 'The first aspect' guided the General Counsel, 'clearly falls outside the Bank's mandate. The Bank's focus is, therefore, on the second and third aspects'.[34] But here, a shift can be seen:

> [B]eyond building the capacity of public sector management, to encouraging the formation of the rules and institutions which provide a predictable and transparent framework for the conduct of public and private business and to promoting accountability for economic and financial performance.[35]

Clearly governance was political, and went well beyond public sector reforms; but the MDBs, at least, had yet to chart a narrower, more technical course around the constraints of their Articles of Agreement. In the event, the World Bank's General Counsel charted a new course without difficulty. In a memo clarifying the 'limits' of governance policy, counsel first identified five aspects of governance lying beyond the World Bank's mandate:

> [The Bank] cannot be influenced by the political character of a member; it cannot interfere with the partisan politics of the member; it must not act on behalf of industrial member countries to influence a borrowing member's political orientation or behavior; it cannot be influenced in its decisions by political factors that do not have a preponderant economic effect; and its staff must not build their judgments on the possible reactions of a particular Bank member or members.[36]

With these provisos, General Counsel held that 'governance' was quite consistent with the World Bank's mandate as it was restricted to the promotion of order and discipline in the management of a country's economic transformation.[37] With this apparently restricted mandate, however, Good Governance opened a breech in poor country sovereignty potentially far bigger than anything since Lugard. Looking back, the road from Lugard's mandate is quite clear. Looking ahead to Poverty Reduction strategies, we are already well on the way to Development as 'security', in the name of 'opportunity' and 'empowerment'.

Good Governance: the tools of a crisis manager

Accountability and transparency became all the more important as the IMF took on the role of global 'crisis manager' and as pressure grew for more security for highly mobile investment capital. In April 1995, finance ministers and central bank governors of the leading members of the IMF committee (the G3) were calling for 'stronger and more effective IMF surveillance of its members'.[38] This shift in the IMF's role towards explicitly advancing the cause of transparency and capital market integration Stephen Gill refers to as 'new constitutionalism', that is a 'move towards construction of legal or constitutional devices to remove or insulate substantially the new economic institutions from popular scrutiny or democratic accountability ... aimed at guaranteeing the freedom of entry and exit of internationally mobile capital'.[39] Events moved quickly at the highest levels of global governance. In June 1995, the G7 Halifax Summit identified the promotion of 'good governance' as an important goal for multilateral institutions and in 1996 the policy-making committee of the IMF Board of Governors added an explicit mandate. Its *Declaration on Partnership for Sustainable Global Growth* stressed the importance of 'promoting good governance in all its aspects, including by ensuring the rule of law, improving the efficiency and accountability of the public sector, and tackling corruption, as essential elements of a framework within which economies can prosper'.[40]

By pushing accountability, management, laws, regulatory surveillance and ownership up front, those building on *Governance and Development* established a firm utilitarian basis for operations that finessed the vexed problem of 'politics'. This was backed up by evaluation results purporting to show a 'strong correlation' between various indicators of this litany and 'satisfactory' outcomes.[41] Here was a classic case of driving political ends that could not be approached directly, through utilitarian technical means. Thus interpreted, good governance could defensibly reflect a legitimate concern about the *efficiency* of the state – what would soon be reshaped as a question of lowering 'transaction costs'. Alongside these indicators, it was still possible to maintain a rhetorical commitment to deal with the *equity* effects of the economic system and the *legitimacy* of the power structure. But these classic concerns of 'territorial' modes of governance would be judged in terms of the adoption (or 'ownership') of these technical, efficiency oriented, good governance precepts: in sum, *institutional* strengthening.[42] It would take time to connect all the dots at an operational level, as we will see, and find a way through to far wider political engagement, but the General Counsel's advice in 1992 provided a durable basis for an unprecedented expansion of international agencies administrative ambits.

Much was still being said about the problems of corruption. But a longer litany of principled statements about transparency and accountability – of the public sector – and the auditing of governments' financial and policy credentials began to turn up in the conditionalities and 'prior commitments' attached to adjustment lending. A new generation of conditionality, labelled 'structural conditionality' in the case of the IMF and 'governance conditionality' in the case of MDBs, was being confidently advanced.[43] So defined, from 1996 to 2000, the World Bank alone initiated over 600 governance related programmes in 95 countries and was underwriting explicitly focused governance reforms in 50 countries.[44] By 2002, governance featured in the bulk of the conditions attached to IFI financing (on average, 72 per cent in Africa, 58 per cent in Asia, 59 per cent in Central Asia and Eastern Europe, and 53 per cent in Latin America and the Caribbean).[45]

In spite of this finessing in the mid-1990s, broader political economy questions about equity, the achievement of population-wide outcomes and the fallout of global structural power in local markets all continued to surface.[46] But these had little operational effect. By the mid-1990s, governance policy was being articulated mostly by a new cadre of public administration specialists recruited to operationalize governance policy. By and large, this reinforced the narrowing of focus. From the mid-1990s there was a palpable sense of technical tautology, with one agency, be it the ADB, the African Development Bank, or bilateral agencies (and none too few NGOs), quoting back on each other's good governance definitions and refining a view of governance that highlighted the conservative discipline of transparency and predictability, accountability to the law, and the 'positive' Liberalism of participation. Refined further, for the World Bank, 'governance' became 'a predictable and transparent framework of rules and institutions for the conduct of private and public business'.[47] For the ADB, Good Governance was 'sound development management' and rested on four interrelated 'pillars': 'accountability, transparency, predictability and participation'.[48]

While IFI's intent in depoliticizing governance reforms is understandable, it is important to understand the effect this had. Some important beginnings notwithstanding,[49] there has been amazingly little public debate on this issue, not least because IFIs are reluctant to publicly pursue it. Treating governance as a mostly technical issue didn't take governance entirely out of political contest and debate. But it did take the political edge off it, in a context where arguably it is political accountability that will provide the most potent fix to the problem. Good Governance provided an illusion that things can be fixed by donor and IFI leverage, mainly applied via threats of aid withdrawal. Recipient governments quickly became adept at playing this game of weak threat and symbolic reform, playing on the need of donors to move the money their way for wider political reasons, aware of the few occasions when aid is actually withdrawn. And knowing they can be seen to be acting by setting up some minor bureaucratic anti-corruption mechanism in say the Ministry of Interior, by new laws that will be honoured only in their breech or selective application, and/or play some donors off against others, or that the Chinese or Japanese will make up any IFI shortfalls. This game playing in itself undermines political commitment, by representing the problem as being attended to, and not requiring significant action. As importantly, making the reforms technical tends to define them as the business of a small group of bureaucrats and ministerial executives, who negotiate with donors confidentially, behind closed doors and via the odd piece of public declaration of commitment. The effect of this is to insulate wider politics from the wrath of the classes that have historically had most success in demanding accountable governance: especially the small, growing middle classes. Some argue, then, that Good Governance reformers would be better to find ways to mobilize class interests, and stringently avoid doing anything to mitigate their anger. This would require a politicization of IFI's own position that would have wider ramifications and risks. Meanwhile, even though IFI's programme of promoting technical consensus backed by weak threats ends up a recipe for compromise, it does keep IFIs in business, negotiating, seeking new kinds of less feeble leverage.

Meantime, other political issues were bubbling under technical governance reform. Declarations by the 1995 *Commission on Global Governance* capped the achievement by reiterating Good Governance's disciplinary security underpinnings and expanding it to embrace all of development's potential partners. But much more was at stake than corruption and mismanagement, or the internal workings of poor countries. A wider concern kept resurfacing, around security and containment of risk: 'the concept of global security must be broadened from the traditional focus on the security of states to include the security of people and the security of the planet'. For this reason governance 'now involves not only governments and intergovernmental organizations, but also NGOs, citizen's movements, transnational corporations, academia, and mass media'.[50] As later developed with far greater sophistication through the High Level Harmonization Forums of 2004–2005, poor country governments were being urged to align to global norms in the interests of global growth and security, and then of course to show their commitment to this responsibility by micro-managing, integrating and disciplining multiple local actors in accordance with these global regulatory regimes. But as early as 1995, globalization of Liberal governance norms based on technical harmonization was evidently the shape of things to come. The ways this global to local reterritorializing and embedding would be driven by crisis were less clear: but not for long.

ASIAN CURRENCY CRISIS AND THE RISE OF THE POVERTY REDUCTION STRATEGY PAPER

Into the crucible again: the 1997 Asian currency crisis

Notwithstanding the fact that the techniques they employed have been largely outlawed by the Liberal juridical arrangements of the WTO,[51] Asian Tiger governments' successes in managing sustained growth over decades is now well established: arguably, their work with managed markets constitutes the only successful model for capitalist development of poor countries over the last 50 years. Nonetheless, the heady days of the mid-1990s when their growth rates stupefied everyone's imagination are now recognized as a classic speculative bubble, albeit with fatal local characteristics. But more broadly, the scope and nature of over-investment in South East Asian economies had political economic rather than governance causes, and needs to be traced to the emergence of the Asian Tigers on the back of Japanese capital and, of course, strategic, security driven state led industrial development in Korea, Taiwan, and Singapore. In particular, growth was fuelled by security linked preferential trading status with the West, and circulation of burgeoning Japanese profits and savings pulled from US bonds (where much of it had gone in the early 1980s to chase sharply higher interest, itself a result of Reagan's security obsessions) and expatriated from speculative bubbles that resulted in Japan.[52]

Although compounded by numerous factors,[53] the eventual crisis principally reversed short-term capital flows. The resultant liquidity crisis sent shockwaves, doubts and fears into an over-committed banking sector and out through a realm that suddenly seemed crucial to wider confidence and systemic integrity, namely the way accommodations were made between corporate capital and domestic regimes. Reviewing mid-crisis statements from Alan Greenspan, Treasury Secretary Larry Summers, and the IMF Director Camdessus, it is clear the Asian capitalist model had unfairly come into the policy cabal's sights.[54] Summers, for example, in the *Financial Times* of 20 February 1998, opined that 'the problems that must be fixed are much more microeconomic than macroeconomic'.[55] Such messages joined to sharp narrow macroeconomic orthodoxy would mean things got a lot worse than they needed to, and what Joseph Stiglitz famously reminded were simple, year one economics solutions to crisis would be notoriously set aside.[56] Together, these politically charged, mixed messages made a short-term liquidity crisis a much more damaging business.

Of the IFIs, the World Bank was by 1997 the furthest along in realizing the need to secure markets, and in lining up the institutional machinery now seen as necessary, in the light of the tragic Eastern bloc experience of marketizing reform.[57] The 1995 arrival at the Bank of Wolfensohn and Stiglitz had begun a significant, pro-institutional reform period, which had already been institutionally pegged out in the 1997 WDR (discussed in next chapter). But for now, the precipitous combination of the all too clear Washington Consensus imperatives (maintain fiscal prudence, raise interest rates), with much less effectively articulated institutional prescriptions (sort out corruption and the weak banking sector, what's needed are unspecified microeconomic institutional and political reforms), ultimately proved lethal in dealing with the Asian currency crisis. This combination powerfully obscured realities, prompted misdiagnosis, sent mixed signals to already panicked markets, and precipitated a ruinous collapse in confidence in currencies, banking sectors, governments and IFIs alike.[58] Most damagingly here, the mid-crisis moralizing censure that the whole Indonesian system (indeed the whole Asian capitalist model) was in need of major governance and transparency reform undermined confidence at a crucial juncture. With few capital account controls, money fled as fast as the IMF supplied it. Old crisis reactions rushed to the breach: stabilizing currency flight apparently demanded grotesquely spiked interest rates. In erstwhile success stories including Indonesia, years of poverty reduction were wiped out in days.

Crises typically induce changes and more contest. In the depths of the crisis, the US was able via the IMF to finally leverage further deterritorializing of some countries macroeconomic policies – most famously, more capital account opening in Korea and Taiwan – in line with a wider agenda of securing other countries' profits and savings capital to compensate for low savings rates and high current account deficits in the US.[59] More spectacular political fallout emerged on the streets of several western cities, prompting a substantial legitimacy crisis for IFIs. Calls for the head of the IMF came thick and fast, and impelled both IMF and World Bank to reposture themselves around Poverty Reduction. As debtor adjustment did its familiar thing, the ultimate crisis culprit (financialized unfettered capital) got off unchallenged. The guardians of the financial system – bankers and ratings agencies – reinforced their hegemony, further narrowed the tolerance for policy difference, and enhanced their ability to get the money out quick, whatever the local consequences. Hence the enhanced terror that Thomas Friedman's 'Electronic Herd' of mouse-clicking capital movers holds for any government. As Friedman had noted in 1996, super-power status had come down to the US and the ratings agencies.[60] After the Asian crisis, states everywhere certainly moved (or were moved) to further stiffen their public positions around capital movement, maintaining a rigid, statuesque orthodoxy for fear of provoking that lawless, unaccountable hegemon, finance capital. In the 'developing' world, those wanting to impose more transparency, surveillance and discipline found their moment to hand. The point was clear, not least to IFI staffers ducking from violent reaction behind police barricades: for the IFIs, Good Governance was a security (and development) imperative.

Thus all these factors made it easier to extend demands for conformity with Liberal conceptions of governance, fuelling the institutionalist efflorescence that constituted the post-crisis 'Post-Washington Consensus'.[61] The key, following the 1997 crisis, was how Good Governance could be directly pinned back, not to the imperatives of a globalizing economy, but to 'the new focus on poverty reduction'.[62] By 1999, the G8 Summit in Cologne had turned the two key concerns, macroeconomic adjustment, fiscal reforms and debt relief measures, to an explicit linkage with poverty.[63] This link was dubbed by the G8's 2001 Genoa Summit as the 'strategic approach to poverty reduction'. Read this as post-Vietnam War 'Defensive Modernization' rebirthed in a global financial bullpen. As former IMF Deputy Managing Director Stanley Fisher remarked:

> [w]e took tremendous heat – unfairly, because I think it was not consistent with the facts – over the impact of the Asian crisis on poverty. That was a tremendous factor in the debate over whether the Fund should get more financing – the perception that we had supported policies that hurt the poor.[64]

With this move firmly underway, the IMF, all MDBs and multi-/bi-laterals were through 'good governance' able to accomplish the full convergence of risk, crisis and security management, all joined to the adoption of slightly more 'inclusive' neoliberal market reforms by what was seen as the unassailable 'moral duty to reach out to the poor and needy'.

Ownership, PPA and fixing the poor in places

As Dollar and Svensson showed in their persuasive retrospective on structural adjustment, the 1997 financial crisis served to reiterate that successful reform depended on poor country government 'ownership'.[65] But the political fallout from this crisis made clear that obtaining 'national ownership' and 'long run commitment' for a reform process needed to extend well beyond the handful of technocrats who had negotiated the conditionalities for Structural Adjustment proformas. What emerged in the late 1990s as the PRSP process was far more ambitious, something that 'must not stop with the government', but must involve a 'broad based consensus' including 'middle management, as well as top-level technocrats, global and local civil society and the private sector'.[66] Thus, Development's recipients, especially the hopeful contenders for HIPC debt relief, were soon being taught a host of process consultation and psychological ownership techniques. After all, they were responsible for initiating reforms, they needed the skills to garner intellectual conviction, broad-based political support, and 'institutionalization', so that they could 'stabilize expectations around a new set of incentives and convince people that they cannot easily be reversed'.[67] Left aside was the pointedly global content of policy, instruments and processes. Poignantly left aside too in the wake of IFI responses to the 1997 crisis was the duplicity of calling these approaches 'country driven'.[68]

Internationally, in both core and peripheral countries, aligning local responsibility to global risk was taken forward in three ways. As the book's Part II cases will show, it is achieved through a resurgent moralizing about the poor's responsibility; second, by innovations that fixed the poor, and poverty, in particular, localized places (most favoured being the 'local community') while, third, being seen to give the poor in these places a crucial voice in legitimating the whole enterprise. Ownership was securely pinned to a renewed emphasis on 'rights and responsibilities' at a very local, individual level of 'real people in (local) places'. Here, the OECD antecedents and parallels are striking. The incoming New Labour government of Tony Blair in 1997 expressed this most clearly. According to Blair, 'the rights we enjoy reflect the duties we owe: rights and opportunity without responsibilities are engines of selfishness and greed'.[69] 'Community' was to provide the main, disaggregated, 'etho-political', domesticated domain that could be invested with moral and other responsibility for achieving pro-poor results.[70] In Blairite terms, 'opportunity plus responsibility equals community'.[71] So locality, along with labour markets, became a preferred container for responsible 'inclusion'. Here was where the vulnerable poor, the child, the family were most visibly in need of protection, and it was at this scale that rights and responsibilities could be brought to bear in firm, disciplining ways. This appealed to adherents of the communitarian Left and Right, and it chimed well with the heartland of NGOs who had romanticized 'community' as the site of moral responsibility for all Development. As we will see in Chapter 4, this also occurred because interventions at the national scale (such as redistribution of wealth and opportunity) were being actively set aside in policy, which was beginning to focus on narrowly targeted, highly disaggregated approaches and techniques for governing poverty. At the same time, then, as the UK government's 1997 *White Paper on International Development* was being resolute about the 'moral duty to reach out to the poor and needy',[72] the categories and spaces that were being marked out for the poor and needy were being more clearly incised, and given a whole new disciplinary edge.

The technologies for inscribing poverty in a localized place had already been elaborately assembled. In the early 1990s a commitment was made to produce Poverty Assessments. These, the 'basis for a collaborative approach to poverty reduction by country officials and the Bank',[73] routinely analysed sectoral, infrastructure and wellbeing indicators, provided appraisals of national economic management and delineated sectoral efforts to 'develop the human resources of the poor' and special measures to ensure 'the extent, reliability, affordability, and cost-effectiveness of the social safety net for protecting the most vulnerable groups and the very poor'.[74] As we will show in Vietnam (Chapter 5), by the early 1990s, it was common to find social, environmental, economic and other endowments being cross tabulated with quintiles of wealth and poverty in ways that framed the poor's immediate social order in a series of disaggregated and localized categories of common interest (gender, ethnicity, parentage, landless, disabled) and by local scale (commune, district). By the mid-1990s, Poverty Assessments were well established practice, with more than 25 carried out in Sub-Saharan African countries by a top heavy mix of international consultants and, in the main, staff from national ministries of finance and planning. As will become clear, the later development of *Participatory* Poverty Assessments involving NGOs and participatory research methods constituted a classic 'inclusive' neoliberal approach to framing poverty in 'local vulnerability' rather than political economic terms. These extended the qualitative reach of Liberal framing and surveillance of poverty, and had the effect of turning potential civil society critics into consensual governing partners.

As noted in Chapter 2, NGOs were projected as efficient, trans-territorial localizers for Development, particularly in fraught political contexts and where the state's reach was doubtful or was being curtailed by budget cuts, conflict or legitimacy crises. The move to PPAs was achieved through the partnership with NGOs that had been contracted to apply their apparent expertise in participatory assessment techniques, known as PRA, or Participatory Rapid/Rural Appraisal.[75] Of wider significance than this active governmentalizing of NGOs (see Chapter 2 on this) was how these PRA techniques promised to give the poor 'voice' in what was being reframed as the wider contest of knowledge and social regulation. Crucial too was the overall legitimation PPA participants unwittingly gave to the wider Liberal approach to poverty reduction, and to the framing of local poverty in ways that suited these forms of governance.

Although PPAs augment the indicators of vulnerability and insecurity contained in Poverty Assessments with local voice, colour and content, they add little wider analytical scope. As we will show in detail in Vietnam in Chapter 5, in the course of a local PPA event community members and their NGO advocates engage with a series of participatory mapping, issue and interest based focus groups, listing and prioritizing events. The usually one-off process can generate immediate excitement. Pakistan's PPA reported a 'local analyst' in Karak District, North West Frontier Province as remarking, 'It is the first time that we have had some government officials here in our village to discuss our poverty issues', and another in Dera Ismail Khan District bemoaning that: 'The policies of our government are blind and deaf. Most of these policies are wrong. Nobody listens to the poor. The policies that are developed without our participation make the situation worse.'

Results from 81 PPAs in 50 countries ultimately appeared in a series of World Bank *Voices of the Poor* publications.[76] These provide compelling, almost voyeuristic reading, and have been both hailed as opening up new dimensions of poverty and derided for their depoliticized vision. Pakistan's PPA reaffirmed the sense of humiliation and shame reported worldwide while highlighting the huge inequalities in the basic distribution of livelihood assets – land and water – along with gender, and landlord power. Debate around PPAs focuses on their political durability and a range of other weaknesses. [77] More salient for this book is the role PPAs played in enabling the administrative, fiscal and political arrangements through which governance became disciplined around PRS. As local poor participate, they identify local issues and, in the very same breath, fill out and naturalize the new categories and prescriptive analytics through which their plight is described. This allows the poor and their places to be presented as spaces available for particular kinds of (governance, service, NGO) intervention.[78] In essence, PPAs frame poverty in classic Liberal terms that have Liberalism's predictable structural, political and historical blind spots. Routinely under-represented are the poor's political organizations such as unions, parties and left leaning (i.e. pro-poor) governments. Left-right politics, and the role of class in the capture of power and the distribution of its benefits, the collusive relations of political elites, local judiciaries and police have been generally absent. Similarly, typically ignored are existing social protection measures, or over-writing them with new, formalistic concepts around 'social capital'. Wider structural issues of political economy are obscured; the structure of core productive sectors, and labour markets in them (i.e. these workers are badly paid and poor for structural market reasons), factors that are clearly effects of market economy development and adjustments (this is a subsistence economy prone to commodity trading shocks), trade, market power and wider access issues including migration.[79]

Where these political economy effects do occasionally turn up in PPAs, and thence as we will see (below) in the wider PRSPs, it is in manageable ways that can be *locally* owned and governed. Thus, the effects of market externalities and fallout from production processes (pollution, child labour, workplace injury) and the responsibilities of global and local corporates, become governable in two ways. First, by being localized as problems of 'overcrowding', 'poor sanitation' and 'poor location of slum buildings beside polluted streams', these issues are framed as in turn reflecting governance issues such as the failure of 'rule of law', 'ineffective local governments' or 'inadequate local tax bases'. And second, shaping governance up in these terms provides a new set of markers against which the performance of governments can be monitored and, as we will see, rewarded or punished in the allocation of concessional aid and other financial flows. This double 'inclusion-exclusion' process will be explained further in Chapter 4. For the moment, its important to note that this aspect of reterritorializing poverty (the focus on local, politically manageable 'containers' of poverty, at the same time establishing markers for inclusion in wider global governance regimes), was to prove indispensable for the 'ownership' partnerships that were being crafted to help poor country governments 'face up to the challenge', and make commitments, soon to be echoed throughout all PRSPs, to be 'in the driver's seat' and 'be ready for the long haul'.

The PRS Initiative

By the end of the 1990s all the elements of Poverty Reduction's three-legged strategy – Opportunity, Empowerment, Security – could be laid out in the 2000 WDR *Attacking Poverty*.[80] The three legs expressed not just the 1990s ideological consensus embodying elements of neoliberal market opportunity, together with the social democratic ('positive' Liberal, Senian) empowerment,[81] and conservative security. More practically, the three legs would become a mantra of PRSP processes wherein poor countries' governance was marshalled around what were now branded Poverty Reduction reforms and, as importantly, used to signal their 'ownership' of the wider reform process. In all this, they also provided a set of ideological 'quilting point's'[82] onto which could be stitched the whole gamut of technical Good Governance's voice-listening reforms (such as PPAs) and, as we will see in Chapter 4, a host of technical devices for planning, financing and sustaining the focus from global compacts – later the MDGs – through to specific, localized actions. Together, under the enemy-less brand of Poverty Reduction, all this seemed to constitute an historic high point of Liberal development consensus.

When announced in September 1999, the PRS Initiative key objective, to 'assist low-income countries in developing and implementing more effective strategies to fight poverty . . . through supporting and sustaining a country driven Poverty Reduction Strategy process in low-income countries',[83] drew plenty of fire, mostly because of the brazen repackaging of structural adjustment policies at its core. It was, as we will see, more than this. But the policy commitments annexed to the first Interim PRSPs were in fact 'lifted from past or existing ESAF and Structural Adjustment Credit agreements and incorporated into the new PRGF and Poverty Reduction Strategy Credit (PRSC) documents'.[84]

For one thing, there was little time for their authors to do much else. For example, Cambodia's PRSP – rumoured to have gone though nine hasty drafts before its Khmer translation – was quickly produced primarily to allow the IMF to convert their impending ESAF into a PRGF. A similar imperative pushed Tanzania from interim to fully blown PRSP in just six months.[85] But the PRS Initiative's principle function in the wake of the financial crisis, and at the end of a long decline in Development's legitimacy through the 1990s, was never to question or even elaborate core market doctrine. Not surprising, then, is that the UN Conference on Trade and Development 2002 retrospective found only two instances in 27 African PRSPs that departed from the mantra that maintaining Liberal trade regimes is beneficial for the poor.[86] Similarly, the consultants and country officers huddled to write PRSPs stuck all to obviously to the opportunity, empowerment, security rubric. And so struggled the dubious, but strenuously maintained claims of a 'no blueprints' approach, 'hallmarked' by participatory process, and 'owned' or at least 'driven' by countries themselves. Inside Development, there were critics of core doctrine, notably the Japanese government that was still smarting over the beating that Asian state developmentalism had taken in US Treasury inspired post-Crisis diagnoses, and who 'did not think jumping from one idea to another every several years is the right way to promote development'.[87] But most agencies, like the progressive NGO, Oxfam, put aside doctrinal worries in the belief that PRSP 'offers a key opportunity to put country-led strategies for poverty reduction at the heart of development assistance'.[88] Remarkably quickly, given Development's Byzantine inter-institutional intrigues, agencies from the UN through all the significant bilaterals and most large NGOs welcomed and engaged the process.

Certainly the process itself needed this scope of consensual support: for its own ambitions were considerable. The PRS Initiative was designed to give MDBs, the IMF and the many UN and bilateral agencies that quickly joined-up a place to fully articulate (and continue to harmonize practice around) all the elements of macro-adjustment, economic and social governance reform that had been cohering over the 1990s. And, through a shared process with national governments (in practice, a small group in say the finance ministry, conducting wider consultation missions), to frame all this up in terms of single country situations, with a view to getting ultimate signoff that would signal ownership, commitment, and other antidotes for Structural Adjustment's earlier failure. In sum, PRSPs set out to become poor country's headline, comprehensive strategic development document, describing its 'macroeconomic, structural and social policies and programmes over a three year or longer horizon, to promote broad based growth and reduce poverty, as well as associated external financing needs and major sources of financing',[89] focussing on the 'whole public expenditure program, to ensure that all foreign and domestic resources are spent well'.[90]

The requirements were not lost on the authors of Pakistan's PRSP, who offered the following checklist as evidence of their own compliance:

> In short, [Pakistan's] full PRSP is a living document based on six principles with in-built mechanism for adjustment overtime: (i) it is home grown, involving broad based participation by key stakeholders; (ii) it is result orientated, focusing on monitorable outcomes that benefit the poor; (iii) it is comprehensive, recognizing the multidimensional nature of poverty; (iv) it is prioritized, so implementation is feasible in both fiscal and institutional terms; (v) it is oriented to build public-private partnerships; and (vi) it is aligned with the Millennium Development Goals.[91]

As we've seen, much that had gone on before was knittable into the new process. NGOs brought the legitimacy of their participation in PPAs. The UN agencies brought their headline MDGs,[92] a set of tough indicators of core poverty related outcomes that would give Poverty Reduction badly needed credibility and accountability to a policy consensus that was so apparently not doctrinaire. MDGs included eradicating extreme poverty and hunger (MDG 1), achieving universal education (MDG 2), and promoting gender equality and empowering women (MDG 3). Developing a global partnership for development (MDG 8) provided a generous goal around which were retrofitted many Good Governance initiatives. As marshalling and legitimating points for all manner of projects, they reinforced the overall ideological impression that the plethora of social service, community development, Good Governance and kindred projects were all ultimately not palliative actions, but powerful devices all coherently joined-up to the common PRS purpose.

If not for the PRS Initiative, the MDGs would probably have languished on the UN's cocktail circuit. On the other hand, it was quite clear that the IFIs' neoliberal policies would not drive sufficient economic growth to finance MDG outcomes, and nor did it seem likely that sufficient aid would be forthcoming to fill the gap. World Bank economists had acknowledged that even the target of halving 'dire' income poverty by 2015 could be met only in the unlikely event that the 1990s average per capita GDP growth rates were doubled. Even then, the number of poor would increase by 345 million, that is, at a faster rate than in the past. Getting close to the MDGs would require $40–$70 billion in additional Overseas Development Assistance (ODA) per year, and this implied that current global ODA levels would need to double.[93] In this context, the PRS Initiative offered two critical ingredients: a set of agreed operational principles, and a way to tap into the expected fiscal fillip of the HIPC initiative.

The key principles – ownership, results focus, and country-led partnership – were adopted from the World Bank's CDF.[94] But whereas the CDF encapsulated a way of 'doing development', it did not provide the kind of pre-qualification regime nor the incentives needed to encourage poor countries to become full graduates of PRS doctrine. For its part, the HIPC initiative provided the resources necessary to link the CDF principles with the spending priorities of PRSPs and through this, a pathway to realization of the MDG targets.[95] Whereas the initial 38 HIPC qualifying countries were required to prepare PRSPs as a condition of debt relief, this was soon extended and made a pre-qualification requirement for all the 80 or so low income, aid dependent countries. Not surprisingly, access to HIPC resources became the key incentive too for poor countries to prepare PRSPs and to participate in the ownership exercise.[96] With HIPC came a comprehensive framework of language, priorities and programmes, and this in turn created some tension with both the 'home grown' and 'locally owned' principles of PRS. But the focus in practice was on completing the PRSP documents and, since the PRS guidebooks were fairly light on how to adapt the Initiative to different country conditions, key sections in country PRSPs were all too often verbatim borrowings from elsewhere.[97] Where they did hybridize with home grown policy (in Vietnam, Uganda), there were interesting deviations, often in the direction of sectoral production and other initiatives culled from socialist territorial planning doctrines. Indeed, across PRSP, otherwise moribund (because unfunded) planning processes now came into ironic, fraught accommodations with neoliberal institutionalism. The result: more planning and PRSP-one party state accommodation, now down to province and even lower levels.

The fact that the PRS Initiative offered a consistent view of what was to be achieved – the MDGs – has encouraged a focus on the operational rather than the doctrinaire in PRSPs. They tend to be preoccupied with where investments need to be made and with creating a set of instruments for directing and monitoring these resource flows. Aside from their routine preamble statements which represent data and surveyed voices about poverty, the bulk of these documents comprise tabular presentations about the overall public expenditure programme and its allocation among key areas, a matrix of policy actions and institutional reforms and target dates for their implementation. The fact that PRSPs tend to comprise these two parts – the contextual and representational preamble book-ended by quantitative commitments – means that actual strategy to reduce poverty tends to be treated very lightly; the principle concern seems often to ensure that strategic statements include suitable tributes to wide ranging constituencies of interest. The taken for granted nature of the consensus and its narrow institutional, technical and political scope, in turn greatly enabled further elaborations.

PRSP's Development beyond neoliberalism?

Thus, the three legs did in their PRSP formulation register shifts beyond the neoliberal Development of structural adjustment. Here, under subtitles, we briefly consider the nature of those shifts. Whereas in the early 1990s, Poverty Reduction stood resolutely on two legs or pillars, broad-based growth in incomes and investment in education and health care (along with, in some cases, recognizing the human costs of structural adjustment, safety nets for those 'unable to participate in growth'), by the 2000 WDR, these had been tremendously elaborated into three-legged versions.

Opportunity

The 1980s and early 1990s economic reform agenda was dominated by macroeconomic structural adjustment; after 2000, at the heart of Opportunity lies a broad, pro-poor growth that 'can be defined as one that enables the poor to actively participate in and significantly benefit from economic activity'.[98] The focus on macro policy constraints – overvalued exchange rates, trade and financial sector regulations, regionally biased industrial location politics, etc. – still remains, but more as a taken for granted precept that is seldom clearly elaborated. But now, by using Poverty Social Impact Analysis (PSIA) for instance, Opportunity in PRSP gives more attention to how structural adjustment impacts on the poor, and to measures believed necessary to 'make markets work for the poor', including pro-poor legal and regulatory arrangements, investment and savings schemes, access to financial markets and business services. Opportunity is still keyed to the scalar of income, but consumption based poverty indicators now include also the imputed value of human, social and physical assets available to the poor and assessments of constraints that effect the 'capability' and 'empowerment' of the poor to access product markets or important factors of production, like land, capital, or small to medium enterprise. Growth, or more consistently, the mantra of fiscal-stabilization-leads-to-growth, then, is valued both in itself and as a means to an end: poverty reduction.

In all this, its hard to conclude that PRS's Opportunity simply carries a brief to serve as handmaiden to Liberal, global capitalism. In fact, the Joint Staff Assessments that must be done on each PRSP by IMF and World Bank technicians commonly lament that while they may be strong on exogenous and endogenous risk assessment, PRSPs are typically weak on the actual economic growth strategy.[99] Indeed, PRSPs often focus predominantly on leveraging public consumption expenditures for poverty reduction and give scant attention to non-expenditure related policies that might enhance or constitute obstacles to growth, such as exchange rate management or tax and revenue policies, or, at least until 'second-generation' PRSPs began to be discussed late in 2004, to public investment in production oriented to economic growth.[100] One indicator of this is that economic growth and financial sector management rank last of ten priority PRSP spending sectors. In fact, finance and private sector development as a share of IFI overall investments dropped from a pre-PRSP share of 40 per cent to under 20 per cent after PRSPs were introduced. Thus, in most PRSPs, after a typically optimistic statement about real GDP growth prospects, the bulk of rhetorical and fiscal space available is devoted to targets and expenditures for HIPC's 'quick win' sectors, health, education, water and sanitation scaled to levels of investment that a country's advisers judge the donor community will accept.[101] That said, we may well be back in institutional overreach and political economic blind spot territory here. However elaborate your definitions and monitoring of 'pro-poor growth', that growth still has to emerge from productive sectors. And here, Poverty Reduction has yet to significantly elaborate sectoral or industry or even trade policy of the kind that will radically boost productivity and international competitiveness in sectors that can most impact poverty. We have more to say about this in Chapter 9.

Empowerment

A primary task for PRSP was to create and operationalize consensus between disciplinary, conservative and social democratic, participatory ethics around the neoliberal market agenda. Thus the early 1990s investments in human capital could easily be recast in terms of empowerment, and thus draw on the popularity of Amartya Sen's *Development as Freedom* precepts about individual capability, and about the need for poor people to participate, negotiate and hold accountable the institutions that affected their wellbeing. Here Poverty Reduction's Empowerment had dual appeal – to the market conservatives, keen to rein in the discretionary ambits available to market-meddling states, and to Development's positive Liberals and communitarians of the right and left keen to build 'individual' and 'community capability' and 'wellbeing'. Thus, as we will see repeatedly in the chapters to follow, in Empowerment, PRSPs can offer a place for the epidemiologist worried about health service standards, room also for celebrants of 'community voice' and the larger Community Driven Development approach to spending service delivery monies, while leaving plenty of space for social democratic public sector reformers and Third Way state re-inventors to promote their various approaches to decentralization.

But positive Liberal notions of empowerment are always sorely in need of sociological caveats. Like other Liberal notions, they present their possibilities in universal terms, and greatly understate the constraints on different people in different territories. When substantive differences are recognized (and they are very good around gender), there is a disturbing tendency towards tokenism: a little programme for these people, another for that community, an Information, Education and Communication programme for those behaviour puts them 'especially at risk'. A little voice here, a little community partnership there. This round table multi-inclusion may feel good at the time, but in practice is not just sociologically, but politically naive. In its full-blown form, it creates a powerful 'inclusion delusion', a perfect ideological smokescreen, legitimating wider Liberal market relations and obscuring the powerful inequalities they lead to.

Security

Security received equally polysemous treatment as PRSPs came to elaborate a wide range of security, risk and vulnerability analyses. Again, these appealed as much in a neoliberal court (first concerned for security of capital) as to either conservative or social democratic judgments about poverty. The low level of security, including vulnerability to shocks which cause a decline in wellbeing, may arise at the household or individual level (e.g. illness or death), the community level (pollution, vagarious rainfall, rebellion or riots) or national level in the form of terrorism, gangland rein, civil war, or other strife; all which may act to undermine Opportunity or Empowerment.[102] The World Bank's *Social Protection Strategy Paper*, developed in 2001, intended to operationalize the risk and security dimension of 2000 WDR. And then 9/11 drove the security dimension all the way home. 'Security' issues in PRSP terms have been framed overwhelmingly within a Liberal, or 'inclusive' neoliberal conception: liberty within the law (and within markets and communities). Here, for a time, security framings were relatively free of wider sphere of influence geopolitics and nationalism, and constituted a policing and vulnerability reducing exercise around the edges of the international order. In this, no doubt, substantive aspects of economic security, the territorial political economy of poverty have been underplayed, and substantive redistributive moves that might enhance the poor's security remain underdeveloped. But at least the risky poor were not submerged under increasingly shrill nationalist and militarist developments: as, if Polanyi was right, may be the case in the future.

Assessing PRSP: what impact in moving beyond neoliberalism?

Cynics have fairly remarked that the PRS Initiative is still something of a headline commitment, a kind of 'one size fits all' ideological clasp at the top of many poor country policy commitments. Its real echo in budget and expenditure management, or in sector investment strategies is often far weaker than rhetoric would have us expect. There's no doubting its rhetorical overreach; in many respects the advent of PRSPs and MDGs is not yet impacting on the core business of how aid is organized or delivered, or on the retinue of consultants and bureaucrats that comprise the public face of Poverty Reduction. Let alone on the substantive structure of the economy. But there is little doubt the new high volume lending instruments launched by the World Bank (the PRSC) and the IMF (the PRGF) have had a big impact on how concessional finance is calibrated to PRSP's thematic points of focus. In the World Bank's case, no country assistance programme will finance sectors that fall outside the PRSP and concessional lending clearly favours countries with PRSPs.[103] Increasingly, bilateral donors have shifted towards PRSP-sanctioned direct budget support systems that augment these larger budget funding instruments, and this, more importantly, has placed the World Bank/IMF even more clearly in the role of being a 'signaller' to others. Table 3.1 tracks the kinds of shifts towards Poverty Reduction's Opportunity, Empowerment and Security that we have outlined so far. Public Sector Governance, before and after PRSP, remains as expected about the same. Opportunity's private sector, finance and economic management, all significantly decline; while the 'Empowerment' financing for Human Development, Social Development/Gender moves from 3 per cent to 26 per cent. Security's Social Protection and risk mitigation measures moves from five to 21 per cent of total IFI investments.

Thus, as we will show in Chapter 4, where government-donor coordinating mechanisms exist, 'the PRSP has increased donor participation in, and coordination of, budget support instruments and the streamlining of their performance monitoring'.[104] Yes, there are cases where the PRSP is apparently well integrated with country budget and expenditure management – as in Uganda, Chapter 6 – but in most poor countries the match is still sporadic. Ethiopia's PRSP is perhaps closer to the norm, here before the PRSP ink was barely dry, key policy decisions, for resettlement, education, water sector reforms, all differed from the PRSP commitments.[105] That said, poor

Table 3.1 Share of IFI investments pre- and post-PRSPs[106]

Thematic focus	Pre-PRSP adjustment lending (% of total)	Thematic focus of PRSCs (% of total)
Public Sector Governance	23	26
Human Development	2	13
Finance and Private Sector Development	40	19
Environment and Natural Resources	3	6
Economic Management	17	1
Urban Development	1	4
Trade and Integration	5	4
Social Protection and Risk	5	21
Social Development/Gender	1	13
Rural Development	1	7
Rule of Law	2	

country governments are now seldom in doubt about the persuasive intent of the parties drawn into the Poverty Reduction consensus. The new IMF/Bank instructions about PRSPs empower staff to 'discuss with the Authorities any modifications to the strategy that might be considered necessary to allow managements to recommend to the Boards that the PRSP be endorsed'.[107] These modifications in public policy and fiscal commitments are stepped out through a rubric that has the IFI's Boards endorsing each step, Interim PRSP, sometimes called a PRSP Preparation Status Report, the full PRSP, followed by a Joint Staff Assessment, then Annual Progress Reports, all linked back to fiscal transfers and, as we will see in Chapter 4, larger gate-keeping announcements about fiscal and political probity, fiduciary risk and security.

Thus in its programming and commitments, PRSP has indeed moved beyond aspects of neoliberalism. But we should not be blinded by the intentions, or even the practices: for as ever, the most powerful drivers of real opportunity, empowerment and security exist not primarily in Development's institutions and programming but elsewhere in the more powerful structures and changing relationships of political economy: markets and societies, markets and territories, markets and capital. In this sense, we might see PRSP not so much as leading innovation, as reacting to and narrowly encapsulating shifting subterranean political economic changes, and framing them up for strategic attention within a strongly Liberal rubric.

Thus in a sense all these PRSP dimensions were new only in their new contexts, and in their apparent institutional joined-up-ness. It's important then to see how much of this was also happening in wider political economic and institutional contexts. And, to be sure, such arrangements had been already coming together in OECD political and governance circles, as Polanyian market backlash joined-up with social democratic and Third Way political process and 'positive' Liberalism to produce another interesting governmental hybrid. So before turning our full attention to how all this was deployed in poor countries (Chapter 4), we need to understand just a little more of the core, OECD country, political developments through the 1990s and early 2000s.

THIRD WAY 'INCLUSIVE' NEOLIBERALISM: GOVERNING AND EMBEDDING THE NEW CONSENSUS

Many years ago I suggested that the first party to occupy the new political terrain that combined the ethics of community with the dynamics of a market economy would be in power for a long time. It would be the equivalent of [New Zealand's] first Labour government because it would usher in a new political settlement that would last for decades.

Steve Maharey, *The Third Way,* 3 May 2002

Through the 1980s and 1990s, neoliberalism introduced the possibility of governing more through the market, and less through the territorial and social. With the demise of Keynesian demand management and the rise of Liberal openness to capital markets and globally networked commodity production, core economies found that a welter of other Liberal governance and policy settings now became plausible. The new policy smorgasbord invited the market and corporate interests into what had seemed natural territorial monopolies of utility and public service provision. Markets or quasi-markets were being created for what had been socially-driven and territorially governed education and health care. Competitive contracting was being encouraged for provision of all public sector functions, in which NGOs or private interests from anywhere might deliver services as one of many contractors operating in the same erstwhile territorial domain.

As we will see in New Zealand in Chapter 8, the integration of production into liberal markets and the emerging, fragmented nature of contracted service provision created both evident inequalities and deep-seated social reactions. In the mid- and late 1990s, erstwhile social democratic parties taking government after long periods of radical and conservative-Liberal reform (as in UK and New Zealand), launched themselves on hybrid 'enlightened reactionary' platforms. These promised both Liberal market orientations and as we will explain, a 'wraparound' array of less certain social and territorial dimensions. Retained was the Liberal commitment to global capital markets that promised fiscal integrity, current and capital account openness, floating exchange rates, a reserve bank focused on inflation, and ongoing, contracted third sector involvement in service provision. But at the same time, these commitments had to be seen to be serving the needs of, if not society, then at least 'people and communities'. Here, we briefly elaborate some dimensions of the re-embedding, hybridizing process, which became known as the Third Way.

These governments, beginning perhaps with the Clinton presidency in 1992, faced the tricky task of being seen to conform more sharply than their conservative predecessors to basic neoliberal macro and microeconomic practice, while responding to popular disaffection about the socially disruptive effects of neoliberal reform. As we describe later (and elsewhere[108]), 'inclusive' neoliberal approaches involved a basic double orientation to both markets and some aspects of social 'inclusion', often framed in quasi-territorial terms like 'community' and 'partnership'. The double orientation was most obvious (some say notorious) in hybrid rhetorical formulations and policy innovations involving for example rights *and* responsibilities, 'social capital' and 'social entrepreneurship'. There was a pervasive emphasis on individual and community responsibility, and on local or decentralized approaches to social programmes. In all, the orientation was towards 'enabling' citizens to be 'included' in markets, especially the labour market, via training, job market incentives and obligations, and support through periods of 'vulnerability' and lifecycle transition.

Opportunity plus responsibility equals community

In many respects, these approaches were reactivating an older Liberal discourse. Now, instead of the 'negative' Liberalism of getting governments out of markets and cutting away regulation and social programmes, here was a 'positive' Liberalism,[109] emphasizing in terms familiar since the 'new Liberalism' of J.S. Mill the need to empower individuals to take their place in markets and civil society. In this positive Liberal frame, government itself could be 'empowered to be enabling', creating frameworks wherein plural (empowered) actors in government, markets and civil society could be marshalled and 'joined-up' to focus together on delivering services that worked to 'enable' and 'include' people. Agents, whether individual, NGO, community, corporate, or governments, were seen as needing capabilities in order to do their job: an educated, healthy and active child, student or worker; a sustainable, strong family, community, or NGO; an efficient and effective state. Here, the vulnerable and at risk were not to be merely cut adrift to fend in the market, but invested in and wrapped around with inclusive support and trained into a mature, sustainable role in the marketplace and workforce. Whereas risk and security fears were now globally pervasive, affecting people and communities everywhere, this programme set out to aggressively manage risk, to the extent of enforcing – or at least highly incentivizing – inclusion in labour markets. If early neoliberalism was sink or swim in the global market, this programme insisted on swimming lessons until further notice. Even a mild nationalism was acceptable, as long as this enabled better, smarter market inclusion for active, innovative subjects and businesses. Thus as in New Zealand's erstwhile neoliberal Treasury vision for an 'Inclusive economy', the three pillars of PRSP-style poverty reduction (Opportunity, Empowerment, Security) appear as:

> Productive Capability (arising from inputs of labour, capital and technology, and productivity), Social capability (arising from norms, values, trust, institutions, networks, human capital) and Wellbeing (arising from consumption of goods and services, family, health, job, security, community, freedoms and opportunities).[110]

Good Governance (or, in Third Way terms, 'modern government') came to signal three things: one, a 'democratic centralism' of tough accountability regimens that featured whole of government targets, contracts and managerial accountabilities focused on service delivery outputs and social outcomes. Second, Good Governance also signalled a disaggregated, market friendly approach to services that empowered all manner of local agencies (governments, NGOs, entrepreneurs, etc.) to participate as partners in creating local 'political markets' that would efficiently allocate resources and competitively contract service delivery in ways clearly linked back to specifically defined mandates at different levels of government. Third and above all, Third Way governance promised that together these plural but joined-up services might provide a 'soft' institutional approach to both accountability and 'social inclusion' (sometimes disparagingly referred to as the 'inclusion delusion').

As we will see in Chapter 8 when we turn to examine how this was applied in New Zealand, under inclusion-delusions like 'the whole of government responding to the whole of community', local partnerships were allowed and encouraged to join up. At the same time, partnerships became ensconced in governmental relations that set out to include individuals and communities in webs of mutual responsibility, often by creating joined ways of working at the frontline of service delivery: interagency family conferencing around youth justice issues, programmes involving several agencies' social workers partnering to plan family support. In this, ways of partnering and creating accountabilities were elaborated unevenly and diversely. As we will see in detail in Chapters 4 and 8, and in line with its emphasis on state 'capability', 'inclusive' liberalism has continued to roll out a number of sharp governance ambits: targets, benchmarks, narrow accountabilities, outputs and outcomes. Crucially, these have not displaced the competitive contractualism and narrow, vertical output accountabilities of recent NPM or NIE style reforms. Rather, they have accreted these with 'inclusive' and wraparound local partnership dimensions, and with awkward conceptions of 'managing for outcomes'. But hard vertical accountabilities and moral social inclusion sentiments and obligations, visited on frontline workers and clients themselves, were hardly matched by substantive horizontal, or interagency accountabilities. Here, rather, as Chapter 8 shows, all this disaggregation and 'soft' institutionalism was a recipe for messy, ineffectual coordination at local levels, with minimal impact on substantial drivers of poverty.

Here, the Third Way (like the Wisconsin-model welfare reforms rolled out during the standoff between Clinton and the Republican revolution) simply reiterated the salient characteristic of what Esping-Anderson has called 'liberal welfare regimes': their reliance on work as the main form of welfare, beyond which are only minimal benefits, retraining if you are lucky, and sanction if you are not. This is crucial: the Liberal welfare regime model prevalent in the 'Anglo' countries of the UK, US, Canada, Australia and New Zealand is clearly salient for how the PRS Initiative has been put together. It might also be expected that its long-term path dependencies, including high levels of inequality, and fraught horizontal accountabilities, might be expected to follow.

Although many observers questioned how far this went 'beyond neoliberalism', Third Way governments billed this as a major departure. It offered a smart, market friendly way through which privatized sectors could be re-regulated to achieve sustainable social (or environmental) ends, and a morally unassailable way through which self-regulating individuals could be educated, encouraged and disciplined into market participation, especially in the labour market. For many directly involved in Development, the embrace of Senian positive Liberalism[111] constituted a welcome move towards to their personal political positions. In IFI dealings with peripheral developing countries there was no equivocation: here was a way to put Development's Humpty Dumpty back together again.

Clearly, while salient for Development, the particular emphases of Third Way 'inclusive neoliberalism' weren't universally felt. They were especially salient for 'Liberal welfare regime' countries,[112] post-conservative, 'Anglo' welfare states experiencing mild social democratic reform. Elsewhere, conservative, nationalist and security driven versions of re-embedding reaction were by 2001 visible in many places: neo-conservatism in the US; resurgent, post-shock-therapy nationalism in the former Soviet Union; growing poor country reaction to WTO efforts; strategic bilateralism in free trading. Nonetheless, the international political effects of more 'inclusive' modes of Liberalism have been manifested in Poverty Reduction, not least because the Clinton and Blair governments powerfully supported such an ethos in the World Bank.

In a speech on 8 September 2001 that addressed his 'Priorities for Fall: Education, economy, opportunity, security'[113] even George Bush could sound inclusive and Liberal. Many such positive Liberal ambits, it will be clear, remain within US foreign policy, and can be seen in the allocation

Table 3.2 Beyond neoliberalism?

Conservative neoliberalism	*'Inclusive' (neo)liberalism*
Getting the state out of markets	State must create institutions to enable markets
Markets deregulated, a law unto themselves	Markets need to be embedded in institutions and community by 'smart' re-regulation
Countries open up and conform to global market rules: sink or swim	Countries 'own' reforms, commit to good governance, enable markets
People participate/are included in markets: sink or swim	'Enabling' people to participate/be 'included' in markets and community, through basic services
State/others as funder of services, market as provider	'Joined-up' co-production of services, central and local state, markets, civil society: 'soft' institutions
Poverty reduction through market integration (structural adjustment)	Poverty reduction through markets, 'enabling' services and empowering partnerships
Market and fiscal discipline: IFI loan conditionalities	Market, fiscal and moral discipline: the obligation to participate, govern through efficient incentives
'Thatcherism', neoclassical economics, Structural Adjustment	'Third Way' social democracy (Blair/ New Labour); IFI/multilateral-led PRSP and Poverty Reduction

criteria of the US Millennium Challenge Fund. Evidently, however, 'inclusive' neoliberalism has not turned out to be the kind of enduring settlement,
linking market to community, that New Zealand's social development
minister envisaged in this section's opening quote. Rising systemic risk
and security fears have added a conservative twist, especially since the
shock of 11 September 2001, and its 'therapy' in Iraq. By his 2002 State
of the Union address, Bush could confidently declare his budget supported
a new three-legged ambit, a long way from 'inclusive' neoliberal conceptions of opportunity, employment and security: 'We will win this war; we
will protect our homeland; and we will revive our economy'.[114] In the
uber-Reaganist budget blowouts that followed, and the increasing gap
between multilateralism and neo-conservative unilateralism in US foreign,
economic and trade policy, the Consensus in Washington could be declared
over. By 2005, *The Economist* would consider the ensconcement of Paul
Wolfowitz in the presidency of the World Bank as evidence that the Bush
administration wanted to capture the Bank, and make it more overtly the
agent of US foreign policy. In all this, the high point of 'inclusive' neoliberalism appeared to be receding from view. What implications, then, for
governance and Poverty Reduction?

4 Local institutions for poverty reduction? 1997–2005

Re-imagining a joined-up, decentralized governance

In sum, governance is a continuum, and not necessarily unidirectional: it does not automatically improve over time. It is a plant that needs constant tending.

World Bank, *Governance and Development*, 1992[1]

Unable to make what is just strong, we have made what is strong just.

Blaise Pascal, 1670[2]

By early in the new millennium, Poverty Reduction seemed to present a comprehensive 'inclusive' neoliberal framework. This came replete with universal policy framings around economic Opportunity and financial Security, that were apparently linked back through a range of participatory devices to achieve Empowerment and mitigate social risks through service delivery. But governments remained the weak link in all this. Still missing were transparent budget and expenditure processes, budgetary and fiscal transfer devices to actually move the money down into places it might reinforce NIE's incentives and give strength to sanctions when behaviour was not PRSP-compliant. As described in Chapter 1, actual practices here were plagued by all kinds of frailty and perversity: underpaid public services, unreliable resourcing, budget skimming, rent seeking and corruption. . . . Addressing this was to be achieved by no short route for it required, in the words of Pakistan's 2003 PRSP, a 'series of fundamental transformations'[3] that amounted to a thoroughgoing reform of governance in its practiced form. This chapter will sketch a history of how the new politics of governing was finessed, and the technical instruments crafted by paralleling the 1997 to 2005 period covered in Chapter 3.

The process we will describe involved familiar dimensions of Liberal governance: using the market, contracting out, and allowing the private and civil society sectors to deliver services. But there was more to it than market privatization, as the first section explains. Here potent, formalistic and normative theories of governance, especially drawn from NIE, were deployed to disaggregate functions of government and reassign them across the board.

Here, in sum, a full-blown formalistic institutionalism got the better of substantive governance. As we will show, governance functions such as policy, planning, audit, financing, interdiction and more were identified and reassigned both horizontally and vertically: down to decentralized, sub-national authorities, including local governments, and across to various mixes of private contractors, consultants, civil society. Here, normative principles and assumptions ruled, especially the belief that better informed, competing providers and governance being 'closer to the people' would result in more efficient allocation of resources, and more accountability. As we will see later, the result was not just the disaggregation, but the fragmentation of governance: moving it away from central political, social and territorial accountabilities (which to be sure, had not always been effective and were being further frayed by structural adjustment), and out into localized, marketized community, partnerships and other decentralized quasi-territories.

The second section explains that these shifts became the hallmark of a resurgence of interest in decentralization. Under the code of privatization, this kind of fragmentation of the state had been underway since the 1980s Structural Adjustment response to poor country indebtedness. This continued through the mid-1990s but was extended to other disaggregations around community ownership and participation, to local knowledge, market-styles of competition and 'partnership' based 'co-production' of services. And, as we will show, to enforcement systems that were 'locally owned' but which all adhered to globally sanctioned norms. This process generated enormous complexity and contradiction, uncertainty and risk and, as we will see in later case studies, precious little confidence that accountability to Poverty Reduction could be achieved. Indeed, the precepts of NIE, that several accountabilities – the result of disaggregating government – would produce better accountability, was assailed by experience from practice. But again, none of this happened because of accumulated evidence that any of it delivers pro-poor outcomes: rather, this was driven by normative politics, in this case reacting to fragmentation, and promising, in the case of Third Way OECD governments whence much of this derived, to bring community back into rampant market governance. Thus over this period, we will show something of a double movement; on the one hand, market disaggregation and deterritorializing state functions and delegating responsibility for setting norms for standards and practices for health, economic behaviour, education, regulation of land, water or other rights and so on to the international stage. On the other, responsibility for determining how these all cohered in 'governance' was in theory devolved to the locality, and various reterritorialized (or, again, quasi-territorial) forms of sub-national governance, that were bestowed with responsibility to bring all this together.

But there was another, perhaps even more Polanyian part of the story. Adherents of these neoliberal and communitarian reform doctrines were compelled also to accommodate the disciplinary, conservative edge of neoliberalism. How to contain the profligacy of national and local politicians, even the much-vaunted poor citizen? How to ensure they in turn delivered on *their* Poverty Reduction promise? A range of governmental techniques were crafted especially around PEM and the budget, and transfers of funding from centre to sub-national government levels, to allay these concerns. Originally created as a reaction to Structural Adjustment fallout and, in the context of HIPC debt relief, to channel the dividend of international concern with poverty and security to poor countries, these devices evolved into means to put fetters on (and incentives around) the way plans, budgets and expenditures were decided. We will explain how these techniques for aligning sectoral expenditures, and what is called intergovernmental fiscal transfer systems for moving resources from the centre to local governance arrangements, were in turn backed by a range of 'extra-territorial' restraints that it was believed would create a 'common accountability platform' and facilitate international collective action. Combined, they would make sure that at all levels governments and donors worked in harmony, and complied with moral, fiscal and political injunctions about what they were responsible for, and with how poverty needed to be governed in particular ways.

The governing techniques described here will be elaborated in the case study chapters that follow. The techniques are legion and, before seeing them at work in particular places, our aim here is to describe a few in the polished and packaged form in which they were made to travel, as part of what James Wolfensohn, World Bank President from 1995, regarded as 'part of an ongoing process through which the Bank and the borrower develop and nurture a mutual trust and commitment as the reform proceeds'.[4] We begin with 1997 WDR, *The State in a Changing World*, and move through 2002 WDR, *Building Institutions for Markets*, to 2004 WDR, *Making Services Work for Poor People* that, for reasons we explain in the third section, may be regarded as the high water mark in neoliberal institutionalism.

FROM 1997 WDR TO THE DECENTRED BUT CAPABLE STATE

As should be clear from Chapter 3, the publication of *The State in A Changing World* by no means signalled the start of Good Governance. But 1996–1997 does seem to have been a watershed: the first impact of Wolfensohn and Stiglitz recent arrival in the World Bank and the election of Tony Blair's New Labour meant new blood and commitments were prepared to seize the unforeseen possibilities that would emerge from the Asian financial crisis. So retrospect met opportunity: looking back eight years, 1997 WDR assembled a much broader range of already current norms, commitments and rationalities capable of supporting Liberal re-embedding. In particular, it drew extensively on the emerging NIE view of governance (Williamson, Coase), both theoretically, and in the expert panel advice of NIE guru Douglass North, and Graham Scott, who was a central figure in New Zealand's radical embrace of NIE, discussed in Chapter 8.[5] But the 1997 WDR and its NIE underpinnings also served to greatly expand the domain of 'good governance' in Development: it made possible 2000 WDR's 'Opportunity, Security and Empowerment', and by 2004, the redirection of the overwhelming bulk of international donor and IFI attention to Poverty Reduction's 'service delivery' ambit. This, while leaving entirely intact the neoliberal prescriptions of global integration and macroeconomic policy that hid in the wings of Opportunity.

Citing the collapse of the former Soviet command and control econ-omies, the fiscal crisis of the welfare state in industrialized countries, the 'important role' of the state in the miracle economies of East Asia, and the 'explosion of humanitarian emergencies in several parts of the world', 1997 WDR announced that the determining factor behind all is 'the effec-tiveness of the state'. Policy now had converged around the idea that there had been an over-withdrawal of the state and the realization that, para-doxically, a strong and capable state was required to implement the neoliberal reform agenda.[6] On page 1 of the 1997 WDR was reflection that 'today's intense focus on the state's role is reminiscent of an earlier era, when the world was emerging from the ravages of the Second World War, and much of the developing world was just gaining its independ-ence'.[7] But in place of the post-Second World War state's extensive engagement in managing demand and production, today's state had to be permitted to take on only what it was 'capable of doing'. While the analysis was typically formalist, and less certain about what 'capability' would prescribe in actual country situations, it was clear in general that a glob-alizing economy had narrowed tolerance for capricious behaviour. Rather, the state's main responsibilities were emphatically bound to making markets work: tax, investment rules and economic policies must be 'respon-sive to' the needs of globalized markets; services need to be unbundled to take advantage of the opportunities provided by technological changes, and multiple potential providers.

A striking degree of unanimity: the Capable State

Both markets and citizens, 1997 WDR announced, have 'come to insist' on transparency in government, and on changes 'to strengthen the ability of the state to meet its assigned objectives'. The 1997 WDR clearly reached well beyond previous structural adjustment fixations, but there remained a 'striking degree of unanimity'[8] about both macroeconomic policy and post-adjustment changes required to ensure public institutions were com-plementary with the market. The state was brought back in, albeit 'not as a direct provider of growth but as a partner, catalyst, and facilitator'.[9] Nonetheless, responsibility to manage risk and vulnerability, the poor and the environment was being shifted firmly back on the transforming state. Macroeconomic stability required that the state also 'keep its eye' on the 'social fundamentals', and this added four tasks to the litany of 'market fundamentals': establishing a foundation of law; investing in social services provision; protecting the vulnerable; and protecting the environment. It was not just that markets and governments were again inherently complementary, or could be made to be: in this marriage of necessity, states that failed to embed market reforms would be marginalized, and the resulting social dislocation and lawlessness would make them even more risky and unstable.

The inference in the 1997 WDR is clear; a veil was firmly pulled over the extent to which problems of dislocation and instability reflect 'frontier issues' of peripheral or emerging economies in the international political economy. Poor states must both conform to global norms, and manage their own exacerbated problems. Thoroughgoing reforms were required. But even though change was to be domestically contained, the role of government was boiled down further, to stability, risk management and discipline – staying the course, and owning the reforms necessary to connect better with globalized market 'opportunities'. This move was intended to avoid a rerun of events following the debt crises of the early 1980s. Policy then had further peripheralized them as debtor states, providers of raw materials and export platforms to core countries – and poverty and inequality had risen within and across countries and globally too, to assume renewed strategic prominence.[10] The urgent need to avoid a repeat of immiserating catastrophe was clear by the mid-1990s, when the IMF's former Managing Director dubbed poverty the 'ultimate systemic threat',[11] and was potently reinforced by the 'unfair heat' the Fund took following the 1997 crisis. The parallels between the 1980s and late 1990s were clear in other ways too. Both crises had reduced states rich and poor to 'competition states', distinguishable from one another only in the degree of their subordination to policies that would signal both their private portfolio capital credit-worthiness, and the strength of their moral fibre as worthy of concessional financing for essential service delivery.

Yet, the post-1997 'capable state' was to be critically different, as the distinction between what came to be called 'first-' and 'second-generation' governance reforms illustrates. Whereas the shocks of first-generation structural adjustment reforms could be introduced quickly, often by a small group of 'competent technocrats', second-generation reforms need time, 'they involve changing institutional structures' through which a different set of rules of the game could become embedded in practice. Second-generation reforms are typically transaction intensive and require discretionary actions by vastly greater numbers of people and agencies operating at every level.

This was exacerbated by policies designed to disaggregate and marketize the state, and as we will see, the plethora of arrangements put in place to re-join the local and national elements in ways that were simultaneously owned locally and compliant with global norms. Obviously, it was all the more critical that actions by the state are well attuned with the reform requirements, lest they increase the transaction costs – a concept we will return to later – of reformed market economies.

We will develop the distinction between first- and second-generation reforms in the Uganda and then New Zealand chapters. But one way to illustrate what this shift to 1997 WDR's good governance and capable state, is in the contrast between 1980s NPE and the later embrace of the NIE. To recap, 1980s NPE basically entailed four things for the state. First, sharp downsizing, privatization and withdrawal, not just from private good producing enterprises, but also, institutions producing merit goods, public goods and those creating large externalities. Second, for those state functions that could not be successfully privatized, introducing corporate market practices such as user fees in health and education. Third, cuts in progressive tax ratios, which resulted in cuts in public expenditure. And fourth,

Table 4.1 First- and second-generation reforms[12]

	First generation	*Second generation*
Main objectives	Crisis management: reducing inflation and restoring growth	Improving social stability and reducing risk; maintaining macroeconomic competitiveness and stability
Instruments	Drastic budget cuts, tax reform, price liberalization, trade and foreign investment liberalization, deregulation, social funds, autonomous contracting agencies, some privatization	Civil service reform, labour reform, restructuring social ministries, judicial reform, modernizing of the legislature, upgrading of regulatory capacity, improved tax collection, large-scale privatization, restructuring of central-local government relationships
Actors	Presidency, economic cabinet, central bank, multilateral financial institutions, private financial groups, foreign portfolio investors	Presidency and cabinet, legislature, civil service, judiciary, unions, political parties, media, state and local governments, private sector, multilateral financial institutions
Main challenge	Macroeconomic management by an insulated technocratic elite	Reorientation of middle and local government to apply global institutional norms in locally sustainable ways

increasing and sometimes wholesale reliance on NGOs and charitable institutions to tackle poverty and social exclusion. NPE was firmly directed at sub-Saharan Africa, but transported virulently elsewhere. The task was not just reconfiguring the role of the state, but dismantling state power so as to undermine the instruments used by elites to accumulate political and material wealth and 'free the peasantry' to engage with a newly privatized economic and political markets.

NPE was initially persuasive. Studies showed that the economic irrationality of states ('neo-patrimonialism', the 'economy of affection') tended to produce the 'vampire state' that was not accidentally but inexorably ingrained in African social formations.[13] The central assumption was that people engage in political and public action principally in pursuit of material self-interest. If duly freed of the oppressive state, political life would guided by a political version of the rationalistic calculus underpinning economic behaviour. Research inevitably exposed the material self-interest that really lay behind public action, such as a reluctance to contract out services, or a preference for tariffs and other means through which rents could be extracted.[14] This kind of analysis put steel into operations designed to curtail corruption and dovetailed nicely with the personal orientations of some MDB staff.[15] But beyond the cuts, withdrawals and moralizing chants about transparency and accountability, it offered little from an operational viewpoint. While NPE provided a useful support to cynicism about developing country states, it had few prescriptions for how to deal with the central paradox of structural adjustment, namely, that the market was a political construct that needed strong states to create the enabling conditions for its effective functioning, locally, or nationally.

The transparent skeleton of the invisible hand: the rise of NIE

Comfortably clear of the iconoclastic but operationally limited tones of NPE, the central ambit of New Institutionalism sounded constructive ('building institutions'): a 'positive Liberal' approach, perhaps, enabling rather than cutting. In fact, it was premised on a mentality of governance that disaggregated state functions into market-like transactions and their immediate contexts. In NIE, 'institutions' refer in the first instance not to the conventional institutions of the state (departments, ministries, bureaucracies), but to the plural elements which together can be seen as 'governing' transactions at micro-levels: formal and informal laws, rules and conventions of exchange (such as markets), and social phenomena informing those exchanges (such as 'information' and arguably 'social capital'). NIE doctrines defined 'institutions' as the 'formal and informal rules and their enforcement mechanisms that shape the behavior of individuals and organizations in society',[16] or, more generally, 'humanly devised constraints that shape human interaction'.[17] Such rules and their composite mechanisms (especially markets) are seen as the transparent skeleton of the invisible hand governing market behaviour, permitting individuals to benefit personally from doing what will serve the material interests of society as a whole. Also crucial is the notion of transaction costs: the observation that exchanges between actors, in the economic or political markets, are seldom costless or fully informed. Here, 'the transaction is the basic unit of analysis and governance mechanisms are needed to create order, mitigate conflict, and realize mutual gains'.[18] A whole world, in other words, could be built on the foundation of the efficient market transaction. This, NIE advocates in Development set out to do.

Broadly, NIE argues that efficient transactions (and, as we will see, 'accountabilities') depend on three ingredients: first, information, in the sense of both expertise and knowledge that can make better informed choices (in markets), second, laws, contracts and their efficient enforcement (supporting markets), and third, contest, in the sense of having competition (e.g. to contract to provide services) between multiple different players (that is, market competition). In short form, these three precepts of NIE could be rendered as 'Inform, Enforce, Compete'. And so they were, as we will see. In these terms, the task of good governance was two-fold. First, it required that obstacles standing in the way of locally knowledgeable but still rational policy choices be removed. The powerful central state is often exemplar of 'obstacle' in NIE doctrine. Second, it was necessary to encourage 'champions' of those chosen policies and create incentives for them to build broad based alliances that would 'embed the voice of powerful interest groups in mutually acceptable rules'.[19]

By enforcing these rules and instigating market competition for contracts to deliver (previously government delivered) services it would be possible to break 'path dependencies' and rent seeking and corrupt capture by central bureaucrats. Central state decision makers in NIE (and classic Liberal) terms were thus caricatured as self-maximizing agents: laws unto themselves, under-informed and anti-competitive, bent on resource and authority capture. Smugly trusting their own unchallenged expertise, they are too far from local situations to be able to know what needs are and who is best placed to meet them. Far better, then, to devolve decision making to local agents, and to keep these honest (and incentivize their information gathering) by making them compete. Thus NIE would deliver more efficient services by manipulating the incentive structures which bear on individual actions and above all, reducing transaction costs by limiting 'the opportunities for corruption by cutting back on discretionary authority', thus 'reducing the scope for opportunistic behavior'.[20] Meanwhile, effective rules, an efficient banking and legal system, neatly informed and unimpeded market transactions would together remove state and other impediments to growth.

NIE thus seeks to 'build institutions for markets' by maximizing accountability and transparency. It does this by disaggregating state functions, assigning them as widely as possible and fostering competition through markets to discipline governance and service delivery. How these efforts to discipline, foster participation, efficient service delivery and account-ability across the disaggregated terrain fare in practice will be of special interest in Part II of this book. For now, it is evident that these doctrines had immediate appeal to those responsible to frame second-generation reforms. They fitted tidily within the Liberal market paradigm, while portending a clear role for the state in facilitating efficient, almost spon-taneous governance. They also facilitated wider consensual crossovers. For instance, the esoteric precepts of economics could be read into the domain of the state and politics; as Joseph Stiglitz remarked 'in a sense, the polit-ical notion of accountability corresponds closely to the economists' concept of incentives'.[21] Similarly, it made clear why civil society should be governmentalized: their 'shadow state' activities could play a role in plural-izing the state and facilitate the voice of its poorest clients.

While the role of NIE in 'enabling the state' was apparent in 1997 WDR, it was not until the 2002 WDR that the NIE rubric became central. Here it was trimmed and packaged to travel in a three part framework: 'inform, enforce, compete', with 'messages' (complement, innovate, connect and – again, compete) which could, the authors urged, 'be applied regardless of the specific sector studied'.[22] Local resource allocation decisions would be based on 'information', with competition providing the incentives for procuring knowledge. Bureaucratic capture would end by cutting back on government discretion; health, education, welfare, all could be improved (and made more likely to help the poor) by making them work in compet-itive market-like manners, providing there was enough information, juridi-cal enforcement, 'complementary' innovation and joined-up-ness. It was a marvellous fantasy, even though it flew directly in the face of the experi-ence of the very few places on the planet where such reforms had actually been tried, especially, as we will see in Chapter 8, New Zealand.

Decentring the State: up, down and sideways

By the end of the millennium, then, it had already become popular in Development to talk of 'reinventing' the role of the state. In NIE terms, this meant re-conceiving the state as a plurality of governance and service functions, then contracting out the services, while both honing the institu-tional governance arrangements, notably its disciplinary, enforcement functions, and disseminating these throughout the social, political and eco-nomic order. This would break down old territorial and bureaucratic fief-doms, and enable whole new spontaneous (but 'informed' and 'enforced') ways of joining global market and juridical norms and techniques to central and to local versions, and sector by sector. The NIE vision, which in the third section we will see articulated in 2004 WDR's three-cornered con-ceptual frame for 'local – local' dialogue among policy-makers, service providers and citizens, thus involved what some have called the 'flattening out' of politics and accountabilities, and the replacement of contests over resources and entitlements with narrowed notions of administration and management.[23]

This had the effect of 'decentering' the state in ways now referred to as involving both vertical and horizontal pluralization (Table 4.2). NIE reforms instigated a vertical disaggregation, as the political, administrative and fiscal functions of the central state were delegated or devolved 'down' to multiple levels of local, regional, state/provincial authorities (and to private sector consultants and audit agencies) as well as 'up' to international frameworks (such as the WTO and its global rules). 'High level' policy was separated from 'on the ground' operations just as funding for services were separated from the provision of services. This pluralization of the state occurred 'horizontally' as well, through which the state's regulatory, policy, enforcement and service delivery functions were variously contracted and delegated to private sector groups, NGOs and a growing array of 'public-private' partnerships.[24] In theory, policy was to become a matter of contest between plural 'voices' (consultants, government officials, the poor, NGOs, policy analysts). Operations were to be the site where freely participating individuals could 'enter' or 'exit' and shift between service providers. All this disaggregating was superintended and supplemented by different levels of accountability: audit, contract compliance, peer review, consumer voice, participatory dialogue that were recognized, legitimated, funded, and often assigned to different groups. A local government agency might deliver services in one place, and a for-profit contractor in another, or both could compete in the same jurisdiction. A local NGO might facilitate workshops around contract compliance, while a global accounting firm audited the books. This was an unprecedented dispersal of responsibility for policy and operations, standard setting and regulation, contracting and co-production, evaluation and surveillance of state functions that was soon felt in reforms of sectors as diverse as the judiciary and police, health and education, municipal services, land and water management. In the new world, accountabilities and functions would be deliberately assigned, across fields of overlapping and indeed competing organizations of all kinds.

In sum, more *accountabilities*, more stakeholders informing, competing, enforcing, more points of voice and exit for clients should mean more *accountability*. Practice would be all joined-up, inclusive and participatory and empowering as never before. But on the ground, it was often a very different matter. At the same time, the state's ability to manipulate local outcomes, or address local dimensions of poverty through systemic central or regional interventions, was greatly reduced. With many of its core responsibilities (and their budgets) contracted out, the state could no longer wield significant fiscal muscle in socially or territorially interventionist ways, or in the redistribution of assets. Once the fragmentation had set in, it created a whole series of perverse effects and path dependencies of its

Table 4.2 Decentring the state

	Pre-disaggregation roles	Functions that were vertically and horizontally disaggregated
State functions	Policy, financing, planning, design, compliance and outcome monitoring, technical and financial audit, regulating, interdicting, service delivery	Policy, financing, planning, design, compliance and outcome monitoring, technical and financial audit, regulating, interdicting
Decentralized/ civil society/ privatized functions	Advocacy/voice, some service delivery	Policy, financing, planning, design, compliance and outcome monitoring, technical and financial audit, self- and contract-regulating, interdicting, service delivery, co-financing, subcontracting, advocacy/voice
Citizen roles	Service beneficiaries, citizen rights and entitlements through representative statutory bodies	Voice, service client, co-funding, participation through facility planning and operations

own. As we will see in New Zealand, putting this Humpty Dumpty back together again would be a task that all the Kings horses and men, tasked only to frame the rules, inform the players and enforce the contracts, would find themselves very poorly positioned to achieve. It would require major political commitment to reterritorializing accountabilities, something which as we will see Third Way and poor country governments typically lacked.

THE GOVERNANCE TECHNIQUES OF 'INCLUSIVE' NEOLIBERALISMS' GLOBAL AND LOCAL RETERRITORIALIZATION

The ascendancy of NIE occurred at a time of resurgent interest in decentralization, that is, the transfer of authority, government functions, resources and responsibilities from the central state to subordinate or quasi-independent local authorities. NIE proved enormously useful. Past experience, the most recent round of decentralizations having made an appearance in the 1970s, taught that decentralization was not automatically pro-poor. It was well known that decentralization resulted in 'local centralization' in which local elites captured public resources and opportunities, ignoring the interests of those unable to command or influence how decentralization's rules were applied or resources allocated. Emboldened by NIE's precepts however, the new decentralization taking shape in the late 1990s involved a more thorough horizontal disaggregation of the local state. Here, a host of quasi-public, commercial and civil society agencies were bestowed with statutory rights to engage well beyond the actual delivery of services. In this new decentralization, this plethora of local authority arrangements would all be informing, competing and enforcing; authority, and powers of exit and voice were invested in planning, budgeting, expenditure control, social and other forms of audit, monitoring and evaluation. And, as explained in this section, they were to be backed and surveilled by far more sophisticated governance techniques to guide and discipline relations between the 'local' and higher orders of the national and global Poverty Reduction consensus.

Decentralization resurgent: in theory, then practice

As we saw in Chapter 2, decentralization has moved in and out of Development fashion since Lord Lugard's Dual Mandate. In its longer history decentralization reaches to John Stuart Mill and de Tocqueville's *Democracy in America*. Its mid-1990s resurgence was inspired by such NIE writers as Mancur Olson (although his *The Logic of Collective Action*, was first published in 1965). Ardent proponents of decentralization argue that incentives for collective, pro-poor action, disincentives for free-rider problems and reduced transaction costs occur best under decentralized, federal or consociational (multi-agency) systems. Alongside this, by 2000 in World Bank and other literature, we find Amartya Sen's *Development as Freedom* also providing populist comfort; decentralization offers a 'smart and dignified' way to 'institutionalize the empowerment of communities'.[25] But whereas many advocates of popular NIE forms of decentralization downplayed the importance of local government and linkages with higher authority, for the World Bank, the links were crucial. For decentralization to become again *operationally* ascendant, NIE doctrine and populism needed to be matched by governmental techniques that would guide and provide assurance that the micro would be firmly linked back to emulate macro interests. We will examine some of these techniques shortly. Thus, although the World Bank's 1992 *Governance and Development* suggested that decentralization might offer the 'linkage between macro and micro', it was almost ten years before decentralization was pushed to centre stage in Poverty Reduction. Until then, decentralizing moves tended to feature only as handmaidens to (or a shorthand for) privatization: that is, as part of generalized disaggregation of the national state. There, it was assumed, decentralization had merit because 'in theory it can lead to significant improvements in efficiency and effectiveness'[26] by reducing the load on central government. Here, as so often before, was Development imagining that localized governance could be cheaper. ... But, as crucially, international donor confidence about decentralization would need its 'successful cases', or at least, cases where decentralized governance was apparently being applied to whole of government reforms around Poverty Reduction. In these cases, as we will see from the Uganda chapter, neither Poverty Reduction nor realizing better allocative efficiency in resource use were at the forefront of the political imperatives driving decentralization. Rather, as we will see, decentralization has everywhere been driven by states, often authoritarian ones, looking to consolidate polit-ical support in local areas and use this to stabilize larger regime interests. This political agenda, as we will see, has usually been laid down precip-itously to political ends, and has typically intersected only at certain points with what has otherwise been a donor driven set of actual good governance reforms.

For the donors part, they had a more limited rationale for decentraliza-tion – doing more with constrained resources, that is, more efficient service delivery. The donors' preoccupation was starkly evident to the Ugandan government when their decentralization programme, built on Resistance Councils crafted during the horrendous 1981–1986 'bush war', became the official cornerstone of the government's reforms in 1992. Academic fashion was by then shying away from the state minimalist positions of NPE, it was still five years before Uganda's decentralization could start to be embraced by the World Bank/IMF. Uganda's donors, in a pattern which again would repeat, responded primarily to the technical, fiscal, service delivery and communitarian rationales for decentralization. The World Bank's 1993 *Uganda: Growing out of Pover*ty stared straight past the avowedly political and territorial, local-to-national state building ambitions Museveni had for decentralization and remarked, almost in passing in its last pages, that its merit may be as a platform 'to support community and NGO initiatives, particularly in light of the fact that the government lacks the capacity to delivery much needed services to most of its citizens'.[27] Three years later, with political and fiscal decentralization well underway, the World Bank's 1996 *Uganda: the Challenge of Growth and Poverty Reduction* still gave only begrudging recognition to that country's decen-tralization. It might 'help anti-corruption' by encouraging 'civil service downsizing' and 'kick-starting' community participation in the manage-ment of vertical programmes for service delivery. Better still, it may foster a willingness amongst the local population to add their wealth to national tax revenues.[28] By the mid-1990s, for Development agencies decentraliza-tion had both the populist appeal of empowerment, and some growing technical allure. But it was still too risky. Thus still, no major financing commitment was made in support of Uganda's decentralization. Museveni's skill, as we will see in Chapter 5, was to foster a rising star in Africa status for his country, and to reach an accommodation with the international community around decentralization in the name of Poverty Reduction that would allow him to consolidate his political project of 'no-party' democracy.

Thus decentralization found its footing in Development's operational policy and practice not because of any demonstrable successes in terms of poverty-reducing service delivery or local empowerment. Rather, the practical application of decentralization doctrine, which by 1997 WDR placed it right at the centre of governance reform, occurred only once worries about the inherent risks of this approach were assuaged technically and politically. Technically, this required development of a range of techniques to contain *local* politics (such that it would be narrowed and restrained as political contest around technically defined service delivery mandates), to ensure *national* public expenditure decisions were focused on PRSP outcomes (through medium-term budget/expenditure management techniques), and to discipline *intergovernmental* relations in ways that would ensure global policy priorities were accountably projected into local spaces. Politically, and as we will see, equally important, were the accommodations reached in particular countries that gave Development institutions prima facie assurance these techniques were locally owned, and recipient country regimes comfort that their larger territorial and stability interests could also be advanced by the flows of resources promised when they adopted these technical arrangements.

Notwithstanding the breathtaking march of assumptions and accountability blind spots ('local information good, central information bad', 'fragmented accountability good, central accountability hopeless'), Good Governance could with these provisos now be recast as the joined-up sum of all these market, law, information and contract-disciplined public and private service providers. By 2001 the World Bank was announcing that decentralization and local empowerment 'is a form of poverty reduction in its own right' quite independent on its actual effects on income or other poverty measures.[29] But that the decentralized, local space could become the site to bestow moral responsibility, to create positive citizen attitudes and behaviors, to ensure local ownership of MDG problems and remedies, and provide a platform for legitimacy and wider stability, all this depended on refinement of interconnecting local to national to global frameworks of discipline. These are briefly summarized below.

National–local governmental techniques for joined-up governance

To achieve the required levels of moral, social and market 'security', the state from local to national needed to be disciplined and to be able to discipline others. But the enabling Liberal state must act not so much directly as through 'institutions' that provided the security of law and fiscal integrity, and empowered and manipulated the incentives acting on individuals, and letting them access the services they choose. The governmental techniques involved, which we describe below, are now judged to be at the heart of delivering on the promises of the Poverty Reduction. We will only briefly illustrate here three that have become prominent; how they were introduced in practice from the early 1990s to mid-2000s is shown in the chapters to follow: medium-term budget and expenditure management frameworks (MTEFs); Sector Wide approaches (SWAps), and third the techniques of fiscal decentralization, which may be seen in the transition from social funds, to Local Development Funds (LDFs) and to fully blown intergovernmental grant systems.

The medium-term expenditure framework

Not long after the introduction of PRSPs, an IMF review remarked that 'the Poverty Reduction Strategy Papers will only become truly effective when the Poverty Reduction Strategy Paper itself is closely aligned with the budget process in each country'.[30] Ambitious, but nowadays persuasive in their comprehensive framing, MTEFs (sometimes called Medium Term *Budgetary* or *Fiscal* Frameworks) are the fiscal complement to the PRSPs. Whereas PRSPs set medium-term policy priorities – detailing sector service delivery targets, for instance, which would see primary school enrolment of female children shift upwards from x to y per cent. The MTEF, complementing this, would indicate how budget outlays for particular sector targets would change in the ways needed to meet this target. The MTEF was developed as a tool for linking policy, planning and budgeting over the medium-term (i.e. 3–4 years). The steps in constructing an MTEF consist of two features. One, it requires a reasoned projection of likely total revenue, from domestic and external sources; this projection is a product of 'fiscal profiling' in which national revenue projections are read off projections about economic growth which, in turn, are calculated on the basis of monetary and fiscal policy measures (tax regimes, expected flows of donor resources, debt restructuring and so on). These calculations are then used to provide sector managers – the head of the health ministry for instance – what is known as a 'resource envelope' or a 'hard budget constraint' within which s/he must plan expenditures, keeping in view the PRSP sector outcomes. Second, in the meantime, sector managers are expected to calculate a 'bottom-up' estimation of resources required to get from existing levels of service delivery to the PRSP desired outcomes. During this process, they negotiate with their finance ministry and raise or lower their targets according to total fiscal resources being made available to the sector. Needless to say, in most situations the basic data and analytic skills required to prepare an MTEF are seldom available, and costly to assemble. But a potent incentive to be seen to be engaged in this arduous process was provided by IFI insistence that completion of the MTEF would be one of the triggers used, as we will see in Uganda, to gain access to HIPC debt relief. MTEFs were believed necessary to ensure that the resources freed up by this 'debt dividend' would be directed only Poverty Reduction outcomes; it was to provide, more to the point, a measurable and globally legible way to bind local resource management decisions to the global and national policy commitments around poverty reduction. Thus, MTEFs provide one, comprehensive device through which global agreements, national inter-sectoral budgeting and highly localized investments are bound together in ways that were never possible through the 'national development plan' and 'national accounts' type devices popular in the immediate post-colonial period. For example, the Pakistan government's MTBF, announced at the same time as the 2003 PRSP, aims to:

(a) provide greater certainty about the level of available resources, permitting clear-cut decisions over what can and cannot be funded;
(b) enhance participation and ownership among line ministries in the budget process as a result of its improved predictability and consistency;
(c) improve management of the overall resources available to the budget, implying both enhanced allocations to priority sectors and more efficient management of funds received; and
(d) greater consistency between macroeconomic performance, policy formulation and public expenditure. Efforts will be made to integrate recurrent and investment budgets gradually under a coherent MTBF. It will require consistency between sector-specific policies, programmes, project selection and the Poverty Reduction Strategy Paper strategy.[31]

The Sector Wide Approach

The MTEF both encouraged and built upon what had earlier and more tentatively been tried through SWAps, or Sector Wide Approaches. By the early 1990s, both education and health were increasingly being redefined in terms of their economic implications. On the strength of this, the World Bank made its debut into international health policy with a 1993 WDR entitled *Investing in Health*, around the same time has it was shifting to Sector Investment Programmes (SIPs) which advocated what became the defining features of SWAps; donor collaboration around common implementation arrangements, locally-owned sector-wide policies and priorities.[32]

The defining characteristic of a SWAp is alluringly simple. It brings together

> all significant public funding for the sector (health, education, roads, etc.) under a single sector policy and expenditure program, under government leadership, adopting common approaches across the sector, and progressing towards relying on Government procedures to disburse and account for all public expenditure, however funded.[33]

Thus SWAps responded to concerns about lack of coordination of development assistance, the need for 'concerted action', focused and well targeted assistance, not least also to find a mechanism through which governments could be held to commitments in the face of a 'dysfunctional public expenditure management system'. The design of MTEFs or SWAps, as might be imagined, is no small feat: it involves vastly improved data systems and surveillance measures so that multilateral, bilateral and major NGO donors can sit round the table with government officials with the books open, and 'partnership' and coordination in allocating expenditure and input rights and responsibilities. Transparent, apparently all enjoined rationality, as we will reiterate in later chapters, only comes at extraordinary price.

Social Funds, to LDFs to Intergovernmental Grant Systems

Social Funds came onto the scene in the late 1980s to provide a direct, easily measured and high profile device through which to channel resources to sectors of a country's population particularly vulnerable to the immediate consequences of Structural Adjustment (the removal of subsidies on food or utilities, public sector downsizing, industry contraction, etc.). This was achieved not by direct income supplements, but usually by providing more money for health and education services, small-scale public works, or local savings and credit groups. Social Funds could be used to 'ring fence' resources earmarked for such pain mitigation and poverty reduction: that is, to keep them separate from the rest of the budget, so they would get to the places intended. They were a separate bucket of money, held at national Treasury level, not a part of recurrent transfers, and administered by special purpose, internationally contracted Project Implementation Units (PIUs), again based in the capital, close to Treasury. PIUs transfer the monies direct to local levels – right to the door of the new school, clinic or other capital investment – and most importantly could operate in ways that cut out the layers of government in between. Social Funds thus parlayed to a popular view that government was an obstacle, a corrupter that needed to be sidelined. Social Funds teams moved back and forth between capital cities and local places where 'social gaps' had been found, and offered direct, controlled transfers to local health and education institutions. Importantly, these were apparently insulated from both local politics, and, again other forms of government control that led to them not reaching local beneficiaries. Between 1987 and 2000, the World Bank approved about 100 social fund-type projects in more than 60 countries with a total value of about $3.4 billion.[34] They remain popular and are the backbone of the World Bank's Community Driven Development approach.[35]

However, by 2000 the ascendancy of Social Funds had been lost to LDFs.[36] LDFs in some respects became popular for much the same poverty-gap filling reasons as Social Funds, meaning that they tended to fund much the same agenda – basic social service infrastructure, irrigation, drains and water supplies, and occasionally some social protection measures around savings and credit, or household income generation activities. But in comparison with

Social Funds, LDFs were tied to relatively small fiscal transfers – they were designed to transfer around $1–$5 per capita into the jurisdictions of local governments; as we will see, in Vietnam, around $10,000 for a commune of 8,000 people, or in Uganda where over three years around $11 million was made available to 2.2 million people in five districts, and then upscaled across the country in much the same proportions. Their popularity lay in their simultaneous ability to invest in the emerging 'empowerment' leg of the Poverty Reduction approach. LDFs were, in the parlance of the mid-1990s, 'policy experiments' for a larger governance reform agenda. Accessing the LDF, the actual fund allocated to local community groups or councils of elected officials (or administrative staff, where councils did not exist) depended on compliance with certain procedures: participatory planning procedures, contracting by tender, with special audit requirements. Thus LDFs had the added attraction of being avowedly concerned also with 'building local governance capacity', something Social Fund proponents were never able to demonstrate.[37] Consequently, it was acceptable to finance from the allocations made to pro-poor investments also the local administrative expenses associated with building this capacity: the costs of councils meeting, participatory planning exercises, audit compliance, buildings for people to meet, to store records, that is, all the paraphernalia thought necessary to make wise use the resources made available for local decisions. In contrast with the Social Funds' 'direct route' to service delivery, LDFs came to typify a longer route to poverty reduction, in which service delivery would improve on a lasting basis through local empowerment. More importantly, though usually piloted in a few communes or districts, it was the ambition of LDFs that they could be scaled up into national programmes of centre to local transfers, all within transparent, accountable and participatory governance frameworks. So, whereas both Social Funds and LDFs are largely concerned to reproduce national (and international) spending priorities for Poverty Reduction, over time, from the mid-1990s, LDF's rising importance as 'policy experiments' for national level governance reform saw them, as happened in Cambodia and Uganda, scaled up to national level transfer systems, and through this good governance reforms could be joined-up all the way from community participation, to more formal moves to devolve powers to local governments.

In their most ambitious forms, as we will see in Uganda, LDFs formed the beginnings of more complex, multi-grant intergovernmental financing systems that are now the backbone of fiscal decentralization across the developing world. Intergovernmental fiscal transfer arrangements – involving complex formulae for sharing resources between levels of government, and multiple systems for targeting grants to areas of special need or performance – have been central to OECD country fiscal systems for many years. In the federal systems of Australia, Canada and the US for instance, the plethora of fiscal grant systems are central to federal, state/province and local government politics and regime stability. Such complex, technically and politically demanding fiscal sharing techniques could only move into Poverty Reduction practice once the edges of what was expected of PEM began to harden and the instruments to do so become more assiduously applied. In a sense, they depend on variants of the MTEF and the SWAp on the one hand, and more especially, on a host of what are referred to as 'fiduciary risk' and disciplinary measures on the other, to ensure that the 'local incentives' are matched by arrangements to enforce 'global accountability'.

As we will show in the chapters to follow, elephantine they may be, extraordinarily expensive, time-consuming and demanding to pull together and sustain: yet the clear aspiration of inclusive neoliberalism (and neoliberal institutionalism) is that all these governmental techniques will become localized in developing countries. These governmental techniques are becoming a day-to-day part of how poverty is governed, an essential part of the maelstrom of NIE competition and enforcement through which Poverty Reduction's solution will be articulated. At which point, as might easily be imagined, the grand joined-up-ness and poverty-reducing potency of the whole scheme starts to seem less emphatic. But there is a final layer to add before concluding our survey of inclusive neoliberalism's governmentalities.

Securing the new order: 'facilitating international collective action'[38] through donor harmonization

With the local apparently being clasped by the ambitious national-to-local governance techniques noted above for directing Poverty Reduction resources downwards and ensuring accountability back up the line, as part of a larger 'donor harmonization' effort from around 2000 the MDBs began to roll out a 'selective' performance based system for rewarding the capable state that had adopted 'a clear definition of their roles as public institutions and their development mandates'.[39] In a high level articulation of Tony Blair's moral precepts to the effect that poor countries will enjoy 'no rights without responsibilities', MDBs quite quickly developed the technical apparatus they believed would create higher-level incentives that would bear down through the governance techniques from national to local levels. The first, and least hard-edged were coordinative frameworks – for the World Bank, the CDF, for the UN, the UN Development Assistance Framework and as later followed by MDBs, in the case of the ADB, the Poverty Partnership Agreement. After being implemented in around 50 low-income countries, it is apparent the 'ownership effects' of CDF's are yet to be seen, but there's no doubt IFIs are serious this will occur.[40] A harder edge was given to poor country incentives to adopt Poverty Reduction by new levels of strategic and operational coordination amongst IFIs and donor countries on how aid allocations would be used to reward Poverty Reduction policy adopters, that is, rather than simply allocate aid according to relative needs or levels of poverty.[41]

From 1999 the MDBs began to tie allocations of their discretionary finance[42] to assessments of each borrower's policy and institutional performance in 'areas relevant to economic growth and poverty reduction' to ensure the international Poverty Reduction consensus was guided by empirical research on 'What works and why?'.[43] A common accountability platform for all developing nations came to the fore, to 'translate the international consensus into action'.[44] Quite explicitly, 'resource allocation is aimed at concentrating resources where they are likely to have the most impact and to ensure consistent treatment among International Development Association (IDA) eligible countries'.[45] In one example, the World Bank's Country Policy and Institutional Assessment (CPIA) criteria include familiar NIE territory: economic management (fiscal policy, debt management, etc.); structural policies (competitive environment for the private sector, factor and product markets, foreign exchange policy, etc.); policies for social inclusion and equity (including equality of economic opportunity, investments in human resources, safety nets, etc.), and public sector management and institutions (which include property rights and rule-based governance, the quality of the budget process, efficiency of revenue and expenditure, and transparency, accountability and corruption in the public sector). Governance measures include accountability, transparency, the rule of law and participation. All IDA countries are scored, with the implication that 'countries with highly unsatisfactory ratings for three of more out of the seven governance indicators may be considered as facing severe governance problems, in which case a downward adjustment is applied to the overall country rating'[46] and a corresponding drop in allocations would occur.

These efforts to harmonize around a common accountability platform had to confront donor perfidity. But contrasting the situation a decade ago, Dollar and Levin's 2004 report, 'The Increasing Selectivity of Foreign Aid, 1984–2002', shows that most donors – with the usual outliers, including the US[47] – are now 'very policy focused' and 'have bought into the aid selectivity model'. Aid allocations now 'favor the better governed (poor countries)' judged according to policies robustly researched and referenced in the outpouring of cross national statistical analysis from the World Bank Institute and similar agencies that have established a closely worked empirical basis for good governance. Dani Kaufmann and others[48] have distinguished six main dimensions of Good Governance:

  i)   voice and accountability, which includes civil liberties;
 ii)   political stability;
iii)   government effectiveness, which includes the quality of policy-making and public service delivery;
 iv)   the quality of the regulatory framework;
  v)   the rule of law, which includes protection of property rights and independence of the judiciary; and
 vi)   control of corruption.

For each dimension, the World Bank website has a wide range of indicators, complete with colour-coded maps, graphs, charts and diagrams which purport to illustrate the cross-workings of these elements of good governance and institutional quality. A ready-reference ranking of country performance is provided for targeting of development assistance and to focus policy dialogue around common reference points for the media, NGOs, technocrats and reforming politicians.[49]

These cross-tabulated incentive structures have been put to work through new performance based systems for aid allocation and even further bouts of harmonization. All this is billed as part of a shift from the discredited 'policy conditionality' of the past towards Poverty Reduction 'ownership'.[50] At the same time, getting this far had to appeal to the disciplinary requirements of conservative neoliberals worried that 'there is no certainty that institutional frameworks conducive to growth and poverty alleviation will evolve on their own'. Poor country governing regimes had been known to waiver in their commitments. So, at the operational level, away from the warm camaraderie of the policy roundtables, it was clear that discretion needed to be curtailed, restraints applied and transparency guaranteed; 'strong central guidance' was needed to complement these 'incentives'.[51] Blair-ite UK's DFID, a leader in the 'like minded harmonization' effort, regularly chorused for 'improved surveillance – better monitoring of the performance of developed and developing country economies'.[52]

Thus to complement the performance based incentive framework, what's termed 'concurrent capacity' is promoted by an overarching framework of 'extraterritorial and international restraints'.[53] This features an array of internationally harmonized fiscal monitoring and surveillance systems, so called 'process restraints'[54] – for accounting, budgeting, auditing, procurement, etc. – that are built around 'codes of good practices'.[55] These come in various guises. One is benchmarking for public finance management[56] and mandatory standards for data dissemination, public access to information and fiscal practices. The IMF backs these with 'Press Information Notices' (PIN), which are publicly disclosed at the conclusion of IMF Article IV consultations. These pose a very credible threat that failure to adopt good governance principles can result in capital flight or investment strikes.[57] Reports on the Observance of Standards and Codes (ROSC) adopted in 1999, document the extent to which countries observe internationally recognized standards in the areas of direct operational concerns to the IMF. Also in 1999, the IMF adopted a range of finance sector analyses to 'help countries assess vulnerabilities in the financial sector and identify the needs for corrective action'.[58] Alongside, the host of fiscal monitoring instruments, Public Expenditure Reviews (PERs), Country Procurement Assessment Reviews (CPAR), Country Financial Accountability Reviews (CFAR), Financial Sector Assessments (FSA), were developed to help PRSPs to embed the reforms in country 'ownership' commitments.[59] And binding this all together, at a series of High Level forums since 2003, the OECD's Development Assistance Committee (DAC) members agreed to adopt these and other instruments promoted by the UN's 2005 *Millennium Report* to harmonize, share knowledge and promote common 'threshold conditions' to 'fast track' aid to countries which have good track records.[60] The state, radically reconceived as governance and a mere bundle of NIE institutional functions, disaggregated and reterritorialized in internationally transparent ways, disciplined where institutionally necessary by formidable budgetary accountable frames, could be put to work on Poverty Reduction.

GOVERNING THE POOR: WDR 2004 AND 'INCLUSIVE' NEOLIBERALISM'S LONG MARCH OF ASSUMPTIONS

Although NIE precepts were beginning to underwrite a host of Good Governance reforms from the mid-1990s, it was not until publication of the 2004 WDR, *Making Services Work for Poor People*, that it became clear how Poverty Reduction's inclusive neoliberal, joined-up ambition could be articulated through NIE's inform, enforce, compete neoliberal institutionalist mantra. For this book, the 2004 WDR represents both a high water mark of the technical and wraparound imagination of neoliberal institutionalism, and an ideal point from which to summarize what's gone before, and point forward to the book's second Part.

Making Services Work for Poor People began with the chilling announcement – already well demonstrated though barely if ever publicly stated in Development's inner circles – that neoliberal economic growth alone would not lift the poor out of poverty. Rather, reaching the poor was to be achieved by a longer route, by investing in human wellbeing through social services – in 2004 WDR this meant 'services that contribute directly to improving health and education outcomes'[61] – and through this add the crucial 'enabling condition' necessary for the poor to participate in market Opportunity. The 2004 WDR accepted the now evident fact that a 'substantial increase' in external resources would be required, but argued that if poor countries could show they were using resources well, that a 'persuasive argument' could be made for directing increased assistance to them.[62] Parked right alongside this announcement was moral responsibility – which governments would demonstrate by financing, providing or regulating services that contribute to health and education outcomes. This compelling moral basis was then hinged back to the facts of economics, that is, to the inevitable market externalities and social justice/equity issues arising with market-led economic growth. And back again to global and national ownership, by targeting the MDGs and referencing the Universal Declaration of Human Rights – to guarantee 'elementary and fundamental' rights in health, education, housing, food and clothing – and to guarantees enshrined in national constitutions. Thus, in these classically Liberal terms – moral responsibility, economic facts, and constitutional rights – governments would be judged for their adoption of WDR's conceptual framework to *make services work for the poor.*

The Accountability Triangle: making services work for the poor

The critical point the 2004 WDR focused on was not dealing with technical or content issues – how much to spend on one input relative to others in vastly different contexts – but how the institutional and political context was to be shaped up to efficiently convert these inputs into pro-poor services; 'The answer' was delivered in classic NIE terms 'successful services for poor people emerge from institutional relationships in which the actors are accountable to each other'. Quite a mouthful for so early in the report, but the very next sentence urged 'Please be patient, the rest of the Report works out exactly what that sentence means'.[63] Recognizing that accountability is worryingly slippery, 2004 WDR distilled accountability to five familiar features: delegation, finance, performance, information about performance, and enforceability, and prescribed ways in which these could be made to play out in 'local organization'. The local organization, the principle site of obligation to convert resources into pro-poor services, was now shaped up in an appealingly simple three-cornered relationship – the 'Accountability Triangle' – between the state (politicians and policymakers); service providers (the managers and frontline workers); and citizens or clients of services.

We will show WDR's Accountability Triangle at work in Chapter 7 in some detail. In essence, WDR proposed two routes to poverty reduction involving governance and services. The direct route, involving providers dealing directly with the poor, would as we will see in Uganda and Pakistan involve sending large amounts of money down the silos of service delivery (especially health and education), where they would for reasons we develop later have quick and demonstrable impact on MDG poverty indicators. The indirect route, also known as the 'long route of accountability', involved improving services and poor people's outcomes by increasing the quality of governance at central and local state levels. Overall, the range of NIE accountability modes described above are invoked, and held up as ensuring that services and their governance are focused on outcomes for poor people. For their part, the poor are active and included in a web of service and governance accountabilities that should lead not just to their empowering, but to their enabling to better participate in markets, and to lead more secure lives. All this would occur in the context of comprehensive integration into global markets wherein they and their territories remained peripheral, vulnerable, and comparatively powerless.

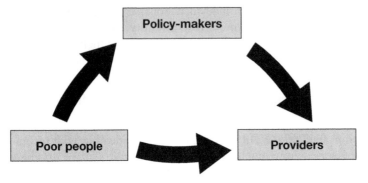

Figure 4.1 The Framework of Accountability Relationships

Thus 2004 WDR fulfilled the crucial function of dot-pointing all the policy learning from the 1990s ascendancy of NIE into contemporary operational terms, and turning this out as a primary weapon in attacking Poverty, enhancing Opportunity, Empowerment and Security. This was at very least a remarkable diversionary coup for 'inclusive' neoliberalism: it served not just to distil and obscure the lack of a robust relationship – observed long before by Dani Rodrik and others – between neoliberal economic policy and economic growth. In fact, the common and profound failure of the model's original precepts was hardly noticeable: that failure too was now firmly associated with governance failure and domestic failures at that. More importantly, perhaps, it finessed a redirection of attention from neoliberal inevitability and its social fallout to a thoroughly pared back focus on national *moral responsibility, technical competency* and *national/local political will* for the *local* provision of social *services.* And, in so doing, it thoroughly bound central government reforms to these three-cornered consequences by requiring nation states to 'fundamentally transform' in comprehensive neoliberal institutionalist governance terms. It is hardly surprising that a wholly different kind of state emerges from these moves, one thoroughly oriented towards its primary function to facilitate the movement of capital across space, while at the same time ensuring that the logics of the market, in NIE, are both articulated into and bolstered by social services, protection and security measures. In retrospect, of course, as we look over the chapters to this point, this is not odd: rather it all makes a certain kind of all too familiar sense.

But now, we need to see what sort of sense it has made on the ground.

Conclusions

As shown in this chapter, it was only once this comprehensive Liberal governance frame had been elaborated, and accommodations made with particular kinds of poor country governing regimes, that Poverty Reduction's Opportunity, Empowerment and Security could be imagined and made to work and, crucially, could the entire enterprise of Poverty Reduction be decentralized with confidence. The net effect, then, as we will reiterate and illustrate in this book's Part II, is a Liberal governance joined apparently powerfully to disciplinary frames, but also disaggregated and re-associated locally, in what we call the quasi-territorial domains of local governance, community, partnership and individual responsibility. We believe that what comes from the plural accountabilities assigned down into these quasi-territories are a quasi-accountability.

Part II of this book will show in particular cases the various vertical and horizontal disaggregations being developed and played out over the decade of the 1990s. What we will show is that in contrast with the past, where power and authority was at least theoretically vested in a national state accountable to protect the entitlements of citizens living within a defined authority, these emerging efforts to 're-join' politics in 'inclusive' neoliberal ways around quasi-accountabilities provide only weak and unstable points for leveraging accountability and even then, only to clients of services. The progression of chapters in Part II show how these efforts were elaborated, step-by-step, from the early 1990s, starting in Vietnam, where PRAs were deployed to create quasi-territorial spaces of consensus around 'community' onto which were pinned LDF financing systems. As a consequence of the weak and ephemeral nature of the accountabilities generated by quasi-territorialities, their form, and the responsibilities assigned to them are continually being adjusted according to technically driven but competing perceptions, for instance, of how decentralization principles like 'accountability', 'subsidiarity' or 'non-subordination' – concepts we will explain in the Uganda chapter – should be applied. Another outcome is that how territories are governed becomes radically uneven. Change becomes the only constant, as assignments of responsibility to and connections between different levels and sites of authority are made to coalesce at different levels, then shift again and reform in yet new ways. In one community, a community development project receives devolved and discretionary funding or NGO driven service delivery, then it stops, and is replaced by new requirements. In another, accountability for delivery of say education or policing is diffused across three or more levels of government, and their component parts dissembled to multiple points for directing, providing and monitoring these services in ways that weaken any possibility of accountability. In some territories, you get all of this at once. The reterritorialization of governance is, in other words, these days a very messy and uneven business indeed. But it is this business on which the Poverty Reduction paradigm has placed considerable expectations.

As we transit from Vietnam to Uganda in the mid-1990s, here again we will see Development's Liberal and travelling formalism at work: framing one place in terms of lessons distilled from elsewhere, developing by analogy, retrospect and overreach. Throughout, we will see the repeat tendency in Liberal Developmentalism for the universal to assert itself over the particular, the travelled over the placed, the technical over the political, and the formal over the substantive. Yet what we will see emerging on the ground will be a much richer, more fraught accommodation between Liberal doctrine and a much broader political economy of place, history, production, market and territorial government.

In all this, wider questions of power, security and stability come to the fore. Poverty Reduction, as we will see, comes congenitally equipped with huge blind spots on how underlying power relations can appropriate its efforts, or act in ways that are at continually destabilizing odds with Poverty Reduction's arrangements for governing poverty. This is not because Poverty Reduction is any hand-servant of naked capital integration. Rather, it is because Poverty Reduction – especially when trimmed down in 2004 WDR terms – includes highly formalist rationalities that seek to put into practice a set of Liberal humanist concerns encompassing a whole range of transformative hopes and ideals. From Uganda, to Pakistan in the early 2000s, and then New Zealand, we will show the agents and institutions of Development overreaching, taking on technical and other transformative tasks that are simply beyond what can be achieved and which have unanticipated political and social implications well beyond what's imagined at the outset. Commonly here, we see some facet or programme or institutional branch of Development creating a delusion that agency can be incentivized to operate independently of political economy, misjudging the potential for change, and making ill-considered and badly timed interventions which are only partially successful and generate reactions and backlashes which cause roll back and all sorts of unexpected complications, prompting another round of interventions.

Our aim, then, is to show how Poverty Reduction's highly formalist travelling rationalities for governing the poor work out in the potent contexts of political economy, history and territory. Thus we have to track them in very different places. But meantime, it will be clear that wider political economy still matters. Reducing poverty is primarily dependent on market outcomes, and on stability and security – things over which Poverty Reduction frankly, has not too much control. In fact, the frailty of Poverty Reduction's central elements against far more embedded structures of politics and history will sometimes become all too evident. But while apparently technical, Poverty Reduction's governmentalities can have powerful political implications, not least because their shifting, slippery application makes easy their appropriation by powerful officials, elites and other actors acting with far less Liberal intent or concern for 'the poor'. As we will see, Poverty Reduction's Liberal economic dispositions co-exist best with authoritarian regimes and, ironically, as we track through Vietnam, and especially Uganda and Pakistan, we will point to its politically centralizing effects at the same time as it moves to empower the poor in their places. This is all particularly poignant in the post-9/11 global security alliance, of course. Since then, active resistance to alternative policy mixes has apparently effectively closed the 'development policy space' available to poor countries. Thus, in the OECD DAC commitment on 'harmonization' for instance, Poverty Reduction has been thoroughly buttressed by the staggering commitment by all partners to 'resist introducing indicators of performance in poverty reduction that are not included in the policy matrix of the national Poverty Reduction'.[64]

This Poverty Reduction frame, then, for all its perverse elaboration and apparent frailty, is also backed by increasingly potent political will, driven in turn by international security fears, all of these visited on the poor and their places. By the time we get to the Pakistan-Afghanistan border at the start of Chapter 7, it should be clear that the perverse plurality of travelling rationalities is also a recipe for brittleness and risk in an increasingly fraught context. This, even as its impact on the poor and their places seems by no means assured. And, as will be seen in New Zealand's experience, its none too clear how Humpty Dumpty can be put together again. But all this, we are sure, needs to be shown, rather than told.

Part II

Cases from Vietnam, Uganda, Pakistan and New Zealand

5   Vietnam
Framing the community, clasping the people

The people know, the people discuss, the people do, the people check a good job is done.

Slogan of the 6th Congress of the
Vietnamese Communist Party, December 1986

Introduction

Vietnam has a long history of governing the poor. From the state's perspective, the poor are 'the people', and this is more than just rhetoric and sentiment. They are, as not just politically correct Hanoi cadres will tell you, the bottom line of legitimacy and national security, and ensuring their wellbeing is the primary function of a 'socialist democracy' like Vietnam. On the other hand, the poor, as many Hanoi mandarins will also confide, are poor in the pejorative sense of being uneducated, bounded in and inexperienced beyond their local *que* or ancestral village. Here, they need to be 'gone down to', governed, instructed, chided, led in 'correct' ways, by those higher in the governing and territorial hierarchy that stretches all the way from the hamlet to Hanoi. From a liberal governance perspective, too, the people are the very basis of liberal governance, legitimacy, democracy. They need to be gone down to in order to have their voices heard: to be able to participate, make locally appropriate choices of infrastructure and services, and to hold unresponsive local territorial patrimonies accountable for local outputs and outcomes. This, it is clear, will mean breaking open the sealed territorial constraints on information and competition, even or perhaps especially where this causes existing, territorialized local governance arrangements a headache or two.

Certainly, viewed in liberal governance terms, the Vietnamese socialist state has been in various times and places as territorialized as can be caricatured: planned national production, extensive market-disrupting subsidies, totalitarian political order. In Vietnam, in this view, there was but one official truth, and it was handed down the hierarchy of territories from centre to commune and village in a monological way that brooked no local voice: one truth for the whole nation, not to be contested bottom up, or with ideas from elsewhere. On the other hand, these same apparently monolithic territories were never extended fully across all of Vietnam, even after triumphant socialist reunification in 1975. And, as we will see in this chapter, they have always been transected, most notably by networks of personalized loyalty and patronage, which have been routinely used to short-circuit or circumvent hierarchical torpor and authoritarianism. For liberal governance, however, these networks too smack of patrimonialism and corruption: personalized, special deals, personal links between patrons and clients bringing impunity and reducing and even blocking formal accountability; lack of contestability in recruitment; no free flow of information, and so on. And, of patriarchy: both the formality and the informal male bonding had powerful gender exclusionary effects.

In fact, as we will see, there are two caricatures of Vietnamese governance: the hierarchy and the network, the formal and the informal, the Confucian and the South East Asian, the north and the south, the closure and secrecy of the fabled northern 'bamboo hedge' and the open flows of the river across the widening southern delta.[1] The stiff official socialist, perhaps, and the populist and personalized: this latter based on *tinh cam*, literally to 'feel the feeling', the sensitivity with which all matters, governmental or not, might and must be managed, at least among men. But of course both are right and wrong: the socialist state was never a totalizing, monolithically territorialized construction, and the populist is anything but informal, and has local limits of its own. But in any case, bringing liberal governance techniques like the PRA events and LDFs described in this chapter into a Vietnamese commune is an amazing 'travelling rationality' experience. As they intersect with traditional socialist governance, both the high formalism and the personalized, informal intimacies of both Vietnamese and inclusive neoliberal governance techniques are brought into intimate contest and collusion.

For Poverty Reduction Liberalism has a reterritorializing logic of its own. Here poverty is framed in terms of a local quasi-territory, the community or village, site of local lack and vulnerability; thus, rather than as a peripheral effect of wider political economic forces. This lack and vulnerability, it has been believed, could be best relieved by markets and deregulated modes of service delivery functioning to increase voice and choice, all within a new set of governing arrangements where again, liberal participation and planning based on local information could open up new empowerments and accountabilities for the poor. Not, of course, that any of this was spelt out to the Hanoi or local cadres involved in participatory events and programmes like the one described in this chapter. These people and the local poor, as we will see, would struggle with the new kinds of liberal openness assumed in the participatory planning and competitive contracting techniques used, and would question exactly what benefits came to whom from these novel, albeit minor and ephemeral liberalizations of local planning.

In this chapter, we describe the governance situation emerging in Vietnam in the early 1990s, and in the context of a particular project in central Vietnam, the ways some of Poverty Reduction's early liberal governance doctrines and techniques described in previous chapters – PRA, LDF – began to make their presence felt in this erstwhile socialist territorialized governance regime. We will see how, in different ways, both Vietnamese and international liberal governance have 'gone down to', framed and governed the poor, both in an earlier, strongly territorialized dispensation, and now, as that strong socialist state is increasingly riven by markets and new modes of more liberal governance.

Liberalization and the socialist state

> In order to take the land from the hands of the landlords, we lost half a century and paid with our sweat, blood and bones. Now landlords are taking back the land . . .
>
> Former soldier, Bac Thai, *Nong Dan*, 5 August 1993

Certainly, Vietnam is by almost any standards a 'liberalization-brings-poverty-reduction' success story. But it is also a 'strong-state-brings-economic-success' story, a place where security and empowerment have a strong socialist ring to them, and where governance is run along powerfully illiberal lines. Arguably, it is the sustained coincidence of both strong, territorialized state and the simultaneous opening to deterritorializing market reforms that have brought Vietnam to where it is now. By 2001, having weathered the Asian crisis, Vietnam would emerge as one of the models for Poverty Reduction, and the exemplary host for the first Asian Regional Poverty Reduction Conference held in Hanoi. Vietnam's PRSP, known as the Comprehensive Poverty Reduction and Growth Strategy (CPRGS) had been inculcated into at least central government planning in ways that other countries simply hadn't. Existing state poverty alleviation and economic planning processes of long provenance had ironically turned out to be quite compatible with PRSP, and had facilitated the engagement of a number of line ministries that had taken it beyond the core Ministry of Planning and Investment. Sectoral dimensions largely glossed in other countries' documents, notably agriculture, were written into the CPRGS in substantive detail, and as early as 2002 moves were underway to pilot the process at regional level, using PPAs and otherwise promoting integrated poverty planning at province level. MTEFs were being put together for the 2003–2008 period in four sectors, education, health, agriculture and transport, fiscal decentralization and a top to bottom restructuring of national accounting and audit procedures was underway. Poverty had continued its decline, from 58 per cent in 1993, to 29 per cent in 2002.[2]

Again, despite ongoing corruption, Vietnam's 'strong state' turned out to be a considerable asset. Limited political contest under a one party regime (a situation not dissimilar to that other PRSP flag-carrier, Uganda), meant the planning environment was already hierarchically endorsed, and consensual and technical at several levels, in which there was much sectoral and centre-province wrangling and even competition over plan ascendancy, alignment and autonomy. Prompted by a government keen on coordinating donors under its own planning ambits, and enacted through an extensive partnership, sector forum and networking process sponsored by the World Bank, most substantive donors came into line with CPRGS orientations. Donors were harmonizing in other ways too: the Poverty Reduction Support Credit, financed by the World Bank, the corresponding agreement with the IMF on the PRGF[3] was being used to signal a new confidence to donors that they could release aid directly into the government budget, thus shifting from project specific controls on the use of development assistance and placing full confidence in the government's own policy and planning, budget and expenditure management systems. Overall, donor contributions have continued to rise. The emergence of Vietnam as a socialist 'model state' for PRSP is not without its ironies, but the socialist orientation has been largely ignored in the lessons drawn from Vietnam.

In 1992, when this chapter begins, these 'model' outcomes were none too certain. The winds of *Doi Moi*, the economic transformation or new change officially announced at the Vietnam Communist Party's 6th Party Congress in 1986, were blowing through Vietnam, but their political and economic effects were uneven; inequality was apparently increasing dramatically and, although it was still hard to check these numbers, there was no doubt public protest about the collapse of services and rising corruption by opportunistic officials was at times flaring into open conflict with the authorities. Most worrying for the government, all this appeared to be undermining the Party's hard won legitimacy.

By 1992, the rise of the market economy had been officially endorsed, state planning had receded into the background, and the household had once again become the basis of economic life. Though still a cub, the new Asian Tiger had quickly come to symbolize the 'Confucian' passion, power and daring for many then holding up Vietnam's bold transformation.[4] Economic growth was impressive. By 1989 Vietnam had begun a run that would last through to 1997 of between 7 to 9 per cent GDP growth each year. Poverty was declining. But the winds of globalization were blowing unevenly. Under pressure to reduce the fiscal deficit, running at 8 per cent of GDP in 1989, a budget-induced crisis impacted heavily on delivery of services provided through commune health stations.[5] Radical deregulation of drug supply worsened the financial positions of commune health stations, and meant that pharmacists' relatives and completely untutored village market stallholders were trading restricted antibiotics over the counter, with or without photocopied licenses. Mothers' fears over fake and ineffective medicines coming over paper-thin borders from China was palpable, and everyone wanted either the state to 'guarantee' medicines, or provide more officially correct information for themselves with which to self-prescribe/regulate in the free market.[6] Medical care standards were radically marketized as privateers of all ilks commandeered local primary health care service provision, and traded on their reputation and the special qualities of their particular medicines. Children's participation in schooling seemed to be declining[7] and alarming reports had surfaced in 1992–1993 that close to 70,000 classrooms had either been destroyed or made unusable due to poor maintenance. Across the country, many were surprised by just how quickly economic liberalization, and especially the parlous performance of a 'withdrawing' state, was opening a yawning poverty gap between urban and rural areas and between emerging classes that was becoming clearly visible across and within provinces.

Wealth was concentrating quickly. Whereas the difference in incomes between the top and bottom deciles was one to two times during 1960–1970 and three to four times during 1970–1980, by the early 1990s, the difference was reported, admittedly on the basis of poor statistics, to have grown to between 10 and 100 times.[8] Life for most rural people remained harsh; inequalities were becoming more obvious and worrisome. Although some officials felt that the concentration of wealth, land and other assets signalled the rewards of entrepreneurship, surveys showed that voters were very concerned.[9] Numerous disputes were emerging, some ending in bitter confrontation. A Communist Party document of 1990 cited incidents of villagers beating and killing each other, engaging in arson, and arousing the police to use force;[10] another study in the same year reported 6,000 conflicts nationwide, disputes over land and administrative boundaries, over seceding from cooperatives, refusing to pay taxes, fees, and compulsory contributions to public infrastructure projects. As a resolution of the Central Committee's June 1993 national meeting warned, poverty still prevails in the daily life of peasants' and 'democracy and social justice in the rural areas are still being violated'.[11]

For the Vietnamese state, change was the order of the day. On the surface it was rapidly, if not transforming, then at least transiting, back and forth in official rhetoric, trying to give a socialist face to capitalist society. But the first rounds of economic reforms did little to change the basic formal mechanisms within government.[12] Among a series of increasingly shrill and reactionary clampdowns, high profile 'lesson-teaching' steps were taken against corruption. Between 1990 and 1992, close to 20,000 people, including seven vice-ministers, were punished for corruption. Official language began to admit the need for 'democracy', 'participation' and, borrowing from the economic doctrine of *Doi Moi*, even political 'openness'. More pressing was the obvious need to find a way through the tug of war between local and central, the people and the state, between official voice 'going down' and the increasingly troublesome noise 'coming up'.

PRAs and LDFs in the wider story of this book

In the early 1990s, the socialist state was strong and stable – certainly when compared with Africa's sub-Saharan state casualties. But new sources of legitimacy were needed through which the economic transformation might be managed in an ordered, socially agreeable way. The government had had limited success combating corruption through existing state hierarchies. For their part, international donors were becoming alarmed at the evident social costs of economic change, as they began to eat away at the primary health and literacy statistics Vietnam had once proudly displayed. They began to introduce new ways to frame up poverty in Vietnam they hoped would simultaneously begin to shift resources to the poor and create conditions for participatory, accountable local governance. PRAs and LDFs, although by no means yet joined-up into the full mechanisms of decentralized, intergovernmental transfer systems for pro-poor service delivery, certainly embodied these plural hopes.

PRAs, described and indeed all but participated in by the reader in the first section of this chapter, are a liberal governing technique par excellence: a set of Participatory techniques for doing a Rapid Appraisal which together frame the poor up as suitable subjects for liberal development and governance. PRAs, as our telling illustrates, involve a set of qualitative, participatory research methodologies conducted at local level, designed to elicit local people's voice and concerns around issues of need. LDFs, as explained in the second section, were 'devolved funding' arrangements, championed in the early 1990s as a means of both funding local social and productive infrastructure, and getting local authorities to listen to and be accountable to local needs that had been captured through devices like the PRA. Both LDF and PRA carried a larger ambition – as we will see in this chapter they aimed to create a 'policy demonstration effect' through successful replication across all villages and communes in Quang Nam Da Nang (QNDN) province and from there up through national policy and legal arrangements that would resonate across the country.[13]

Seen against later, sophisticated decentralization arrangements discussed in Uganda and Pakistan cases, the LDFs and PRAs shown in this chapter seem now quite limited in their ambition. And, frail and naive in their attempt to create Liberal governance arrangements in distinctly illiberal environments. Nevertheless they are a good place to begin to show the dynamics of decentralizing liberal poverty governance we want to pursue in this book. In the early to mid-1990s, these techniques were typical of the instruments then used to articulate global commitments to ameliorate socially disembedding consequences of market led transformation. LDFs promised to get resources quickly into palliative social services and basic community infrastructure and to do this while introducing 'participation' and new international codes of competitive contracting and enforcement. PRAs provided a legitimate way of framing up the recipients of aid in manageable, liberal subject terms, as needy but active individuals keen to participate in planning and implementing service delivery. Both provided systems to alternately bypass local leaders (who were at best seen as getting in the way of delivering self-evidently needed results, at worst, opportunistic corrupters of the public interest), or to make them listen to local voices whether they were keen to or not. As importantly, both techniques are part of the process of linking distant, peripheral places with metropolitan cores then just getting underway in poverty reduction: in this illustrative case, we see the New York headquarters of the United Nations Capital Development Fund (UNCDF) being joined with a commune nestled on the Lao border in western QNDN province in central Vietnam.

For UNCDF officials, these devices authoritatively represented poverty, and allowed the projection of governance and service delivery solutions into the middle of them. They also did this in ways that could be transferred and replicated from one periphery to another, and used to both exemplify and legitimate their work still elsewhere, as we will see in the next chapter. For the locals, their framing in this way enables certain resources to be transferred, providing certain strange orthodoxies are complied with. For both sides, there is a need to capture, to clasp, to package for another context, often in quite narrow and formalized ways. In this project, PRA provides the clasp, connecting the new local containers of a political economy that is carried by the LDF. It enabled the whole project to represent itself in terms of long-standing development enchantments: as putting the last first; hearing the voices of the most marginal; representing communities in globally legible and consumable ways; offering a place for community at the top table; presenting poverty as something 'inclusive' liberal development could fix.

Here, the links between a technique like PRA or an LDF and Poverty Reduction were by no means comprehensive, or even sometimes all that substantive, especially given the amount of resource invested. But what was immediately patent on the ground was the extent of the need that drove their implementation at the beginning: the sheer, shrill obviousness of the 'get the data, need-participatory voice, then do it' linkage. This need was especially important at the beginning of a project, and the legitimacy it conferred would be drawn on all the way to the end, and at sites far removed, even where community needs had moved on, or where the project has failed. Beyond this, the long-term viability, or 'sustainability' of these orthodoxies in relation to local practice is in our analysis called into account. As we have often seen on the ground, these highly situation dependant liberal project orthodoxies might set people up to overstep existing boundaries, take unrealistic positions, expose themselves in unexpected and surely unintended ways to a backlash.

In this chapter we want to show – rather perhaps than tell – the ways Poverty Reduction tends to frame up its subjects and places. The story told here is a hybrid, though based entirely in its more prosaic sections on a leading early 1990s decentralization, poverty and governance project in Vietnam: the Rural Infrastructure Development Fund (RIDEF) project, initiated by UNCDF. What the reader will see here, and initially in the Uganda chapter following, is not Poverty Reduction in full barrel: but rather, the parts of the paradigm emerging, coming together, being worked up to travel, at the same time as they exert tremendous efforts to embed themselves on the ground. What the reader will recognize in RIDEF, its PRAs and LDFs, is an embryo, an experiment. In the Uganda case, in the second half of the 1990s, we will see this very approach explicitly ramped up and bedded down, to the point where it ran into and then colonized the more elaborate PRSP framework, to form a wider set again of structural linkages and programmatic reinforcements.

Again, showing, as much as telling: the italicized sections appearing soon in this chapter show a PRA event, again, drawn eclectically from a number of such we have participated in in Vietnam, and told here with no direct connection to RIDEF. Later case studies, disconcerted readers should be consoled, are much more standard and prosaic treatments of programmes and political economy . . .

PRA: BINDING BINH LAM COMMUNE TO NEW YORK

Clearly the situation into which the PRA is travelling is, as throughout this book, not a *tabula rasa*. In 1993 Vietnam, the fabled socialist/territorial/ totalitarian control over foreign access, the traffic of information, and the representation of 'the people' and 'the nation' were not fading and liber- alizing, so much as becoming fractious. On the one hand, foreigners could legally travel almost anywhere in Vietnam, and talk to and visit who they wanted to. On the other, longer-term foreigners could stay only in approved locations, and their activities were subject to all sorts of scrutiny and reporting processes. Foreigners doing Development, doing PRA or NGO- type project research crossed an increasingly ill-defined border into the highly formalized yet subtly managed domain of official information, where the power of representation and knowledge was all too well known and respected. They crossed this line without being organizationally or culturally locked into the formal and informal constraints of official management of information, and so became potentially loose cannons. In the apparent absence of stable, clear official guidelines and past experi- ence, there was still a powerful uncertainty and edginess on the part of cadres at all levels about what you should be told and allowed to see. This was tied up with an unfamiliarity and suspicion over the methods used to research and talk-up plans for future projects, and fears over who would take on the responsibility if anything went wrong.

Sensitivity, checking-in, the correct way

On the way down to the commune, it was more than just courtesy to 'check in' at province and district level, to show your face and ritually seek permission. While authority from a higher level practically guaranteed some kind of access to lower areas, taking anything like that for granted was a travesty of respect for local authority and even security. 'Checking in' with and 'seeking permission' at each level from local cadres showed 'sensitivity' and 'consideration', and was not just highly esteemed, but provided safety for intermediate level cadres, who more than anyone else would bear the brunt of any criticism following the visit. The formality of checking in was always balanced by the hospitality of at least a cup of tea and biscuits, though meals and even feasts were not uncommon. Reciprocally, local hospitality ministered directly to a visitor's 'sensitivity' to the local cadre's situation, and built a powerful, often explicitly stated obligation to 'sympathize' with them when it came time to 'report'. It also helped by using up a good deal of outsiders' time on the ground, getting them drunk and prone to the compromises in 'cuddle bars' and karaokes which gave locals leverage.

> Centre, Province, district, commune, hamlet: a hierarchy of territorial space and governance that you, your project, and just about anything else travelling down to 'the people' must traverse. The place of hierarchy, though, and its close yin-yang relation with other cultural patterns of intimacy and informality and networks, was still an intrigue, something you had often argued up with Dung and others from the program. On the face of things, the hierarchy ruled: it was the most obvious form of social capital on show, linking things up, correctly, and so on. Governance-wise, central decisions were classically implemented in similar ways, down through the hierarchy, 'correct according to the plan', 'following the decision of the party and the state'. This was the 'correct' way, the *chinh* way, referenced in all the governmental words in Vietnamese that combine *chinh* with other words: *chinh quyen* (authority), *chinh sach* (policy). Hierarchies of patronage and other authority articulated in various ways with these up and down linkages. Directives issued by central government and endorsed by the legislative assembly were handed 'down' through the hierarchy of province-district-commune-hamlet, with each level being required to 'report' (*bao cao*) to the next level up on the typically 'successful' implementation of the edict within their territory. Processes of control and scrutiny follow, where a central, province or district official would come to visit the locality to inspect and to write another report. Written reports had a sinister power in Vietnam, expressed in the saying, 'Write it down, and the chicken dies' (*But ra, ga chet!*). Dealing with outsiders coming to write reports had become a well-practised art in provincial Vietnam.

Everyone blamed the communist government for this state of peripheral subjugation by officialese and formalized governance. But it wasn't entirely so: Gourou[14] writing in the 1930s of the 'immense platitude' of Red River delta geography, described personal freedom and movement in the Red River Delta as bound by 'the complete supervisory control of public opinion over the private life of the individual'. But true, neither Stalinism nor the dirigism and discipline of 30 years' war helped, as dissident poet Nguyen Chi Thien wrote, 'The party holds you down and you lie still', and 'If Uncle and the Party, let's suppose, allowed free movements in and out, Grandfather Marx's paradise would soon become the wilds where monkeys roam'.[15]

The troubles arising from all this staged, spaced hierarchy were popularly fabled over, for example in the well known story of project and other funds being like a big ice-cube leaving Hanoi, travelling in a hot climate. By the time it had traversed the cluttered roads of the provinces, districts and communes, and every level had taken their 'percentage', there wasn't much left for 'the people'. But then there was the counter analysis, that pointed out that resource flows, and especially taxes in many provinces actually reverse this flow altogether: the commune collects taxes and haggles with the district over how much gets passed up, and the district does the same, and so on. So the hierarchy cut both ways, and created mutual obligations coming and going. A tonal rhyming poem bespoke an uneven contest of power, extending hierarchy and referring violence all the way down from Hanoi to hamlet gender and domestic violence relations:

Trung uong tuong tinh	The centre pursues the province
Tinh chinh huyen	The province hauls the district over the coals
Huyen kien xa	The district accuses the commune
Xa na thon	The commune puts the squeeze on the hamlet
Thon don dan	The hamlet rounds up the people
Dan dan vo	The people thrash the wife

But as Hanoi and all points south knew, 'the people' were a sleeping giant, who governance corruption might alienate to enormous political effect. The thrashed wife had already caused the adjustments of *Doi Moi.*

In the long committee room, round the long polished table, Dung from the Province handles formal introductions with the lines of men in formal dark garb, and the couple of women at the far end. Apologizing for lateness Khoi joking that the hospitality and 'sensitivity' had been overwhelming at the province and district. Dung eases gently into introducing the participatory methodology. He picks up on the province engineer's use of Ho Chi Minh's aphorism, 'The people know, the people discuss, the people do, the people evaluate and check up'. He pronounces each part slowly, rhythmically, in his clear, Hanoi diction: *'Dan biet, dan ban, dan lam, dan kiem tra'*. He pauses for small effect, and then catches a smile from his ministry counterpart. Khoi snorts out a laugh, and adds out loud *'Dan phai nghe'*: 'The people had better listen.' Dung can't help himself, and soon everyone bar the commune guys is laughing out loud. They smile, not knowing exactly where to look. You too: that's exactly the trouble with using officialese, every single one of these shibboleth formulism has been parodied mercilessly by everyone, including the cadres charged with their deployment. And now, with 'openness' and all, who knew exactly where the boundaries were anymore?

The PRA event itself was bound to be a bamboozler for these commune government guys: it felt plenty odd to you, a liberal, participatory governance tool imposed from great height in a sharply defined local domain, asking people to set aside years of deference to hierarchy and formality, and participate in some chatroom, as if everyone's ideas mattered like anyone else's. As your Ministry of Planning and Investment counterpart chortled, 'You're going to ask those people to tell you what they really think, in public? They're not that stupid'. And to what end? What would all this information and apparently deep democratic process be used for, other than legitimating the whole frame-it-up, slap-it-down process ... you couldn't guarantee anyone the wish lists you'll generate would be funded ... Or that anyone would ever do PRA with them again. Or, yep, that they wouldn't get in trouble for talking frankly and not 'objectively' in this round ... So then, what exactly would the net effect of this intrusion be? But you're here now, the five day process is in swing, and you'll press on, them too.

Clasping community, framing the project

PRA: participatory rural appraisal, participatory rapid appraisal, participatory relaxed appraisal. Always distanced by PRAs many purists and partisans from RRA, rapid rural appraisal, a crude, extractive form of 'data mining', pulling information from local people to serve higher authorities, and from variants like, participatory learning and action. But all are basically qualitative research techniques designed to rapidly engage local communities in identifying their own needs, and identifying solutions and priorities over the course of a PRA programme, lasting from a few hours anywhere up to two weeks. The origins of PRA lie in agro-ecosystems analysis, participatory action research and adult education, applied anthropology and farming systems research in the 1970s.[16] The 100 or so techniques accredited with being 'PRA-compliant' have enjoyed enormous currency in the past 15 years, expanding rapidly from their NGO base to be embraced across the Poverty Reduction community.

Thus PRA techniques have travelled into the world's remotest corners, and emerged intact, apparently validated, widely enjoyed. Much impetus has come from academic consulting centres, especially from the Institute of Development Studies at Sussex University, where Robert Chambers has championed PRA as a methodologically sound means of giving voice to the poorest of the poor.[17] NGOs' enthusiasm for PRA stems too from the much-practiced realization that local, indigenous project officers and their public sector counterparts can be trained in the techniques. 'Doing a PRA' (or doing twenty of them) has become a right of passage for NGO folk, and for a time at least, any credible form of development assistance project.

The relation of PRA to 'inclusive' liberal orientations should be clear enough. On the one hand, PRA is a move to expand participation in process beyond the state, to strengthen governance by getting civil society to contest, discipline and inform, and to open up local planning dialogue to expose and avoid nepotistic, clientist arrangements. In this regard, PRA has also in general proved remarkably untroublesome as a populist technique married to major infrastructure and other projects. Community needs identified through PRA have in general sat easily alongside wider poverty alleviation ends, and the techniques for whatever reason tend not to throw up particularly powerful popular dissent or critical objections. At face value, it's surprising that an avowedly populist set of techniques like those of PRA have not raised more the hackles of powerful governmental elites in developing countries. There've been hiccups, like when the Government of Malawi in 1999 temporarily banned all PRAs countrywide, although it was more outraged about PRA's costs as much as potential for dissent. But for the most part, the critical voices – and as will become evident there have been some important ones – have tended to come from within the institutions practising PRA, or from jaded former practitioners.[18]

PRA processes occur within the heavy editorial framing of development projects, wherein almost everyone knows up front that certain things can be done, and others aren't on the agenda.[19] There is in PRA too a strong bias to the consensual, as if most local interest differences can be overcome and needs agreed to on a community basis. Overall, however, the process is awkwardly depoliticized, and on the ground often tries to get around the 'distorting' influence of power relations and leaders by treating everyone alike, imagining it can set aside hierarchy and exclusion. PRA would become a level playing field of unconstrained, ideal speech. As a technique that produces apparent consensus from fraught local contexts, its depoliticization reinforces a wider depoliticization, consensual and technical bias evident in PRSPs. Here, the poor get a voice, and what they say they want coincides closely with what 'inclusive' liberalism says they ought to. What PRA tells us they really want is micro-change, particular services and basic facilities, and good governance around these. It is this apparently, that keeps them poor, rather than for example wider structural conditions of trade or existing distributions of power/property. At the same time, what they say legitimates so much more. The notorious 'Voices of the Poor' documents,[20] published after extensive international PRA and PPA processes and much editorial grief by the World Bank in 2000, have had a remarkable legitimating effect on the whole Poverty Reduction process, and, by association, the institutions supporting it. We have to wonder whether the people who contributed their Voices about poverty knew to what wider systemic ends their accounts of their own misery would be put.

The commune officials were sitting in rows in the main hall of the people's committee building, men in the front, women in the back, most leaning forward, hunched over, conical hats on their knees, listening to the run through the processes of the next few days. The Commune chairman stands, walks across to the front.

'Today I speak on behalf of all you members of this commune who are fully determined to realize and implement the decisions of the party. And who want to hear the advice of those higher up and with better education than you. Now, we have explained the decision of the party about development and infrastructure, and the steps of decentralized planning, and the needs of this commune, especially for the development of the roads and other infrastructure'. Dung leans forward again. 'The people had better listen', he hisses.

'So let's discuss this! We must go deeply into the issues, following the spirit of the central decision, and we need to express our opinions about this directive. You should participate in the discussion according to the issues I have emphasized. Should we develop this commune, and combat social evils or not? It's 1993 already, so we shouldn't have to remind you all of everything again! But some people nowadays are getting lazy when it comes to participating, to making their contribution to the development of the commune. Should this commune develop or not? How can we guarantee the happiness of our families, without development and participation? We've heard the way things have been reported so far, so now I propose we participate with our opinions. Let's discuss! But first, some of you younger men, you must be ashamed of yourselves! These days there are many social evils. A road is not just for "going outside". Apart from that, you sisters who were complaining last week about the day rates for carrying the new road materials, you had better consider who is paying you that money, and who is going to feed your children. So, participate, express your opinions about how to implement the decision for the best. But before we discuss it, I would like to invite the vice chairman to give his opinion.'

The vice chairman did, repeating most of what the chairman had said. This he followed with an overview of the principles of participatory planning, almost word for word what Khoi had said yesterday at province, district, then here.

'Now it's time for you to speak! As the chairman said, some of you people really are awfully demanding, but we have to have everyone's participation in this project. Whoever wants to express themselves, go to it.'

The vice chairman steps down from the rostrum, comes outside, and notes you are still there. He asked again for sympathy that the people aren't yet in a position to respond or report according to the principles of participatory planning you require. Everyone has been mobilized to participate, but the people will have to 'study' a bit first, so that no-one is saying stupid things. People here have not had much opportunity to study. It is true, he says what they say in Hanoi, in Da Nang. There are many stupid people in the countryside, who will not speak diligently, but will say whatever comes into their minds. He asks not to treat them too harshly if their comments lack objectivity.

Khoi tells him not to be concerned. 'All we've have heard so far sounded quite correct, he had studied well. And we are very enthusiastic, and would like to sit in and hear the rest of the people study and participate.' He looks to the province guys for a let out, but they shake their stony faces. The discussion consists of speeches from the older men, descending a ranked hierarchy. The formality is high, though not unusually so, considering that 'long noses' and VIPs are conspicuously listening. The rhetoric is nearly poetry, that lovely, lilting oral culture thing, seeing it this way and then that, showing easy grasp of the key principles, turning them over once or twice to create space for inflection, nuance: 'Firstly, we must develop the commune in a participatory way, not without listening to the experience of the older ones, or disregarding the hopes and dreams of the younger. Neither set determined in old ways, nor bending too far with the wind. Secondly, the principles are new, but not really that new. Thirdly, we must work together in a spirit of sensitivity and objectivity to form a comprehensive, integrated plan for the future of the community, a beautiful and correct plan from top to bottom, to link with the higher ups and those below. Fourthly . . . In sum, we must participate gladly in the PRA activity, explaining diligently and correctly our objective situation, and not letting the subjective needs of our individual families cause or speech to wander off into areas that were not for the good of all the people. This would allow the cadres and foreign experts to report the progress the commune is making in its own freedom, autonomy and happiness, so that the much needed roads and infrastructure could be built . . . As Uncle Ho said, "the people know, the people discuss, the people do" . . .'

The commune People's Committee and Women's Union reps are waiting in the stucco walled yard at seven am. Shoes and plastic sandals stomping up and down in the wet clay, lips curled with the chill fog.

Khoi is back, shoes shiny and unmuddied, bright laughter and stainless steel smile to match. The women are bonding too, standing off to one side, around Mai. Cherry, your Canadian NGO-gender counterpart, has latched onto the serious Hanh, in a kind of marvellous Yin Yang partnership. Cherry is yang as all hell this morning, loud Canadian accent and red curls bouncing with her tongue, busting to get out there past this 'official bullshit' and among the 'real people'. She hopes to stir a few things up among the women, knows you won't . . . You break into three groups, Dung, Mai, Hanh with one group each, a mix of central, district and commune guys, you and Cherry floating between the groups, with a commune guide to take you from place to place. Three groups, so different transect walks to start things moving. Then some group activities, mapping, focus groups, various timelines, income ranking . . . Your group includes Dung, Khoi from the province, two district reps, the Commune vice chair, and a young woman from the Woman's Union, dressed formally in a slim, traditional ao dai, and a wide-rimmed Hanoi style hat. 'At least we got the pretty one', Khoi enthuses, loud so Cherry hears. He's teasing again. Tomorrow, when things have settled down a bit, you'll split on gender lines.

You start the transect walk in one built-up corner of the commune by the road, and proceed in a straight line, crossing through houses, round fishponds, past the school. The broad aim of a transect walk is the taking of an introductory cross section of the whole place. Like most other PRA techniques, on one level it's a simplistic, reductionist frame-and-fill-up-the-map device. But on another it's something that can be used to cross borders, to break the normal frames of roads and pathways, and to sample for difference. By proceeding across the whole breadth of the community, you see literally all sorts of things: housing and other indicators of wealth and wellbeing in the better and worse off parts of town, water and sanitation arrangements, transport, livelihood, public amenities. These you sketch in on a series of lines drawn in across your field notebook, and the transect becomes a part of your 'been there done that' documentation. But maybe even more importantly, the things you see give rise to topics of conversation, and, for later discussions and interviews and maps, to gather some sense of what local content is actually being discussed. The walk has a certain informality that enables casual observations to be picked up or let slip in conversations, depending on local reactions. Overall, though, it's safe enough for the first activity: they get a chance to talk, you don't target anywhere controversial, you can slowly move the conversation towards tricky issues, and show them you can be trusted to hear their point of view on them. Locals will usually take cues from somewhere on the walk to tell you of things they personally or as leaders would like you to pay attention to, whether that place is on the transect or not. You too will use whatever you see to push certain things, no doubt: and no doubt they know it too. The need for local voice in the province planning process, for example, is pushing your gaze to look for resource misallocations.

Getting out in the thick of things on a transect also raises local awareness (of what you're doing, if not of what they're doing), and sometimes generates greater levels of participation in subsequent activities. The transect as a formal device makes it harder for interested parties to pull the wool over your eyes, either by showing you or keeping you away from parts of town they have a parochial interest in, or which might generate or harm sympathy. At least, it might make you imagine you aren't being shepherded or blindsided. Khoi from the province keeps up an intrusive commentary all the way, the province did this, they put that there. The others trail off behind, stepping edgily around puddles, looking bored brainless. Not much they haven't seen on a transect walk, you suppose. But who knows: city people never cease to amaze in the blunt scorn and basic ignorance of their appraisals of country folk.

Most people are already well and truly about their business, out in the fields, on the road. You stop and chat where you can: a carpenter, a rice miller. The vice chair is nervous in his introductions, but the banter with his constituents is cheery, sometimes gently teasing. Everyone has a ready laugh, once the first amazed boggle is over. They're steady on their feet, coping kindly with the tall Westerner speaking badly toned Vietnamese. Sometimes, you have what you take to be the small success that no-one starts off their reply with 'I would like to report that . . . firstly . . . secondly . . .' The carpenter, pausing from his attentions to an elaborate bedhead, is looking forward to the road being built, but would in fact prefer it wasn't because of the likely cost. All would depend on the shares of the funding, how much province, how much local, how much voluntary day labour.

You cross paths with Cherry, who's just finished her focus group and is grumpily happy all her worst predictions are coming true. 'He just dragged the women in from the fields', she said. 'pulled them into his sister's place. They were lined up against one wall of someone's house, cowering in the corner while he gave them a lecture on what the village priorities should be. They nodded along, and the more they nodded, the more he lectured. Then the district guys saw their chance, and did it all over again. Each one saying how much their level contributed to the local needs, and how well they understood local conditions. Focus group one. They loved it. Now, they can't wait to get to the next one.'

But back to practice. In most PRA events the transect walks are followed by smaller group activities, participatory mapping of relatively small areas designed to show up its issues and potential developments, its areas of poverty and poor infrastructure and relationships between these issues, potentials and needs. At heart, then, PRA provides a handy set of techniques to address enduring questions in all development practice: how is production and consumption organized; who decides how this happens, what are the costs and benefits, and for whom, of how this happens now and might happen in the future? The techniques usually involve group discussions and activities sitting around large sheets of white paper, where PRA's signal ritual of 'handing over the pen' ensures that local people enact their own PRA representations. There are participatory techniques for discerning relative and absolute poverty levels, community issues, activities and priorities, relationships between various community organizations and service providers, as well as recent history of crop, weather and seasonal cycles. Other techniques help sift through different options to deal with problems identified, and to reach consensus and common commitments to how resources and responsibilities will be assigned to bring them into practice. A good PRA operator will have a working grasp not only of the central techniques, how to work the group to make it engage, but will also have tricks and techniques of their own, ways to overcome some of the biases inherent in the method.

PRA events, then, are to their partisans explicitly not merely tools for extracting information in as short a time as possible from local people bound together by common administrative jurisdiction. They are infinitely superior to any of the previously relied on methods of project visits and other forms of 'development tourism'. Nor are they just a way to frame up for public and institutional consumption something that in fact wasn't there before you started, a 'community', for example, with consensus priorities and ways of making them known. Much less are they merely a shibboleth, a gesture at community participation designed to appease the wet side, the die-hard Left element occupying community development or gender desks in funders' capital city offices. Though of course, a cynical person might see reason to believe that was the case. No, a *real* PRA, with *true* participation would not be being done at this eleventh hour, when the red ink on the big official stamps signifying approval of the project proposal was already well dry, and the province and district officials, who have so much riding on this, were eagerly expecting the LDF project funds to be devolved any week now. But that was how it was going to have to be. The funder, facing a rising chorus of doubts about the role of provincial and district officials, fearful of it all degenerating into unprogrammed social security for the locally well-connected and influential, worried about people's voice in the real prioritizing of issues, about the need for other distant officials, in Hanoi, New York, to raise their comfort level during this radical departure from existing project practice . . . all these things are the burden of this PRA.

It's been a frustrating session: the cadres dominating the focus group, making it interrogative question and answer. You let it go, knowing that tomorrow everyone would be more relaxed, and you could really get down into the participatory stuff. You look out the door at the cadres, chatting among themselves, smoking, the vice chair gesturing out over the fields, talking up plans. On the other hand, with cadre interest now well and truly on higher level networking, now might be a great time to press in a little closer on issues around local representation in decision making and resource allocation. It would certainly be great, this early in the piece, to be able to get a sense of how these men experienced the resource allocation processes. You ask a series of leading questions about where the recent roads have been built, which direction they go, and why people might use them. One of the older men starts drawing with his finger on the floor, and soon another is alongside him, arranging little household objects to show the bridge over the canal, next to the market. When a third goes to join, you ask if they would mind if all this could go onto a piece of paper, you happen to have in your bag. There's a brief hesitancy, the vice chair nods, and the butcher's paper is laid out on the ground, two metres by one and a half, held down by six or seven hands. You produce a packet of crayons, 'hand them over', and soon several of the men are discussing, sketching in the main features of the commune, roads and public buildings first, pathways and stalls and other locally significant places next. All this, especially the very local bits, is done with much discussion and involvement from nearly everyone, and is achieved with an accuracy and detail that enables you to move quickly to the next stage. Now, more pointed questions, using the map to draw things out, asking them to show you on the map how and where these things worked. Which roads had been built in which years, and why those ones? Which ones seemed to be the most beneficial and which less? Who decided the priorities, and how much local contribution (tax, labour) was made? They ask you to sympathize if they do not speak correctly, and hope you will find nothing to criticize in their words, and if you do, please tell them, and not someone else. And they discuss, and draw, and discuss some more. What did they know about the overall cost of the roads, and the proportion of local contribution? Which roads seemed to be holding up better than others, and why? Which ones had better quality, or adequate materials, and how did anyone know? What consultation has happened, at what levels? The debate and the map are getting lively. The map now has different roads ranked from most useful to least. Some of the least useful were sealed in the last year. The cadres outside are showing some more interest, and as they return, the discussion seems to hesitate. You wanted to go further: Why was this road here sealed, it seemed to fill very little purpose? Whose idea was that, and what sort of returns might come on that kind of construction project, for whom? You don't dare to ask, now: someone will say something, taking advantage of the moment, and there could be repercussions.

At the end, there's a multicoloured, detailed map on the floor, a kind of a mini GIS of what looks like most of the commune. The map has several different sets of symbols, and related keys. The commune and district roads are colour coded for dates of construction or renovation, and amounts of contribution per capita. There are estimates, most carefully debated, of numbers of trucks and carts and bicycles travelling from the commune, and guesstimates of numbers travelling through the commune to elsewhere, picking up or not picking up loads of people. There's a list of things different roads will be used for, and by whom. Costs, uses and benefits of the roads are much more diverse than you'd imagined. It turns out a lot more people will see something from the roads than you might have imagined, even if it's just another place to dry rice, or to sell produce, or better all weather access for everyone from province officials invited down to see the local situation, to bicycles and bamboo pole porters carrying things to the local market. But it's also interesting how fatalistic people are about resource allocation, and their own ability to get things moving. And how little they would say about any of the processes around prioritizing or getting attention to things that were worth doing. The map, then, got some kinds of discussion going, and in other contexts, it might have gotten more. On one level, participatory mapping seems to have worked its magic again: people participating in ways they wouldn't otherwise, and the map giving rise to all kinds of questions and differences in representations: a unique insight, at best, into how local people see their world, its salient and significant local features, what matters in what relation to who, the place of the local in the wider scheme of things. In the right wider context, a full on map-based debate between cadres and locals about costs and benefits and priorities around roads would have opened up some amazing possibilities. But here, where odds were there'd never be another PRA done, and where people would have to live with the consequences of what they said, in a far from liberal place?

On a separate piece of paper, you ask them to list the ten most significant infrastructure needs. This done, they rank the different possibilities in terms of desirability, benefits, and also of costs and potential problems. It's a classic wishlist: more classrooms for the school, a new building and better quality, state guaranteed medicines for the commune health station. Well, this time at least, some of it might get funded: but then what? Their list gets recorded in your notebook, basic elements of the report. Then, another ranking exercise, a bit tougher: you ask them to list the ten most significant issues to come out of the map: questions they have, issues, things they are concerned about. They talk about the need to 'control and audit' (*kiem tra*) the amounts of labour and money people contributed to the costs, and for someone to obtain information about whether these were inflated compared to elsewhere. So, something. But how much?

It's late for lunch, as everyone's body language is making clear. Most of the cadres have bolted for the People's Committee building. Khoi, though, lingers, taking one or two participants aside. He asks a few quiet questions of his own, and receives fairly frank answers. The vice chair interrupts, enthusing over the list of priorities and damning the map with faint praise. Back at lunch, Cherry is gloating over more horror stories: she'd decided to use her 'divide and rule' tactic, splitting off different groups away from the officials, sending them off with the younger researchers, while she kept the big men occupied with officialesque, and her 'flirt and flatter' cadre-capture routine. The commune chief, though, figured out what was going on, and had gotten himself more and more worked up rushing from group to group, correcting participants. In the end he had stood in the middle of the compound, and screamed at the top of his voice that people had better speak diligently or there would be big trouble!

DATABASES, FILTERING AND SENSITIVITY: GROUNDING THE LDF

As Scott Fritzen recounts, a Ministry of Finance department chief affirming in late 2002 that Vietnam was 'rigorously pursuing decentralization' could also report that decentralization of control over poverty reduction funds to the commune level 'can never happen' particularly in mountainous areas, since people there 'are too poorly educated and cannot objectively assess their own needs'.[21] Until early 1996, there had been no official statements showing interest in decentralization, in fact, 'grassroots decentralization' (*dan chu hoa tai co so*) did not become the idiom of choice for government or donors until decade's end.[22] Much debate was however occurring around official Circulars on budgeting and financing that would later be promulgated as part of the 1996 Budget Law. These pushed out the boundaries of what province governments could do, in relation to infrastructure provision, socio-economic development plans, sectoral coordination – in this sense promoting 'decentralization' – but at the same time they also pegged local discretion back to budget and expenditure norms that tried to make sure that real local discretion remained just out of reach and in Hanoi's control wherever possible. It was not until 1996, when the World Bank sponsored a high level seminar on decentralization, that any true sense of discourse around 'decentralization policy' got underway.

That said, in the early 1990s a mood of experimentation could be felt, both in UNCDF's New York offices and in Hanoi counterpart official agencies. The ethos of policy experimentation hinges on the belief that particular times and places present opportunities to experiment in a high profile way. Policy windows are conceived as short periods during which external and domestic circumstances join to create an opportunity for specific, preformulated proposals to rise in prominence in a form that allows action to be taken that can have a wider, more lasting salience.[23] Typically these windows open, or are forced to open, in times of political or economic crisis, something we will see repeatedly in following chapters. For reasons made plain in the first section, Vietnamese authorities were keen to experiment with an international donor promising in QNDN province what would then be the largest externally financed development project in the country – which is not to say government and UNCDF had the same policy experiment in mind. In fact, their attention focused on the fact that UNCDF had agreed to provide close to $20 million capital assistance for power supply, irrigation and road network development, concentrated in Dai Loc district – then judged to be a 'chronic food deficit' area with 'severely underdeveloped infrastructures, but having the relative advantaged of close proximity to Da Nang urban/industrial area'.[24] Officials in Hanoi and Da Nang were prepared to experiment with limited privatization of service delivery agencies – irrigation, infrastructure, electricity departments – and introduce internationally standardized competitive contracting procedures, so long as the capital intensive, regional infrastructure project was captured, and directed to this politically important district.

Creating a 'window of opportunity' for policy experimentation

As often happens, domestic concern to chart a course between economic transformation and social stability coincided with the Vietnam's turn to a re-embedding phase that would at minimum put a human face on Structural Adjustment. Thus donors' litany of 'growth with stability', 'community and participation', public sector reform and bottom up, consultative planning were easily cobbled together with Communist Party rhetoric. To the members of a visiting UNCDF Programming and Project Identification mission in February 1990, it seemed that they could use their commitment to the Dai Loc investment project to lever open another 'policy window' through which could be thrust their experimental LDF. To the mission, the LDF policy experiment offered a decentralized, local participatory planning system tied, through donor funds, to rapid delivery of small-scale infrastructure; this, to areas of poverty that had been prioritized objectively through a comprehensive local database. What's more, this special form of a social fund carried the firm objective of being institutionalized in ways that would provide a long-term discipline to province and lower level authorities. In this respect the LDF appealed to officials in Hanoi anxious to pass 'corruption' off as a problem of ill-disciplined local officials, and to donors worried about widening inequalities, and keen to find a direct course through Vietnamese bureaucracy. The LDF also promised more transparency, greater accountability and a competitive bidding process for infrastructure contracts that spun benefits out to local private contractors.

As we will see, officials at province and lower levels were not quite so convinced that corruption was a problem, nor that more transparency and participation was required.

What follows is not a case study (or far less, an evaluation) of the various permutations of the RIDEF, the particular name to the Vietnam policy experiment, then being promoted by UNCDF in five countries.[25] Rather, we want here to describe how the LDF travelled into QNDN province and down through the layers of district, commune and hamlet authorities. We want to show how it mapped poverty, and tried to open up debate about how public funds should be used for poverty reduction, and how it set out to create a new, participatory and decentralized system for this process. How did Vietnamese officials deal with the new rationalities of a liberal project then travelling in on the backs of donor technical missions? We move down through the comprehensive 61 variable database on which the LDF rested, down past the intergovernmental systems of planning, financing and contracting to clasp together with communes who's needs had been framed up by PRA style participatory planning events. Along the way, we pass episodes in which the technical apparatus of the LDF met with a far more complexly structured Vietnamese polity to peal back to what we think is the essence of the policy experiment carried by this re-embedding liberal project. We will have ample opportunity to look more closely at other aspects of this paraphernalia of decentralization, as we track across to Uganda and then to the lofty ambitions of devolution in Pakistan.

Thus the Dai Loc operation was a beginning. Vietnamese officials could agree, so long as the LDF was presented as being a special kind of social fund, a stove pipe through which funds could be channelled to them from Hanoi and then used at selected sites in the province to finance small scale infrastructures – farm to market roads, clinics, schools, small irrigation and water facilities, new markets and power connections. This was appealing. Less so was the fact that the LDF came with a proviso: LDF resources, set at about $20,000 for each commune, were to be allocated in a totally new way that had two aspects; a technical, data-driven planning process backed by a participatory planning procedure based on PRA. Allocation of these resources – at the time, amounts many times local development budgets – was determined through a technical assessment of needs generated from a database that would produce an index of needs, rank communes, rich to poor, and target the poorest 50 per cent. In these poorest communes, it would be up to commune authorities – the smallest administrative unit in the state governing apparatus – in consultation with local people, to decide which particular investments were to be made. Provided this choice matched technical indicators of need registered in the database, investments would be implemented.

Not surprisingly, the province authorities were not convinced that communes had capacity to 'objectively determine' needs, and took a dim view of any funds allocation and planning process beyond Province Construction Management Board's control. UNCDF understood that officials were battling nationwide charges of resource misdirection, and that they were sensitive to charges that favouritism and outright corruption reinforced growing inequalities. So they pressed hard. 'Creating new systems for public finance allocation and spending was urgent, the old systems had no credibility.' 'Without radical change, how could you expect donors to help get resources to finance local needs?' And: 'look how keen local people are to participate in the new Vietnam.' At the time an unprecedented number of new rural organizations were springing up – an estimated 30,000 by December 1992, which were 'essentially voluntary and organized by the population directly . . .',[26] for example to build and maintain irrigation systems, provide mutual aid, construct roads, markets and other infrastructure. With no initial agreement, decisions about who, in which level of government, would decide on how resources were allocated were temporarily set aside. It remained important to find a way to quickly channel resources to local levels, but attention focused on the larger action in Dai Loc. Meanwhile, assembly of the LDF's local planning database and the first rounds of analysis were permitted to proceed.

As this happened, UNCDF's missions were increasingly convinced that provinces had a high degree of fiscal autonomy and a widening range of responsibilities. The political rhetoric was written into back-to-office reports: the country *was* opening up, the LDF experiment could create a demonstration effect and assist opening political space all the way through province and district to the commune. Besides, it was clear that for provinces like QNDN, their budgets were rapidly expanding – doubling in real terms between 1990 and 1993. Provinces had money and increasing discretion about how it could be used. UNCDF reps were blowing hard to inflate any evidence that political and fiscal decentralization were underway, and that the time was right to experiment, for the first time, to get real people's voice into participatory planning. Not seen was that at the same time Hanoi was introducing measures to curtail provincial government spending powers that required the province to clear all capital spending decisions with the State Planning Commission in Hanoi.

Rather like Vietnamese policy on decentralization, the LDF policy experiment was never 'designed' in one go. Until 1995 at least, it was scattered in various mission documents and reports at the margins of reporting on the larger Dai Loc operation. On the other side, what was believed to be Vietnamese policy had to be read through the framing of the external documents. Meanwhile, one of the greatest opportunities seen through this 'window' in 1991 was the availability of an experienced local Vietnamese agency, a local champion, with whom a contract for professional services was made, namely the CERPAD of the Ministry of Construction. CERPAD's planners, architects and engineers were tasked to establish the planning database for all fourteen districts and 240 communes in the province. Data depicting basic needs and existing endowments of infrastructure and services were to be represented through a set of thematic maps on existing infrastructures and disparities in the province. The next task was to define targets and criteria for allocating resources to small-scale infrastructure projects. In response, CERPAD prepared a 'development infrastructure index' drawing from the database, to provide a technical basis for choosing locations for investments. CERPAD then had to design 'community-based planning and implementation methodology' and train personnel from institutions at all levels in the province on its application.

What CERPAD constructed became politically sensitive, and quickly brought official attention back from the largesse being distributed through Dai Loc. QNDN was comparatively resource rich. It was then one of only two provinces in the south central coast region where revenues exceed expenditure (largely as a result of state enterprises and cross-border taxes on trade). In other respects it was typical of provinces in the region, with its 1.9 million people concentrated along the main north-south highway, in the rice producing lowlands and the large coastal city of Da Nang (450,000 pop.) But CERPAD's early work revealed great disparities in social, ecological and economic conditions, against every indicator: production, trade, market location, access to irrigated water, health, education and other facilities. All this highlighted great disparities in how the benefits of public finance decisions were distributed. The Development Infrastructure Indices were calculated from the 61 variables in the database to reflect a composite of endowments in education, health, market facilities, agriculture, electricity, and roads.[27] These Indicator Scores were the basis for a composite, weighted Development Indicator calculated for each of 216 communes, as follows:

$$Ik,\ Ek,\ Pk = \frac{\Sigma\ Bi^*\ Ti}{\Sigma\ Ti} \times 100$$

Where P = population, I = infrastructure, E = economy.
$K = 1, 2, 3 \dots n$, where n = the number of commune, Bi = Indicator Score, I = number of Indicator, and Ti = weight of the Indicator number I.

These scores (from low to high) clearly showed the patterns of relative wealth and access to social and economic infrastructure – as in Figure 5.1. More than 60 per cent of communes in the mountainous region recorded index scores of less than ten, and many were close to half this score, which indicated that infrastructures either did not exist or were in a parlous state. Most revealing was that only 17 of 216 communes in the province, less than 10 per cent, had index scores greater than 14 and all (excluding one anomaly) were nestled on the coastal plain area close to the political centre of Da Nang.

27.00 QUANG NAM DN

1. Tp DA NANG
2. HOI AN
3. TAM KY
4. HOA VANG
5. HIEN
6. DAI LOC
7. DIEN BAN
8. DUY XUYEN
9. GIANG
10. QUE SON
11. THANG BINH
12. PHUOC SON
13. TIEN PHUOC
14. TRA MI
15. NUI THANH
16. HIEP DUC

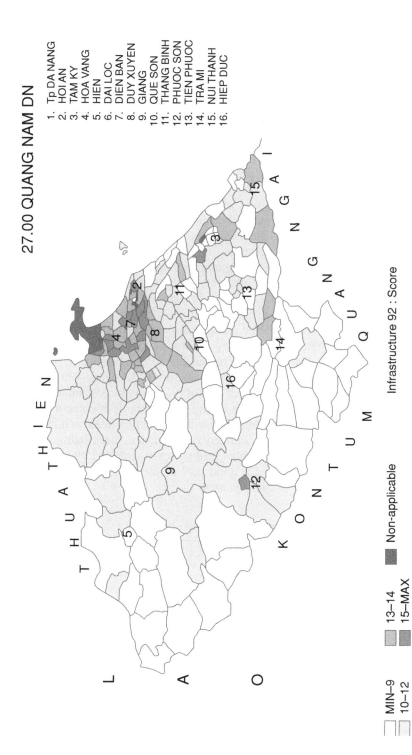

Infrastructure 92 : Score

MIN–9
10–12
13–14
15–MAX
Non-applicable

Figure 5.1 Development Infrastructure Index

These results reinforced claims by province officials that much external support was needed for local investment. But they were quickly also aware that these maps were showing just how biased their own investment decisions had been, clearly favouring very few communes over the large majority. It was not just that communes with scores above 13 had good facilities, but that these had been the beneficiaries of province capital investment decisions that had been in place only since 1991.

CERPAD's work also highlighted another uncomfortable fact, namely the extraordinary volatility in district and commune level spending arising from a highly political budgetary process at the province level (from where districts and communes derived the bulk of their budget) closed to all but a small coterie of province capital officials. CERPAD's team presented data from selected districts to make their point.[28] Pushing their analysis to build the case for introducing the LDF's local planning database process, they were able to show the devastating impact the politicized and highly volatile resource allocation was having on planning and delivering services. Table 5.1 shows the increase or decrease in revenues by source and in total over 1991–1993 against a 1991 base year set at 100. Clearly the pattern for every revenue source and in total was completely irregular and the increase in total revenues varied greatly amongst the districts. The 1993 budget for Que Son district increased by a factor of 3.3 over 1991, whereas that for neighbouring Nui Thanh increased by 2.1, and for poor, mountainous Hien district increased just by a factor of 1.6 over 1991. The consequence was obvious. With such volatility, it was almost impossible for commune or district governments to undertake area-development planning, even if they had the skills and capacity to even maintain minimally stable administrations. Much of their energies went to negotiating deals with province officials, for the months leading up to approval by province authorities of their budgets, and a good part of the year afterwards was spent in actually securing budget releases.

Table 5.1 Revenue flows in selected districts, 1991–1993

	Revenue source	Taxes	Levies	Grants	Total	Annual increase (%)
Nui Thanh District	1991	100	100	100	100	–
	1992	125	347	184	208	108.0
	1993	146	166	348	210	1.0
Hien District	1991	100	100	100	100	–
	1992	356	79	105	135	35.0
	1993	211	46	208	156	16.0
Que Son District	1991	100	100	100	100	–
	1992	145	169	326	326	87.0
	1993	226	151	524	524	75.0

Source: Interviews, district officials, June 1994.

Most districts relied heavily on grants from the province government to balance their books each year – ranging, in any year, from 39–70 per cent of their budgets. Pushing the analysis further, CERPAD showed that per capita transfers to districts varied from 14,100 VN dong to 53,600 VN dong, and that there was never certainty that budgeted funds would actually be made available to local governments. It all, apparently, depended on constant visiting by district officials to the province capital.

The local planning process

'So, then, let me summarize the three stages of the local planning process' Khoi pauses, consults a typed sheet of paper 'which will apply an approach which combines both "bottom-up" needs identification with a regional analysis for prioritizing and selecting beneficiary communes, and prioritization of projects that is consistent, transparent, replicable and sufficiently flexible.'

Stage 1, Planning and Programming, has seven steps, using the database and successive rounds of 'screening' to identify a limited set of communes that have both below average infrastructure development scores and have indicated a 'readiness' to participate in the later project.

Stage 2, Feasibility Study and Design, has a further six steps. Here we match the requests received from communes with the database to see that what you've requested matches with objective measures of needs. So, if you ask for better market facilities, or a clinic, we look at what the database says you already have, then do an on-the-spot check.

'If its approved, in Stage 3', he looks closely, as if for the small print, 'we will allocate resources, then there will be a competitive bidding, tendering and contract award process, through supervision and monitoring to completion of the investment project.'

Khoi stiffens up against the audience again, looking pleased its taken just 45 minutes to run through this time. He had gotten off lightly: those doing all the steps on the ground would be engaged for considerably longer periods of time.

The RIDEF programme then set out to introduce a procedure for allocating resources that depended on technical information, and to make subsequent transfers more transparent, timely and predictable. This, in place of capital spending procedures controlled by the Province Construction Management Board, provincial officials whose main administrative benefits and personal incomes were derived from the various service charges, fees, and levies on the design and implementation of public construction contracts. This was not just a matter of resource management efficiency, but political imperative. In many rural areas, relations between peasants and officials were deteriorating, due in no small measure to the plunder and ostentatious wealth accumulated by officials as a result of privatization of public assets and enterprises. Powerful officials were self-evidently using their positions in the bureaucracy, the party, military and police, to monopolize agricultural credit and commodity trade, to take a cut of rural development and infrastructure funds, and to extract payments from citizens for their own good. Official policy statements, issued from Hanoi, attributed this to corruption amongst the lower reaches of government. RIDEF then was presented as part of efforts to go around the local official political economies, in a sense to evacuate local politics from investment decisions, and to ensure that the channels – international-Hanoi-Da Nang and down through province, district and commune authorities – were connected with local priorities in more legitimate ways.

By August 1993, projects had been proposed for nine communes, including micro-hydro, road upgrading and construction, construction of markets, irrigation systems, classrooms, clinics and rural electrification. Four of these were selected for 'first-generation' investments. An indicative allocation was made, and some work had actually started. Meanwhile, preparations were made to shift to a larger, more ambitious phase, with refinancing arrangements that upscaled the project across the entire province. So the contest began in earnest. The proposal to restrict access to RIDEF to commune authorities was fiercely resisted by the province representatives. They argued that the Commune Development Boards (CDB) created under the auspices of the project, but nominated by the Commune People's Committee, were 'not strong enough' to guarantee 'efficient management' of resources. UNCDF and the Province agreed an additional 'modality' so that district authorities could also access projects, larger 'inter-commune' projects technically beyond the capacity of CDBs. This suited the province, which had just pre-emptively issued official Circulars that reduced districts to administrative spending units of the province, thus pre-empting RIDEF's delegation of executive powers to these bodies. And just as the Hanoi Ministry of Construction was issuing Circulars consistent with the principles of competitive tendering and contracting, the Province People's Committee was simultaneously countermanding these back and forth with administrative orders reinforcing conventional practices.

A UNCDF Review mission was fielded to deliver a few ultimatums about the need to finalize the design. But tidying away outstanding points of disagreement was a big ask. Province officials were arguing for greater oversight powers to be assigned to them, especially actual budget decisions which would enable them to calibrate the 'confusing database which no-one understands' to their objective appreciation of local needs. At the same time, NGOs and dissenting staffers in UNCDF's New York office argued that this RIDEF data and plan driven process was 'too mechanistic' and that people's needs were being subordinated to the LDF model. The Review agreed that the database was indeed too complex, it needed to be stripped down to 'essential variables' to ensure it could be understood, and accepted, locally. But in a spike at the LDF's designers, the Review concluded 'local people's knowledge should be incorporated more fully, via participatory planning techniques, to increase their sense of ownership'. The present system 'empowered' the community only to choose specifics of the investment project, all else was controlled by the planner. The Review mission prescribed that PRA techniques would be grafted on to the local planning process, not just to placate distant critics at headquarters, but in the anticipation that a good dose of PRA could allow commune officials to create an upwards pressure on province government: this, by demanding exactly the kind of allocation and planning process the LDF offered but which was being blocked by the province.

Then, after a final round of negotiations in late 1994, RIDEF was agreed as a 'special facility' of the QNDN province administration. The project would have two specific purposes. First, 'to alleviate poverty through investment in small scale social and economic infrastructures in relatively poorer and under-equipped rural areas'. And second, 'to increase the opportunities and responsibilities of lower level local governments and community groups to plan and manage the development of these infrastructures'. The province's Planning Department would manage $7.7 million to be made available over five years. But before all this could begin, the November 1994 Formulation Mission laid bare demands to ensure 'participation' was centrally introduced to all aspects of the project. Participation would, said the Mission, 'promote transparent, public decision making, drawing on local knowledge and experience'; 'help the poorest in the community to organize to meet their own needs, that is, needs that cannot be met by government'; 'help the community to learn to be self-reliant, by learning to mobilize local cost contributions'; 'lead to a sense of ownership and responsibility for the investment project'; ensure that 'local needs will be met rather than needs as perceived by higher levels of government' and, what's more, participation 'will be essential to creating responsible government, that is, it is an essential aspect of local government empowerment'.

But how to do this? First, it was agreed that district staff and community members were to be trained in 'facilitation skills' so that investments would reflect a 'thorough consideration of the impacts of the proposed infrastructure, equitable distribution of the benefits, and involvement of the beneficiaries in planning for operation and maintenance'. Second, international experts in participatory techniques were contracted to test and apply PRA techniques, first in a few villages, and then, so the contract said, in 200 villages across the province where LDF investments were to be made.

Conclusions: beyond PRA and LDFs

200 village PRAs, it will be clear, was a big ask, an excessive investment, some felt, in the temporary liberal legitimating of what remained a narrow menu of local infrastructure choices.

In time, the overreach and over application of these techniques would be somewhat curtailed, but not before they had been deployed in simply extraordinary numbers in virtually every Development setting on the planet. Their long-term legacy, nonetheless, remains an open question: in Vietnam, they were one tiny element among many in exposing cadres to liberal techniques and governance values. Doubtless in many contexts, they were like water off a duck's back: never repeated, almost immediately lost in both the routine business of territorial governance-as-usual, and the failure of these minor liberal institutionalist modes to actually deliver much more than a very, very few items in response to the local wish lists they facilitated. Certainly the wider liberal transformations they envisaged – empowering the people to own their own knowledge and increase their voice into wider areas of development context – are often remembered now even by their staunch advocates with something like an embarrassed smirk. And so they remain: vivid memories, great Development fun, but a kind of unrealized, incomplete dream, left, as in this chapter, in mid-course. . . .

Yet as discussed in Chapter 3, much has since been added to these instruments as they have been integrated, as part of PPAs, into the 'comprehensive' PRSP frame. However, the way they were used to represent the poor as embedded in local community with ideal typical needs, accessible and understandable outside the local political hierarchies, still resonates at the core of Poverty Reduction. Indeed, in the IFIs' own PRSP process review, PRA-style participation was upheld as the primary distinguishing characteristic of the new approach. This style of participatory democracy continues to legitimate Poverty Reduction, to frame hopes for service delivery and political accountability, and instantiate notions of voice and empowerment in wider governance. LDFs and PRAs, as told in this chapter, played a crucial role in evacuating the messiness of politics from the locality, framing it in new, cooperative ways that clasp together a consensus about needs and appropriate responses. It could open up politics in ways that could be acted upon by international Development, before local hierarchy, history and politics closed in again. Quite quickly, as we will see in the next chapter, LDF's were 'upscaled' into wider schemes of decentralization, then into increasingly complex systems of intergovernmental fiscal reforms in which SWAps were assembled into MTEFs, and in turn enabled the PRSCs, that is, large whole of budget financing instruments, now popular in all the poor countries in this book. But to see all this coming together, we need to travel, with the UNCDF, LDFs and all, to Uganda, arriving around 1995.

6 Uganda

Telescoping of reforms, local-global accommodation

Just the hum of Landcruiser tyres on the hot tarmac. Michael's said nothing for the past hour, face set on the road ahead, hands firmly gripping the wheel, speeding us back to Kampala. Either side of the road is a smoky haze welling up from fields of maize storks, blackened by fires lit ahead of the rains. Big, wide road ahead.

You try again, pacing your words, not sure how it will go this time. 'OK, so they were all pretty pissed off with us, not surprisingly. But six months ago, it was all looking so sweet, those same councillors who harangued us today, they thought our project was the best thing they'd ever seen, they said as much.' You wave as if you had the evaluation report in your hands, ticking off the achievements.

Evidence from the District Development Project (DDP) 1996–2000 has shown that:

✔ Allocative efficiency has improved, thus local government investments are more directly in tune with local priorities and they invest resources consistently in national PRSP priority areas.

✔ Political accountability has improved; as a result of discretionary budget support, Local Councillors can engage in meaningful participatory planning and local investment management.

✔ Horizontal accountability has improved. Local Councillors are more assertive, thus improving the performance of Local Government administrative cadre.

✔ Financial accountability has improved. Despite interruptions in funds flow from the centre, financial reporting and compliance of the local governments has improved.

He turns, eyes narrowing: 'And you, typical, you had to blurt it out that they'd now failed just about every one of the performance measures and then tell them that it was the same story everywhere. It was so bloody humiliating. I thought that woman councillor, you know, Gladys, I thought she'd just about throw you out of the hall. "Don't hold us responsible."' Michael starts to mimic her tirade, '"you set us up for this, you Kampala people come up here, get us all worked up about accountability and performance measures and joining up with these new complicated planning and budgeting whatsits and when it doesn't stick quite like you said, well, you just tell us that we've all failed." '

He laughs and pushes back on the steering wheel. 'That old guy at the back had our number though. Remember what he said, "You know what's failed? You've failed to put together a system that will help us defend our rights. Our women can't even go to the well for water these days. This system has done nothing to help us defend ourselves against this damn war, these army guys on one side, on the other, those fellas coming across the border at night and burning our houses, taking our children away. Now, we seem to have no time to ask these questions, about the real sources of conflict and poverty, about where the real money is going, and who's getting it." '

'You look at these accountability things one way, and they're all the same, and joined-up. You look at them another, and the links are as frail as a paper clip.'

Does he see it that way? 'Nah, you always imagine things to be more joined-up than they are. And besides, we're Africa's Shining Star, at the moment it doesn't matter. Just you see.'

Introduction

> Weak, patronage-based 'quasi-state' regimes face geo-political pressures and opportunities that compel rulers to experiment with administrative innovations, including war, to consolidate their power, control markets and manage rivals.[1]

When Uganda gained independence from Britain on 9 October 1962, Tanzania's President Julius Nyerere is said to have whispered to Prime Minister Milton Obote: 'You have inherited the Pearl of Africa, don't spoil it.' From then till now: Uganda's history has been one of remarkable transformations, marred in the international imagination by a degree of state and military violence against the civilian population that is almost without parallel in Africa.[2] Uganda had become synonymous with pillage, violence, state collapse and misrule. By January 1986, when Yoweri Museveni and his guerrilla fighters captured power, the country had undergone five years of infamously brutal 'bush war'. Museveni had forged a revolutionary government in waiting in 'liberated zones', and to much dismay he upscaled their territorial governance into what looked very much like a socialist state. Yet in a short while, Uganda was being held up by IFIs as the best performer, not just in macroeconomic performance, but also for resuscitating service delivery, and innovative Good Governance. Before PRSP, its national PEAP was already well ahead of the pack. Uganda, a model of reform ownership and process, lent PRSP the plausibility of one of the continent's rare Development success stories. In short, Uganda had once again become the Pearl of Africa, a shining example of the best results of neoliberal transformation.

This mutually advantageous relationship with the IFIs was crucial to Uganda's transformation. We begin the story as they emerge from a stormy relationship to reach an accommodation around stringent macroeconomic adjustment, a PRS and decentralized governance. This accommodation would yield both unprecedented flows of aid and debt relief and an equally unprecedented period of regime stability. Central to this reform (both politically and technically) was the Resistance Council system, crafted during the 'bush war' and then successfully tied up through a renowned decentralization programme, linked to globally attractive public sector reforms, sector planning approaches, medium-term budget and expenditure management techniques and, of course, to the bellwether PEAP. In Uganda, following closely on early 1990s Vietnam, we show all this being worked out largely *before* it became PRS orthodoxy, and before Uganda's experience, retrospectively preened in documents like the 2004 WDR, became a preferred reference case used to encourage and discipline others.[3]

By recounting how international and local imperatives were woven together around decentralization and poverty reduction, this chapter illustrates one of the book's core themes, namely, the formalized 'travel' of development doctrines, and their continual overreaching and upscaling. In fact, Uganda provides an uncommon opportunity to explore what happens when these international travelling rationalities are laid out on fraught and contested peripheral terrains, and then scaled up into elaborate national governance mechanisms. We show how Liberal and Territorial governmentalities overreach, then telescope and get out of synch with each other, and create crisis on the ground as their protagonists struggle to deal with the unintended consequences of their running together. Thus we will show that in these reterritorialized localities of decentralization, outcomes are mixed and fraught: plural programmes working at some odds with each other immerse local actors in a thicket of complex funding, service provision and governance arrangements, mixing diverse incentives and delivering uncertain outcomes and accountabilities.

This chapter also begins to elaborate another of this book's core themes, namely, how these new forms of institutionalism can be used to marshal local governing arrangements in ways that appear to privilege the citizen and local empowerment, appear to be joining up global and national efforts around pro-poor outcomes, but which are entirely consistent with, and tend to be used also to promote the interests of authoritarian, illiberal regimes of governance and outcome. Thus, we show that this travel of international orthodoxy is never a one-way street, just a matter of a global re-embedding process working on a passive, compliant surface. Rather, we show an active, complicit local reception for travelled practice: skilful local actors, from the President, to technical officials, to local councillors, dealing with waves of globally inspired reforms, attracting considerable resources, talent and patronage in ways that maintain a coherent political project.

Over the period 1986 to 2000 Museveni's regime was able to manipulate global actors and norms to pursue his domestic designs by exploiting their anxieties about the need to limit political disorder and state collapse in poor countries, (and to avoid humanitarian disasters and hasty arrangements to write-off unpayable debt) and by presenting Uganda as a place where the post-structural adjustment appeal of local empowerment, accountability and pro-poor service delivery might be played out. This was achieved whilst sustaining a one-party state apparatus, an unjustified civil war with people in the north of his country, frequent breaching of IFI fiscal agreements about military spending, and large scale predatory military adventures in the neighbouring country of Congo, activities dubbed elsewhere as 'criminalization' of the state.[4]

In this chapter, and as taken further in Pakistan to follow, we will see how decentralized approaches to governing poverty do involve framing up poor localities in ways that recruit and discipline the poor to perform crucial legit-imating tasks for *global* liberal orders. But here we show how they can play an equally significant role in embedding and legitimizing recipient country governing regimes as well. As we will see, decentralization in Uganda was designed at the outset to appeal directly to a newly enfranchised, reterritori-alized constituency. Building on liberal notions of citizenship, its remark-able elaboration of devolved, PRSP-linked funding mechanisms provided a political and technical system that proved crucial for legitimating both Museveni's regime and the wider IFI-PRS approach. What emerged was a set of governance arrangements that, despite their heady over conception and apparent lack of joined-up comprehensiveness, both delivered services to poor locales *and* served to mask deeper questions and issues.

The first section of this chapter recounts the building of a global-local, IFI-Museveni alliance in the years after 1986. The political imperatives bearing on both parties would see Museveni jettison policies reflecting a distinctly territorial mode of governance in favour of the neoliberal pre-scription. We will look at the role of decentralization in this accommoda-tion.[5] The second section tracks how Uganda's DDP borrowed from the Vietnam RIDEF project, but was then grounded in Ugandan political and territorial governance process. It then shows the DDP's early results and eventual upscaling into a nationwide schema for fiscal decentralization. The third section charts the rise, fall and subsequent return to fame of the DDP, and how the entire 'joined-up' and 'donor harmonious' apparatus of national governance presented through this project could be used to play a crucial regime legitimizing role.

FROM THE POLITICS OF REVOLUTIONARY COMMAND TO THE TECHNICALITIES OF GOOD GOVERNANCE

When Museveni achieved power in 1986, Uganda had suffered a severe social collapse. It typified everything bad about African governance: excessive state market interference, unbridled misrule and corruption, widespread human rights violation. The incoming government faced a liquidity crisis, runaway inflation, an insolvent banking system, and foreign exchange reserves to cover only two weeks of imports. Around 7 per cent of the population was displaced, and over half of Uganda's private wealth was held abroad. The economy had suffered astonishing contraction. Large amounts of aid were needed to avoid major humanitarian catastrophe.

Reaching a global-local accommodation

The legacy of economic incompetence, war and human rights abuse also meant that Uganda's standing in the international donor community was at its lowest ebb. While IFIs were ambivalent about the 'revolutionary' Museveni's capture of state power, Uganda was evidently a country ready for the 'counter-revolutionary' prescriptions of the period; macroeconomic stabilization, and urgent funding for social protection and security, delivered with stringent measures to discipline corruption.[6] Museveni's talent, in retrospect, was that he could in a short time appeal to a range of international interests, from the radical to the disciplinary conservative.

For adherents of the radical agenda of NPE, he argued that his National Resistance Movement (NRM) offered a new, genuinely post-colonial form of governance, one that would free citizens from the pernicious effects of traditional elites. He began the infamous 'bush war' in 1980 after claiming that the rigged elections showed 'once again, a minority, unpopular clique was imposed on the people of Uganda' and he had 'no option but to take up arms in defense of their democratic rights', and topple the government.[7] Partially decentralized Resistance Councils were crucial to Museveni's control of newly liberated zones. These were loosely based on the neighbourhood committees organized in the liberated zones of Mozambique by FRELIMO in the late 1960s, and were ideologically informed by radical practitioners-cum-scholars like Eduardo Mondlane, FRELIMO's founder, and Amilcar Cabral, the founder of PAIGC in Guinea Bissau.[8]

The Resistance Council system created a pyramid structure of elected committees with revocable members, accountability of civil servants, election of magistrates, devolution of judicial and legislative powers, and extensive local autonomy. This appealed to local people used to a brutal and violent administration; Resistance Councils could be used to discipline fighters and where unpopular chiefs had been expelled, civilian administrative and judicial responsibilities were turned over to them. Within one year of office, the NRM gave the councils legal protection through the

Resistance Councils and Committees Statute (1987) and, by conferring administrative powers to local communities, the regime was able to consolidate its territorial grip over the country, while underwriting it with the popular commitment to 'people's power'.[9] As we will see, this already decentralized governance mode would ultimately be articulated in the DDP, and then upscaled as a primary means of articulating Uganda's PRSP. Thus, Museveni's revolution offered NPE adherents the prospect of breaking old powers' hold, and to Ugandans and 'Africanists' the hope of breaking the familiar litany of *coups d'état*. In the Resistance Council pyramids, he offered a new form of locally legitimate, territorially integrated governance, with strong links to central organization, both separate from and yet potentially part of a new state that had consciously expelled the legacy of an old and sick society.[10]

Notwithstanding this, it took Museveni four years to build a partnership with the then market-orthodox ways of IFIs. He was deeply suspicious of IFIs and, schooled in the ways of dependency theory, was not prepared to cede authority to what he said was 'an imperialistic imposed policy package'; instead of 'the invisible hand of market forces', he preferred the 'visible hand of government'.[11] Thus, the NRM's first step was to draw up an emergency relief programme and seek international support for a $160 million budget built on the territorial governing style of price control and barter trade (with Libya and North Korea), fixed exchange rates, price controls, deficit financing, and state monopolies over external and internal trade.[12] Not surprisingly, the IMF was immediately alarmed at this 'serious retrogression', and announced that the economy was 'out of control', and sent signals around the donor world that this socialist governance model was a landmark of policy failure.[13] As the IFIs laid siege to Kampala's Ministry of Finance 'to break pockets of resistance to reforms', the government in May 1987 announced a radically different economic package. At IFI urging, the Economic Recovery Programme demonetized the Uganda shilling by a factor of 100 and devalued it by 76 per cent, and hiked interest rates.[14] Much later, Museveni reflected that this turn-around would not have happened had he not been convinced of its virtues, but the fact was his regime desperately needed donor support.[15]

From 1987 to 1992, the IFIs provided $1 billion through 25 policy-based loans to the Economic Recovery Programme. But this resulted in an alarming blowout in debt[16] and the accumulated effects of structural adjustment – the removal of price and wage controls, the abolition of subsidies and devaluation – hit the poor before the rich. This prompted critics, including many within his own Movement, to contrast Museveni's much applauded empowerment politics with the increasing grip of Western creditor nations.[17] As one supporter of the regime said, 'the establishment of the Resistance Council system of governance meant the opening up of the political arena to the Ugandan people, including the women, youth and workers. But the pursuance of SAP has meant the exclusion of the majority from the economic benefits of the NRM revolution'.[18] Thus by the early 1990s, the government was greatly pressed to give macroeconomic policies a 'human face'.

Decentralization did not feature in the *Ten Point Programme* the NRM announced shortly after taking office, but it soon became the central means used by Museveni to 'reach over the heads' of discredited political party and central government administrative elites he could not easily control, and build stronger local political ties.[19] In doing this, he was reviving a long history: Uganda had experienced administrative decentralization during the colonial period, but all this occurred within Lugardian structures of indirect rule, more akin to 'decentralized despotism' than to the political enfranchisement and citizenship Museveni espoused.[20] The 1987 statute had already established a five-tier system of local government from village to district levels that persists to this day. Village councils (Resistance Council 1) were comprised of all residents – typically around 100 households. Elected, nine member committees managed local affairs, and at the next three levels of councils and committees – parish, sub-county and county – comprised members elected by the tier below. The district council, then called the Resistance Council 5, drew two persons elected from each sub-county plus two women representatives from each county and municipality. This was an impressive representational structure for a total population of less than 18 million, with around 400,000 people serving in 45,000 village councils.[21]

Decentralization, a contemporary necessity

In October 1992, in the lead up to the Constituent Assembly elections planned for 1994, the NRM saw that decentralization was a way to bolster electoral support. Museveni took confidence from the high profile Odoki Constitutional Commission, convened in 1987 to build a new constitution for Uganda. Although criticized for being stacked by Movement adherents, its work over the next six years was an extraordinary achievement in public consultation.[22] Decentralization, it said, chimed with tradition and people were unanimous in support of 'flexible local arrangements' for governance. Decentralization, it concluded, 'is a contemporary necessity' as it promised to 'empower the citizens at all levels to be masters of their own development'.[23] Reinforced by these findings, the Local Government Statute 1993 maintained the existing Resistance Council structure and promoted classic devolution: chiefs and administrative staff were made accountable to elected councils; many central ministry functions were devolved; and legal guarantees would ultimately ensure the transfer of around 30 per cent of the national recurrent budget to local governments. Fiscal decentralization, to transfer resources and revenue raising powers in accordance with these new expenditure responsibilities, got underway immediately. Initial results were promising.

Although they were keen to support Uganda, in the main donors remained publicly sceptical about decentralization. Certainly, by this time the World Bank and bilateral agencies were revising their state minimalist positions – and in some respects their confidence in the Ugandan regime encouraged them to abandon this ethos well ahead of the commitments later made in 1997 WDR. Views were shifting, but this was not happening nearly as fast as Museveni wanted.[24]

FROM THE DISTRICT DEVELOPMENT PROJECT TO THE LOCAL GOVERNMENT DEVELOPMENT PROGRAM

The preparation and early implementation of the DDP from 1995 to 2000 was achieved in the face of donor orthodoxy and suspicion around decentralization. Initially here, we view the process of rolling out the DDP through the lens of the officials engaged in designing and implementing this programme. As we proceed we will outline a number of core principles of Uganda's decentralized governance, and set these against the difficulties of more direct Poverty Reduction approaches, which involve direct centre-to-local transfers through the use of ring-fenced social funds. In the third section of this chapter, we step back from this closer technical focus, and look at decentralization in the wider PRS frame and in terms of Ugandan political economy in this period.

During May 1995, five Ugandan officials visited the RIDEF project in Vietnam. The invitation to visit Vietnam was part of a concerted effort by UNCDF to sell the approach to Uganda and through this, to ensure this small affiliate of UNDP with barely two-dozen professional staff would be able to cement a deal with the World Bank. If UNCDF could demonstrate the effectiveness of the LDF methodology in a pilot project in Uganda, the World Bank, still sceptical about decentralization, would upscale these efforts through a nationwide programme. UNCDF was playing for high stakes. In 1994, UNCDF had been put on notice that they must in five years demonstrate relevance or face a collapse in donor financing. Relevance clearly meant relevance to the World Bank's operations. Needed was a working demonstration of UNCDF's products in the shining star of sub-Saharan Africa.

For their part, the Ugandan delegation felt not much could be learned from a highly stage-managed visit to Vietnam. They had a clear sense of where their decentralization was headed and they bridled at any suggestion that the LDF's local planning process, the 61 variable commune database, the Development Infrastructure Indicies, the investment menus and PRA results would simply be shipped across to Uganda. But the visit went well, and eventually, Kampala signalled that UNCDF could go ahead. What then was Uganda's interest? At one level, the need was plainly obvious. Fiscal decentralization – transfer of recurrent and development budgets to local control – was well underway, but so far included recurrent budget financing only. Decentralization would not be complete until both recurrent and development budgets were in local hands.

Larger strategic incentives were however lining up to shift development budgets down to the Resistance Councils. National debt, raised in the 1980s and 1990s to reduce poverty, had failed to budge poverty indicators. More palpably, presidential elections were scheduled for 1995 and to secure victory the NRM government knew it had to 'reconnect' with its IFI-sceptical power base in the Resistance Councils. As Mahmood Mamdani, the former chairperson of a government commission appointed in 1987 to study local government remarked in 1995: 'The Resistance Council system increasingly came to reflect two tiers: one local, the other central; one on the ground, the other at its apex. The higher one went up the Resistance Council pyramid, the more watered down was the democratic content of the system.'[25]

The problem was that the over-whelming bulk of development resources were provided by foreign donors who channelled resources into local service delivery predominantly through special purpose local projects that they could brand as their own and that worked largely outside the government system – Social Funds and their variants described in Chapter 4. They were averse to putting funds into national budget/expenditure systems and having them transferred, unguarded, to local politicians. The designers of Uganda's decentralization argued that local politics would become fully animated only if local councils were allowed to allocate development resources in response to local needs. Here, LDF systems, specifically the systems we saw in Vietnam, promised a better solution.

The LDF systems offered a direct transfer of development funds to councils. They thus promised a quick way to get resources into the nexus between NRM local government politicians and voters. At the same time, they could bind the two tiers of government – central and local – together, giving them common cause and closer funding relationships. This approach would thus generate local legitimacy in ways that were politically far superior to the vertical, donor-controlled project arrangements. Although donors could see that the plethora of local projects had the effect of baulkan-ising, rather than national-izing politics and made national public budgeting and expenditure management difficult, they were far from appreciating the immediate imperatives faced by Museveni's NRM. At the same time, it was hoped, a successful roll-out of LDFs would help the regime deal with a reluctance to joining up locally within the big spending ministries, health, education, public works, and break their monopoly control over government spending. Line ministries themselves knew that decentralization was the NRM's objective, but they argued that local politicians were 'not ready', they would 'simply run off into the bushes with the funds', and that existing service delivery had to be protected.

The donors, scarcely attentive to Museveni's wider political ambit, had a range of other reasons to support the programme. UNCDF was already convinced, and if they could be successfully supported, perhaps they would attract the World Bank to decentralization and pull bilateral donors along in their wake. This would help government achieve the larger prospect of harmonizing donors around one national fiscal control system joined to local, decentralized accountability.

In June 1995, when the first UNCDF mission was invited to visit Uganda to talk about the LDF concept, Ugandan officials frankly argued that they wanted a quite new approach, one that would 'test the boundaries' of the ambitious new Local Governments Bill.[26] Their aim was not to replicate UNCDF's 'policy pilot', but to help elaborate a new 'accountability framework'. Its worth here outlining this framework for two reasons; one because it was evidently a version of the 'three cornered' accountability framework that was not fully elaborated into orthodoxy until eight years later in the 2004 WDR – as explained in Chapter 4. Second, it is of interest because it helps later explain what happened during the five years during which the DDP was designed and implemented.

This accountability framework had three dimensions: downwards, horizontal and vertical accountabilities.[27] In formal terms, *downward* accountability, the accountability of local politicians and service providers to citizens was critically important; Uganda had for too long struggled with the legacy of colonial systems of indirect rule not to appreciate this. Downward accountability was essential, many said, in shifting from a system in which people were the administrative 'subjects' of government, to one in which they were politically enfranchised 'citizens' of a state operating within the rule of laws built on their new constitution of citizens' rights.

But they were aware that accountability did not end here, and it certainly was not to be achieved by the 'participatory planning' methodology of the kind seen in Vietnam or promoted by proponents of the 'short route' to more effective service delivery. This approach favoured direct, participatory relations between planners and communities, not as 'citizens', but as 'beneficiaries' of a technical, administrative process prescribed from the project document or administrative order. A balance was needed, *horizontally*, between the technical and the political, between experienced and articulate local administrative staff and newly elected local politicians quite unfamiliar with the wiles of plans, budgets and, in essence, with directing service providers in ways that matched their constituents priorities.

Table 6.1 Three accountability dimensions

Vertical accountability	Accountability between local governments and national politicians, line ministries and donors
Horizontal accountability	Accountability between local councillors and local council staff
Downwards accountability	Accountability between local governments and citizens, clients, participants, beneficiaries

Decentralization was, after all, a political project, and service delivery was both a means and, ultimately, and end of this project. *Vertical* accountability recognized that decentralization was a two-way street. Local councillors and their staff were entitled to receive funding in accordance with the responsibilities defined in the new laws, but they also had certain obligations, not just to show they didn't 'run off with the money', but that they had due regard to national (and global) priorities in how they allocated these resources to competing local priorities. It was hoped that the upcoming Local Government law would eventually provide a 'common accountability framework' in which global, national and local leaders would respect the primary status of the citizen in Uganda's new Constitution. For these reasons, Ugandan officials made it clear they'd brook no shortcuts, and demanded a participatory formulation process for the DDP; as they said, it would 'be written from the ground up' and not borrowed from UNCDF's approaches poised to travel in from Vietnam.

Embedding the DDP: planning and consultation

One consequence of this commitment to build an accountability platform 'from the ground up' was that preparation of DDP from June 1995 to October 1997 was unusually lengthy. Here we cover this period briefly, however, stopping only to explain some highlights, notably, the incentive and sanction system that became central to the project's design.[28] This kind of system has all the hallmarks of 'institutionalism', that is, the effort to induce local actors to behave in predictable ways, not through the discredited, state-centred command and control systems, but through what much later became popularized NIE precepts of 'inform', 'compete' and 'enforce' discussed in Part I of this book.

Three steps were mapped out for preparing DDP beginning with research into how services were currently planned and delivered, to negotiating with local officials over key features of the mechanism to put the accountability framework into practice, through to defining corresponding institutional roles and responsibilities. First was a step called 'Local Council and Community Profiling and Capacity Assessment'. The plan was to set aside 'form' in favour of 'substance', that is, to find out first how services were currently planned, financed and delivered. For three months villagers, local councillors, contractors, NGOs and Community Based Organizations (CBO) in the five pilot districts[29] worked with the DDP team to construct 'service decision trees' by talking through, in a structured way, each step in how local communities and authorities did business.

The DDP team was struck by the range of existing governance practices. Local planning, budgeting and service provision was a highly charged political process, itinerant and in constant tension with the administratively defined, forward moving linear process depicted in laws and regulations. It was not anarchic, practices were not just being 'changed at will', but it was clear the mix of local history, politics, skills and traditions were crucially important in how local governments, community organizations, informal leaders, contractors and so on actually produced services. Clearly, this would not be dealt with only by reasserting the formal system of planning steps from identification, appraisal, design, through to execution of investments. How then to lay down a transparent, comparable and governable process across this kind of terrain?

This meant that Step Two in designing the DDP, 'Definition of Planning, Allocation and Management Systems', had to try to build visible, predictable, accountable procedures out of this diversity. New regulations, rules of business and guidelines had to be prepared so that the provisions of the Local Government Bill could be applied and tested before it was enacted. But it would not be easy to make the three-way accountability framework durable on the ground. For instance, the principle of subsidiarity in decentralization holds that decisions about allocating resources to competing priorities should be taken at the lowest possible level, where the costs and benefits of those decisions were best understood, and where decision makers could be held accountable.[30] Fine, but while local governments were 'non-subordinate' to higher authorities, vertical accountability also required that these decisions must be 'integrated' at higher levels, so that the implications of local decisions could be matched with higher level commitments.[31] For instance, small, local water supply systems often need to be hooked into trunk water lines that cross local jurisdictions, from one parish or sub-county to another. A village decision to build a school would only improve services if district authorities had agreed to provide teachers. Similar debates around non-subordination and vertical integration occurred to answer a host of questions: what kinds of investments would be eligible for support – a narrow range, or the full set of local government responsibilities?[32] How would funds be channelled: entirely through the official budget process, or should NGOs get guaranteed access? After all, in some districts NGOs ran the bulk of health, water supply or education services. Who should be responsible to plan, to make budget decisions, operate accounts, to sign off at each stage: what would be the role of administrators, or politicians, should elders be involved? And what should be the formula for allocating the resources among local governments? Thus technical principles of subsidiarity and non-subordination had to gel with the integration principle and both needed, by design, to be resolved through a political process.

A transparent, easily understood formula was needed through which development block grants (the Local Development Grant [LDG]) would be made to district and sub-county councils who would then be able to invest in development services they were responsible for according to their priorities. But how to ensure the rules were observed? What was needed was some system of political incentive and sanction to motivate performance across all three accountabilities. Enormous political and technical energy was put into agreeing minimum requirements to access DDP Funds that realistically matched what people were capable of doing and, most importantly, were politically durable and would be legally sanctioned when the new Local Government Act (LGA) was promulgated. These are listed in Table 6.2.

Alongside these minimum requirements were set 'performance measures' and a scoring system, with pass marks and public score cards to encourage councils to 'go beyond' basic rule requirements. Achieving these standards qualified councils for additional resources. Were councils posting information about their budget and allocation decisions in public places; on their office walls, at schools, markets, etc.? Were they honouring commitments to participatory planning; were funds spent according to approved plans and budgets? There were strong incentives to encourage, though not

Table 6.2 Project access requirements

District and municipal councils

- Three Year, Rolling Development Plan approved by council as per LGA Section 36
- Functional District/Municipal Technical Planning Committee as per LGA Section 37
- Linkage between the Development Plan, Budget and Budget Framework Paper as LGA Section 78
- Draft Final Accounts for the previous Financial Year (FY) produced as per LGA Section 87
- Internal Audit Function working in accordance with the LGA Section 91
- The local government own revenues have not decreased in nominal figures from the previous assessment year to the assessment year
- Three year capacity building plan and Budget approved by council in place
- All co-financing (10 per cent of the LDG) for the previous FY made, budget for co-financing in the current FY and 10 per cent for the first quarter on the LDG account
- LDG and Capacity Building Grant Account established.

Division, towns and sub-county councils

- Three Year, Rolled Investment Plan approved by council in place as per LGA Section 36
- Functional Technical Planning Committee as per LGA Section 38 -3
- Linkage between the Investment Plan and Budget as per LGA section 78
- Draft Final Accounts for the previous FY produced as per LGA Section 87
- Internal Audit Function working in accordance with the LGA Section 91
- The local government's own revenues have not decreased in nominal figures from the previous assessment year to the assessment year
- 10 per cent co-financing in place. All co-financing for the previous FY made, budget for co-financing in the current FY and 10 per cent for the first quarter on the LDG account.
- LDG Account established.

---

stipulate, observance of priority sectors set out in Uganda's PRSP, the PEAP – primary health care and education, feeder roads, water supply, support for agriculture and local productive activities.[33] For instance, local governments were free to allocate resources according to local demands, but if their decisions put 80 per cent of the LDG into national PEAP priorities, their share of funds would be increased by 20 per cent. If less than half of the measures were met, their subsequent allocation was reduced by 20 per cent. Penalties were also agreed. For example, the incentive for sub-counties to maintain record-keeping and financial management systems is eligibility to receive a LDG. Non-compliance meant exclusion. This was all very different from normal practice, but was intended to create strong pressure on councils to shape up: from above and from below, from within the bureaucracy, elected councillors, and their community. Financial incentives would create political incentives within the council system for better administrative and political performance.

By June 1997, design was complete and a party of officials headed off to New York to have the DDP approved by UNCDF and World Bank officials. It was a festive event of seminars, videos and stirring speeches. Kiwanuka Mussisi, then chairman of the Uganda Local Authorities Association remarked: 'We appreciate the fact that this is the first time that we have projects which are from the people – not imposed. We were involved right from the project design all the way. This is increasing our own sense of commitment and ownership. It is a great opportunity to promote sustainability as people use language such as "our projects". Leaders take a risk if they mess with their [people's] projects.' Francis Lubanga, then Permanent Secretary of the Ministry of Local Government and who had contrived to bring UNCDF to Uganda, declared:

> The strategic importance of the DDP cannot be over-stressed. This is a key project for Uganda. Our difficulty in the past has been trying to find a partner that is willing to take the risk. UNCDF was prepared, and it is paying off. UNCDF was the one prepared to create the bridge for us to take this next step in decentralization.

The Fund begins

On 12 August 1997 approval was granted for the DDP, a $12.3 million project to be implemented over three and a half years aiming to 'support the efforts by the people of Uganda to eradicate poverty in rural areas by improving the inclusiveness, efficiency, effectiveness and sustainability of the delivery of public goods and services'. Around $11.2 million was to be allocated to local governments through an elaborate sharing arrangement between districts, sub-counties and parish councils, such that the 2.2 million people living in these jurisdictions would be eligible to receive around $5 per capita funding over three years.

How did these new accountability systems begin to bed into local governance arrangements? The first evaluation occurred during November–December 1998.[34] Initial results were very encouraging: at the start only 20 per cent of local governments had met the minimum requirements, but barely a year later, 75 of the 100 participating local governments qualified. With some surprise, it was revealed that all investments were exactly the same as PEAP priorities – as Table 6.3 shows. Though the evaluation found no impact, yet, on service delivery, the political impact was already impressive. For example, the evaluation said, 'it was striking that a community would assemble for an impromptu meeting to discuss just fourteen desks that they had been able to install in a classroom with local development funds'. The information campaign was having impressive effects: 'a very positive finding is that councillors and technical staff at sub-county and parish levels are well aware of LDG and their precise entitlements . . . about amounts, dates of transfers and actual transfers'. Accounting and expenditure management was improving, the law was being followed in local procurement, and local councillors were

Table 6.3 DDP: use of investment funds by sector and main activity, 1998–2000

Sector	Allocation (%)	Examples
Education	43.7	Classroom construction Teachers' houses Desks and furniture School library
Roads	14.8	Opening of small roads Culverts
Health	27.7	Construction of health units at parish and sub-county level Mattresses, beds and furniture for health units Staff housing (grass thatched huts)
Water	8.5	Gravity flow schemes Protected springs Borehole rehabilitation Rain water harvesting for institutions Institutional latrines Water for cattle
Production	4.1	Cattle markets Improved seeds/crops for multiplication Improved livestock for multiplication Environmental protection through tree planting
Other	1.2	Sub-county office blocks Cash safes for sub-counties

Note: DDP raw data on actual outputs in five districts and their sub-counties. Actual allocation of LDG in FY 1999/2000. Total budget allocation was UGS 3.9 bn.

demonstrating new kinds of accountability. Similarly, enforcement of the rules was evident: 'No cases were reported where amounts received varied from amounts expected – i.e. the common practice of diverting, delaying or misappropriating funds in transit has been avoided'.

The contrast over 12 months was breathtaking, right down to the procedures of accountability. Most of the improvement was linked back to the political process introduced through incentives and sanctions that created a competitive environment among administrative and elected officials and their constituents. Cases were reported where the electorate had gone as far as threatening physical harm to the sub-county leaders if they did not qualify to access the LDG, in other cases they were run out of town. Citizens were pushing their complaints about leadership through the system too, travelling great distances to request removal of non-performing chiefs, administrators and councillors. 'Our sub-county chief spends a lot of time drinking, it seems he has not put enough effort for preparing our sub-county to meet the minimum requirements, we cannot afford to lose this money, we would like to have someone else who can perform.' And so it went on: district planners, accountants and engineers remarked 'we have gained a great sense of accomplishment and recognition for having meet all the minimum requirements and qualifying for an additional allocation'. For administrators and politicians, community surveillance, backed by regular evaluations, demanded greater compliance. 'This DDP is strict' said one sub-county chief 'you just can't go straight through. There are conditions, for this and for that, and other things. It used to be that Council would just decide and go'.

Evaluating the flagship

Then, the much awaited donor 'corporate evaluation' of UNCDF got underway, and the evaluation team arrived primed to see UNCDF's flagship project in Africa. The project was hailed a success, and glowing reports went back to the donors with 'clear evidence that it is achieving something new and exciting'.[35] The evaluators found that the core elements of the DDP approach – unconditional funds, transparency, minimum conditions with rewards and penalties – should 'command considerable interest in all quarters'. Their report ended with a score card (Table 6.4) much appreciated in New York, for it became a key plank in a positive donor evaluation of the whole organization, and in Uganda, where the government was soon showing the results to keenly interested donors.

The excitement, then, was palpable: this was one of those fairly rare, enchanted moments in local government/development/projects when success is celebrated, when almost everyone feels great about what's happened and their having been associated with it, and substantive, progressive change feels possible. More funding and an expanded influence and on the ground effect seemed inevitable. The project had been taken up across

Table 6.4 Rating of the actual or likely future performance of UNCDF in Uganda

	Effective	*Efficient*	*Participatory*	*Sustainable*
1. Poverty Reduction	3	4	4	4
2. Strengthening local government	4	4	4	4
3. Strengthening civil society	4	4	4	4
4. An innovative approach within the country	4			
5. Impact on government policy	4			
6. Piloting and replication by government or donors	4			

Scoring: 4 = Excellent; 3 = Good; 2 = Indeterminate; 1 = Poor; 0 = Very Poor.

complex situations and multiple players, the systems were working as envisaged, abstract ideas had taken on a day-to-day life of their own. Other places, other donors line up to congratulate, begrudge or compete with this benchmark. By contrast with the pall over a failed project, this outcome was infinitely preferable.

THE FALL AND RISE AGAIN OF DDP: TELESCOPING REFORMS

Time for the programme to figure on wider stages

By early 2000, the World Bank was about to approve $64 million to upscale the DDP pilot into the Uganda-wide Local Government Development Program (LGDP). Bilateral donors too were about to march on board and accept DDP's new systems. But then it seemed to unravel. To everyone's surprise, the December 1999 annual evaluation of compliance with minimum requirements and the performance measures reported that *more than half of the Local Governments had failed to comply with even the minimum criteria to access the LDG.* None had performed well enough to receive the incentive of additional resources.[36] By most measures, reporting, public disclosure, bookkeeping, the matching of plans with budgets and expenditure decisions, accountability was deteriorating. So soon after the publicity of the earlier successes the entire effort appeared to be in jeopardy. Why did this backsliding occur? Evaluation teams were quickly put together to find out what had happened. They discovered that local officials, councillors, administrators and informed citizens were quite consistent in their assessment that the *DDP processes had been overwhelmed by another system – the Poverty Action Fund (PAF) –* that by then was channelling huge resources down to local governments. In one district, an exasperated woman councillor expressed the difficulty:

You government people, you come down here looking for answers to *your* project, but you don't realize you're not the only thing going on here. Don't hold us responsible, you set us up for this, look beyond yourselves for once and you'll see the world is larger than you think.

HIPC, PAF and the Uganda PRSP

The councillor was pointing to the effects of telescoping of reforms, and it is these complexities and their unforeseen effects we want here now to unpack. The DDP was always a cork on a larger sea dominated by bigger Developments: including the HIPC debt relief initiative; the widening adoption of the MTEF; and, crucially, centrally controlled but locally focused pro-poor or 'social' funds (in Uganda's case known as the PAF). As so often in Development, a close operational or project focus initially crowded out attention to these wider, but crucial contextual developments. Fortunately for the DDP/LGDP, it was not too late to react and re-position. In the following sections, we show the interactions among these developments. It is a complex story, but once the general picture is grasped, it should be much clearer how current practices of PRS-style budgetary control, transfers and service delivery do not necessarily enable joined-up and accountable local capacity to address poverty. Once seen here, similar dynamics in Pakistan and New Zealand will be easier to track through.

Uganda, the first country to qualify from a list of 41 countries, benefited handsomely from HIPC. HIPC delivered a 'fiscal dividend' of concessional loans, aid and debt relief sufficient to finance as much as half of the official budget and 80 per cent of development expenditures from 1997 to 2001.[37] But gaining this dividend required compliance with stringent fiscal and other policy reform along with a heady architecture of financing, transfer and surveillance instruments. HIPC eligibility combines two sets of familiar provisos: first, global alignment and integration, through adoption of mildly nuanced Washington Consensus macroeconomic management. Second, local services: the HIPC 'debt dividend' must be channelled to support 'pro-poor' public sector investments (overwhelmingly social services and infrastructure), particularly local health and education, and, crucially for the DDP, ring-fenced and delivered from the centre to the local through social fund-type arrangements of the PAF.[38]

The HIPC was in many ways the core technical innovation that triggered the broader PRSP process. By qualifying for HIPC, Uganda signalled it had fully harmonized with global poverty policy and had restructured national and local arrangements so as to be accountable to this policy commitment. Uganda's bellwether PRSP was published in 1997 and was 'expected to enhance country ownership of HIPC's economic adjustment and reform programs'.[39] As we explained in Chapter 4, governments are bound to back the PRSP with financing instruments like MTEFs to show they can maintain budget discipline and transfer resources directly to people and places identified as 'in need' and for which measurable results can be demonstrated through social and financial audit reports and other legal compliance instruments. As we will see, this was crucial for DDP. The MTEF was first introduced in Uganda in 1992/1993 at a time of fiscal crisis; the actual design of the MTEF got seriously underway in 1995/1996 in the lead up to HIPC.[40] In the persuasive management 'tool box' language typical of MTEFs, in Uganda:

> [T]he objective of the MTEF is the design of all public expenditure by a clear analysis of the link between inputs, outputs and outcomes, in a framework which ensures consistency of sectoral expenditure levels with the overall resource constraint in order to ensure macro-economic stability and to maximize the efficiency of public expenditure in attaining predetermined outcomes.[41]

Uganda's MTEF was the national budgetary complement to the PRSP, together providing one, comprehensive device binding together global agreements, national inter-sectoral budgeting and highly localized investments. Figure 6.1 depicts how this was intended to work.

The PAF provided the final piece of this policy-fiscal management architecture. The PAF, created in 1998/1999, was a vehicle *within the MTEF* to channel HIPC and other debt-relief funds, donor budget support and government's own resources into activities prioritized in the PEAP. The PAF's three key characteristics exemplified its ambitious global-local orchestration. First, it was a presentational device; government could point to the PAF and give donors comfort that parts of the MTEF had indeed been ring-fenced for poverty reduction priorities – health, education, water supply, etc. Second, the PAF helped to verify that promises to put HIPC savings into social services did in fact amount to 'budget additionality', that is, they did result in increased spending on social services and did not get absorbed elsewhere. Donors policed actual expenditures by closely monitoring transactions in the MTEF, and by a third feature of PAF, stringent reporting accountability from the local level. Monthly and quarterly reports, annual plans, budgets and expenditures for PAF were received from local governments and centrally approved. Then, through regular 'round table' reviews donors could check compliance with HIPC's conditionalities and, where differences were evident, could insist on enforcement.

Uganda's PAF enabled a staggering increase and rapid switch in spending patterns. Public expenditure as a share of GDP grew from about 17 to 25 per cent from 1997/1998 to 2001/2002 and allocations to PAF priorities grew from 17.5 to 35 per cent of the budget. Overall allocations to health, education, water, roads and agriculture increased from 39 per cent of the budget in 1997/1998 to 47 per cent a year later, and the share these budgets going to PAF-privileged services increased from 43 per cent to

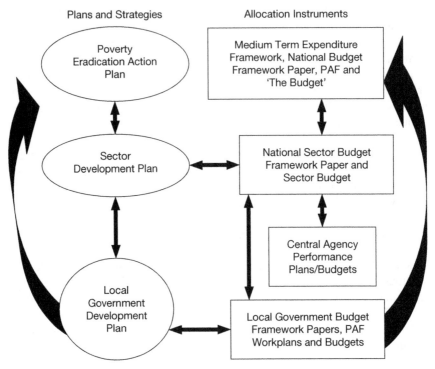

Figure 6.1 National–local framework for planning and budgeting

Source: Government of Uganda, 2000a, 17.

66 per cent. The total resources coursing through PAF budgets reached around US$290 million in FY 2000–2001, almost 33 per cent of the total public spending. In short, the PAF did indeed transfer substantive resources into local level service delivery. In general, we should note that this expansion of service delivery funding in areas where poor people have evident needs is perhaps the strongest claim the PRSP approach can make for being a direct mechanism of poverty reduction.

What implications did this have for DDP? All this PAF money, along with its narrow, vertical accountabilities, had major implications for *local governments' ability and incentives to action the accountability process being created through DDP* and wider decentralization. This dramatic escalation of PAF resources, and the accompanying special purpose budgeting and expenditure controls and reporting requirements, occurred at the same time as the DDP was supporting the transfer of wide-ranging responsibilities for planning, management and accountability for delivery of the same kinds of service by elected local governments. But the PAF and the DDP worked at cross purposes when it came to local government performance and accountability and the PAF, far larger in scope and volume of resourcing, had knock-on effects for DDP. Because PAF funds could be used only for pre-defined packages of investments, transfers were made to local government as 'conditional grants'. These were frequently tied to particular, centre-defined investments and compliance around processes related to these. Of course, DDP's untied LDGs also had vertical account-ability strings attached, but these were designed to be subservient to the two other aspects of the accountability framework mentioned earlier, namely, what the Ugandan's referred to as horizontal accountability and downwards accountability.

Before explaining this contrast further, some additional features of the PAF are relevant. As PAF resources ballooned, so too did the number of conditional grants. From around four conditional grants in 1995/1996, local governments were by 1999/2000 receiving about 80 per cent of central transfers through 26 different conditional grant transfers – of which the conditional grants ring-fenced by the PAF were the most significant in number and volume of funds. In 1999–2000 there were 16 different PAF conditional grants as listed in Table 6.5.[42]

The councillor was quite presciently attributing the dismal performance of her council to the effects of the proliferation of financing and control systems that were quite different to DDP's arrangements. At the time, these observations were unpopular – how to criticize a positive trend in financing for pro-poor services? – although three years later, after a thorough review of PEAP and its financing, one of PAF's architects acknowledged:

> [t]ension is emerging between Uganda's highly decentralized local gov-ernment system and the centrally driven SWAp [Sector Wide Approach] processes where sector service delivery targets have been established at the national level. This has been combined with excessive and

Table 6.5 Poverty Alleviation Fund conditional grants

Health:	Roads:
1. Primary Health Care – wage	10. Rural road grant
2. Primary Health Care – non-wage	
3. Primary Health Care – NGO	Agriculture:
4. Primary Health Care – Development	11. Extension: wage
	12. Extension: non-wage
Education:	13. PMA: wage
5. UPE – Capitation	14. PMA: non-wage
6. UPE – Salaries	
7. School Facility grant	Non-sectoral:
	15. District development grant
Water:	(7 districts; Netherlands funding)
8. District grant: urban water – recurrent	PAF Accountability
9. Rural water and sanitation development	16. Monitoring and accountability grant

Note: UPE: Universal Primary Education, PMA: Plan for Modernization of Agriculture.

increasing central control over inputs through a large number of tightly earmarked conditional grants.[43]

Here, with an eye forward to the Pakistan chapter, we need to consider the consequences of this PAF/DDP telescoping in terms of the three-way accountabilities Ugandans built into the DDP's design. First, technical and political dimensions: it's important to realize that the proliferation of PRSP/PAF conditional grants was a direct continuation of a long running tradition in Development to create arrangements that aim to achieve *technically* what is believed to be very difficult through local official *political* processes. The intent of tightly defined vertical programmes and conditional grants is to put fetters on local political leaders, to discipline them, to create administrative arrangements to make sure they act in predetermined ways – that is, reproduce global/national MDG service delivery commitments in every one of the 1,000 or so local governments across Uganda.[44] Ironically, because they are so complex, *these technical arrangements come at considerable cost in terms of accountability*. Uganda's MTEF included 350 projects in over 100 budget votes; thus it was difficult at the local level to determine which investment was being made by what programme or agency. Confusion and bewilderment was one effect. Another consequence was that the plethora of mechanisms for channelling resources quickly outstripped local capacity to manage, monitor and report on what was going on.[45] In short, as the local political and administrative space became plural and fragmented, all three dimensions of accountability very quickly got blurry.

The main worry for DDP's advocates was that the spectacular early gains they had prompted in political accountability were undermined because of what they called the 'PAF effect' – decisions about priorities, plans and investments were increasingly being made by the administrators, central and local, responsible for managing the PAF's vertical grant programmes. The essence of the original Resistance Council system was that *vertical accountability* – between local and central government – was intended as primarily a *political* relationship that is, of non-subordinate local governments acting as corporate bodies within a unitary state. But with the PAF systems, local governments became increasingly *administratively* subordinated, as district heads of administrative departments (producing, on average, well over 100 technical reports each year to central ministries) became in effect agents of the central government. Crucially, this had major consequences for the disciplinary arrangements that were fundamental for *horizontal accountability* to work. The accountability relations between elected leaders and administrative staff in local governments were in this early stage of decentralization fragile, to say the least. But these emerging lines of reporting between elected councillors and administrative staff – critical horizontal accountabilities – were quickly shunted aside as it became more important (to the end of keeping the flow of funds going) that administrators were now in the driving seat, reporting upwards, and thus re-establishing lines of central – local patronage that Uganda had struggled so long to displace.

The impact on *downward accountability* was also quickly felt in local councils. The ability of local councillors to respond to the needs expressed by their constituents, and being seen to respond, became more, not less constrained. Conditional grants and other vertical programmes provided politicians discretion, at best, only to decide on the location of sector investments like water supplies or schools. Unconditional grants, as a proportion of total transfers were declining at the same time as tied conditional grants were increasing in number, volume and as a share of the total budgets available to local governments (Table 6.6).[46] This placed the local politician to the side, able to influence the process only as a 'spoiler' – thus reinforcing views that local politicians are irrational, and that a further round of ear-marking was required by higher authorities to bring them into line.

The PAF arrangements privileged the administrative executive. This weakened a sense of local ownership then emerging for decentralization as a political project of locally enfranchised *citizens* acting through their elected representatives to bargain and compromise at the local level about how all resources, local and centrally transferred, should be allocated. Instead, under PAF systems, DDP officials felt that citizens were becoming again the *subjects* of administrative procedures for where best to site pre-packaged infrastructure investments. Thus, the idea that the DDP's LDGs would animate local politics, first around small-scale investments in

Table 6.6 Transfers from Central Government to Local Government (UGS bn)

Grants	1997/1998		1998/1999		1999/2000		2000/2001	
Unconditional Grants	54.3	24%	64.4	23%	66.8	17%	79.1	15%
Conditional Grants-Recurrent	168.4	75%	202.1	71%	275.2	71%	321.3	63%
Conditional Grants-Development	2.2	1%	18.8	7%	45.0	12%	107.8	21%
Equalization Grants	0.0	0%	0.0	0%	2.0	0.5%	4.0	0.8%
Total	224.9	100%	285.2	100%	389.0	100%	512.2	100%

Source: Ministry of Finance, Planning and Economic Development/Ministry of Local Government, 2001.

Note: Conditional Grants-Recurrent include both PAF and non-PAF recurrent conditional grants. Conditional Grants-Development includes the PAF capital grants, the PAF District Development Grant (Netherlands financed, seven districts) and LGDP.

services, and then, as was hoped, encourage local councils to embark on other, more complex ventures like supporting local entrepreneurship and production, was dealt a nasty blow. Similarly, underpinning the DDP's push for local control was a larger ambition to link local accountability for how resources were used with lasting participatory democracy in Uganda. As a result of the PAF, in 1999–2000, local councils were 'awash' with what they saw as easy money. Some councillors were concerned at their status as mini 'rentier states', dependent on external sources of revenue, as a consequence of the abundantly resourced national PAF. But councils had few incentives to collect taxes, while citizens felt little obligation to their local councils. Local, own sourced revenue collection all but collapsed.[47] In short, the would-be local territories of DDP governance turned out to be frail indeed in the face of transfer vehicles that were scaled to direct global-national debt dividends back into local sites of investment.

What to do? Overreach further

The problem, in fact, was not uncommon: the unravelling of DDP was a classic case of development project telescoping. Telescoping happens frequently in Development, as one reform is laid down on top of another, only half achieved one, resulting in mess and confusion. Its sometimes compared to a train crash where the carriages are telescoped into one another; the imperatives of one travelling project rationality colliding with another and create an uneven, messy set of development relationships in an already frail local situation. Here, sitting triumphantly in the station junction was the DDP, a *politically* driven project of the NRM, attractively presented in the global doctrines of decentralization supposedly 'growing up from below' via upscaled LDFs. Suddenly, it was being battered from multiple directions by the PAF's fiscal deluge coming down through the *technical* lines of multiple conditional grants and disciplinary devices. The Ugandans responsible for the DDP understood clearly what was occurring, and they arranged the political backing to fix the problem. What looked like nicely framed project rationalities – the DDP's inventive fiscal reward and sanction systems, for instance, crafting up a new local politics – turned out to rely on a long march of unrealistic assumptions about the context in which they were working. Or, as one programme officer quipped 'the right hand column of the Logframe as come completely unstuck: we're awash with unmet assumptions'. Nevertheless, DDP officials acted: once the extent of the PAF's potential threat 'from above' for their worryingly small, localized DDP became apparent, they set about to expose this, playing to both the obvious technical imperative for 'harmonization' and 'coordination', and to the political appeal of 'local control' and 'empowerment', 'subsidiarity' and 'non-subordination' that underpinned the NRM project.

In fact the DDP had enough scale and momentum, and was well enough plugged into both political and IFI priorities not just to survive, but to turn the tables on the PAF. The initially spectacular results of the DDP were enough to stay with the original plan to upscale the DDP through the $64 million LGDP. By December 1999 LGDP had become a 'firm' loan, and LGDP was duly approved; for UNCDF, 'national upscale' had occurred, DDP had succeeded. But government officials were concerned that the 'failed project' message of the December 1999 DDP annual evaluation would be made public. At the same time, donor missions were in town and, in the manner of donor missions, 'jostling' for attention to propose additional conditional grants – for agriculture, for another donor road sector SWAp, and the biggest attention grabber of them all, the first Public Expenditure Reform Credit (PERC), a $400 million World Bank loan that, it looked likely, would consolidate, perhaps proliferate more conditional grants.[48] Moving quickly, officials in the Ministry of Local Government, DDP's parent ministry, set their sights on the 'colonizing' the PAF and after this, the PERC, and the rest of the budget. A large, multi-donor supported technical study was conceived with tasks that rang all the right bells. It was to help *'streamline* and *strengthen* fiscal transfer modalities between national and sub-national governments in order to increase the *efficiency* and *effectiveness* of sub-national governments to pursue PEAP *goals* within a *transparent* and *accountable framework* and in accordance with the Local Government Act, 1997'.[49]

By careful stage-managing, this joint-donor-government team report was endorsed by government at large, including the Ministry of Finance, and the large spending ministries with major commitments to the PAF system, health, education, public works and water supply. Its central conclusion was clear:

> If present trends continue, with local governments increasingly becoming the local implementers of national sector programs, the scope, role and justification of decentralized locally-accountable service provision, as envisioned in the Constitution and the 1997 Local Government Act will be progressively undermined.[50]

Most importantly, it rang bells too at the heart of Uganda's politics:

> The key role accorded to decentralization in the country's political settlement and system of governance specified in the Constitution (and hence in the preservation of peace and stability) is poorly internalized in the planning and financing of service delivery systems. The inherent tensions between sector strategies, SWAps and decentralization are largely unrecognized within government and donors. Initiatives to increase awareness of the need for service delivery mechanisms to support Uganda's chosen institutions of governance and the local government legislation, and to review current and pipe-line sectoral

initiatives and parallel social-fund style implementation arrangements would be desirable.[51]

The Report duly recommended a new system for transfer of development and recurrent financing to local governments that firmly entrenched the DDP/LGDP's status as a 'common accountability platform' across which donor, MTEF, local government and other dimensions of three way accountability could be supported. The DDP/LGDP standards for local government access to fiscal transfers, for rewarding good performance, for giving local elected leaders discretionary powers to allocate resources, along with recognition of the perverse effects of proliferating conditional grants, all this increasingly became part of Uganda's official – government and donor – discourse about decentralization.

DDP to LGDP, with SWAps, the MTEF and PRSP: a joined-up solution?

Stepping back a bit, it's also clear that PAF and DDP had much in common. Both fitted with the emergent Liberal paradigm that pushed certain kinds of responsibility and accountability down to local levels, and transferred or withheld funds according to performance. And both depended for credibility on being part of a joined-up set of governing arrangements for poverty reduction.

DDP took this Liberal governance notion of local accountability further than PAF by explicitly trying to build within the local power capsules of Councils a local politic, firmly territorial its national–local accountabilities. DDP typified what later be referred to as PRSP's 'empowerment' leg, expressed through an amalgam of local responsibility, devolution of power, representative participation. But in this respect, DDP squared up far better than PAF with the need for Museveni's no-party NRM to put significant resources into these local capsules of NRM power.[52] Although the PAF did this only obtusely, DDP reinforced the overall idea that 'three way accountabilities' for poverty could be contained at local level. Both DDP and PAF, in short, built their elaborate accountability and good governance castles on liberalism's travelled ideals and the shifty, technical grounds of NIE in practice. Both PAF and DDP allowed accountability for poverty to become locally focused and disconnected from an historical accounting for the forces that produced skewed patterns of trade, market access or debt risk liabilities, or from the internal civil conflict that was then afflicting a dozen districts on Uganda's western and northern borders, or from the costs of regional military/political moves Museveni was then also engaged in elsewhere.

The DDP/LGDP common accountability platform was successfully used by key government officials to pull donors into a national to local financing system that successfully 'reached over the heads' of NRM-competing national political and administrative elites.[53] But, stepping back, does this also indicate that donors had bought into a new governance approach extending all the way from the globally popular PRSP, through the MTEF and SWAps down to local, pro-poor activity? In other words, did this harmonization really hinge on this extraordinarily high cost apparatus of politics and technicality? And was poverty or poverty-related accountability the central question for Uganda's Museveni government anyway? Yes, these frameworks were used to orchestrate a major switch in development spending. But the more substantive link between governance, services and poverty reduction? The increase in resources for service delivery in Uganda is unchallengeable.[54] Reports did begin to insinuate a link between spending on service delivery and the remarkable decline since 1992 in the proportion of people living below the poverty line, from 56 to 35 per cent in 2000 – although less remarked is the steady increase in inequality since 1997.[55] The data, quite frankly, is too weak to say.[56]

But how much effective joining, bounding and binding does this larger framework really represent? The legitimacy of the PRS is precisely its promise of providing joined-up systems that represent the interests of the poor, building on PPAs and such like and then marshalling the entire apparatus of governance in accordance with the 'voices of the poor'. This is why 'common accountability platforms' are sought after, preened and invested in so heavily. Recall the joined-up-ness of Figure 6.1 – with the PEAP and the MTEF book-ending Uganda's resources at the top, informing budgets, sectors, SWAps, the PAF and down to their local equivalents, the Councils' rolling three year plans, the corresponding budgets and reporting systems. Since credibility rests so heavily on the integrity and effects of this great joining up, what substantive reality sits behind the glossy formalism of these mechanisms? Do they join effectively, and create linked accountabilities to for example poverty outcomes? Does what they generate justify their cost? This has seldom been evaluated, but in Uganda, one review of all this in 2002 noted, 'it is difficult to pinpoint where evidence on performance has influenced the choice of policies and strategies' for although the PEAP describes how each intervention should impact on poverty, in fact 'outcome, output, and input indicators are used haphazardly'; it is difficult to ascertain relationships among them. And, although 'sector plans are usually comprehensive and fully costed', the 'stark conclusion from the costing of the Poverty Eradication Action Plan is that, in aggregate, sector programme targets would not be realistically achievable'.[57]

In fact, it all appears to be so far ahead of what can be reasonably thought through, financed and then wrestled into practice. The PEAP's objectives are clear, but lines of causality and accountability are not, just as it's unclear about which institutions are responsible for achieving what results. The Education Sector Investment Plan, the first SWAp in place since 1998, (another developing country 'first' for Uganda), had no explicit linkage to Poverty Reduction outcomes, no intermediate sector outcomes and no explicit financing plan. The Health Sector Strategic Plan 2000 argued a strong linkage between health sector performance and poverty, but apart from being strong on indicators – with 311 'output indicators' – and a clear claim for resources, it didn't say much by way of justification for how, say, maternal-child health interventions will link back to the larger outcomes. The much vaunted *Plan for the Modernization of Agriculture 2000*, a key plank in President Museveni's 2001 election pledge, shaped up nicely at the rhetorical level, with well articulated goals and purposes, commitment to multi-sectoral, whole of government approaches and participatory management, but had no specific measurable targets or indicators, was largely uncosted and lacked effective ownership by either the Ministry of Agriculture or local governments. In no sector plans are there significant signs of any rigorous *ex-ante* impact analysis of the effectiveness or impact of the policies being adopted.

Conclusions

Under Yoweri Museveni's leadership, and backed by competent technicians aligned with his Movement's political project of building government and consolidating power in the Resistance Council structures, Uganda achieved the political stability of a 'no-party' state that remained hugely popular at the turn of the millennium.[58] But, in his own terms, what local political accountabilities have ensued? Early enthusiasts of decentralization held that it might 'create a politically accountable institutional structure that could ... manage local services and serve as a basis for local autonomy'.[59] But later, others noted that decentralization's designers might have made 'unwarranted assumptions' about the public's eagerness and ability to participate in public affairs, and willingness by individuals to 'do community service' in fact, 'citizens seem disinterested in what is going on in councils'.[60]

What we see here is the rolling out of a technically elaborate, but ultimately bewildering range of participatory and accountability mechanisms, each aiming in its own terms to clasp issues of local governance and poverty together, but singly and together failing to achieve substantive leverage over wider political and economic forces. Given the expense and efforts involved in putting them together, this is a perverse outcome indeed. But it is also clear that both the design and deployment of such mechanisms is not so much viciously imperious as just over-conceived. Like previous promising Development techniques their persuasive power is in this over-conception: they throw up a technically intriguing, multifaceted and pan-dimensional web of governance, empowerment and integration – and, as we have seen, resources follow, sometimes famously. This web may not be all joined-up, it may not create a reliable proxy for the interests of the poor, the partially territorialized local government systems may ever be only a rough approximation of the three-way accountability framework. But the fact remains that these approaches to governing the poor, especially when energized by the PAF-sized transfers of resources can generate more attention to the poor and basic outcomes than any other on offer at present.

What's also clear here is that none of this happened because of a lack of professional skill or application. Those involved worked innovatively and effectively to craft and cut circumstances most often well out of their control to local advantage. A few officials pulled UNCDF out of Vietnam and embedded it powerfully into an entirely different situation in Uganda. Travelling form was disciplined by local substance. And they used DDP to align a sceptical World Bank through LGDP with a new financing system for poverty reduction that helped to consolidate fiscal power with the central Ministry of Finance. They contrived an unprecedented buy-in by donors to a 'common accountability platform' with obvious national benefits. Donor financing to local health, education, roads, market development and so on, was increasingly mainstreamed in national budgets and transferred to points of investment under terms increasingly of Museveni's own choosing. Through this accommodation with IFIs, Museveni did retrieve a wholly untenable international situation, wresting his country from 'an unusually severe social collapse' and continuing instability. He was able to orchestrate HIPC to provide almost 40 per cent reduction in national debt and to take a resounding majority with his no-party democracy through two national elections. This political stability, whatever its perversities, is often considered a genuine public good. As we will see in Pakistan, regime consolidation is often one of decentralization's hoped for 'deliverables'.[61]

But the cost is that the narrow focus on the mechanisms generates blind spots of its own. For all the immediate advantage the mechanisms may bring in global acceptance, and the remorseless light they throw on the technical norms and process instruments of practice, they crowd out crucial wider factors. Bigger picture poverty drivers, the political economy of land, elite and tribal politics, trade and commerce, militarism, debt and donor dependence, were masked and de-emphasized, both nationally and locally. Undeniably, and in several crucial ways, PRSP and decentralization in Uganda reinforced the overall territorial ambit of the one-party Museveni government. During the period of this story, attention given to the PRS kept the international spotlight well away from an unrelenting civil war in the northern part of Uganda, a war that continues, even today, to baffle observers.[62] All through these poverty-related reforms, the NRM government maintained military expenditure in support of an obviously criminal adventure in eastern Congo well above the 'hard budget constraints' of structural adjustment agreements.[63] IMF officials in 1999 set 1.9 per cent of GDP as the upper limit for military spending before loans were delayed or halted. In defiance of this, Uganda maintained spending 'well over 2.2 per cent' during 1999.[64] Protected by private military service companies, army officers traded diamonds, gold, stolen cars and coffee, 'criminalizing the state' by using government resources and administrations to pursue private gain.

It is one thing to note that the PRSP/MTEF to local governance connection is seldom joined-up in practice. Its quite another that its massive fiscal dividends continued to escalate. Despite the Congo intervention, despite release of a UN report that criticized Uganda's army for commercial activities in Congo, despite the untimely release of a World Bank report on widespread corruption in Uganda's privatization process, donors pledged $900 million in 2001/2002 for development assistance and the Bank awarded a $150 million PRSC. In total, this was $2.2 billion for the ensuing three years, on top of initial HIPC provisions. As Reno argues, and as we will see in the next chapter, 'regimes in the most volatile regions may be relatively more immune from creditors' political conditions than are more assiduous followers of creditor prescriptions in peaceful areas, since creditors need to preserve official interlocutors who will recognize the state's sovereign obligations.'[65]

Is it reasonable to case all this as merely a 'Trollope Ploy' of mutual beneficial deception? This term comes from recurring scenes in Victorian writer Anthony Trollope's novels, where a marriage-hungry maiden takes imprudent gestures such as a squeeze of the hand as a proposal of matrimony, regardless of actual intent, and both she and her suitor play along with the fiction? In Uganda, as in Pakistan, it was surely much more; there was talent, commitment and a determination to avoid the litany of coups and unrealized expectations. And the poor are getting some services. But we would stress that within the overall PRS framework, open markets, services and more accountable governance are just about all the poor might get. The market, for its part, will deliver uncertain outcomes, winners, to be sure, and losers, eventually. Under PRSP, funding for services has been shovelled down to localities, but with limited practical accountability and uncertain poverty outcomes.

Similarly, the claims for empowerment generated by local governance reforms need close scrutiny. Ugandan experience shows, and we will see again, that such reforms in their current 'joined-up' forms generate few substantive accountabilities, especially in relation to wider social outcomes like poverty reduction. In fact, as the Uganda story shows, (PAF) service delivery and (DDP) empowerment aspects can even act at considerable odds with each other, all the while reinforcing the basic liberal governance precept: local accountability for problems deemed a local responsibility. Here, to repeat in closing, services and their narrow, technically joined-up accountabilities crowd out a more searching political accountability about poverty; all the while making it look like poverty is being addressed. All this, we will elaborate, at enormous local and national expense, in both DDP and PAF contexts.

7 Pakistan

A fortress of edicts

The tyrannical ruler now is well-versed in power
Builds about himself a fortress made up of edicts;
While falcon, sharp of claw and swift to seize,
He takes for his counsellor the silly sparrow
Giving to tyranny its constitution and laws,
A sightless man giving collyrium to the blind.
What results from the laws and constitutions of kings?
Fat lords of the manor, peasants lean as spindles!

Muhammad Iqbal, 'Divine Government',
The Sphere of Mercury, *Javid-Nama*, 1932[1]

A hope, pinned on a hope

Masood Zafar is an imposing man, with big hands that catch your attention. He's warming to his point, the audience right with him. 'I don't think you realize the destabilizing effect you people are having in Pakistan. You talk about local democracy and empowerment, the separation of powers, reminding us, don't we all know it, that this has eluded Pakistan since Partition in 1947. And then you support this military dictator.'

Zafar leads the opposition in the Province Assembly elected in October 2002 after three years of military rule. He commands the audience with his eyes, then turns on you again. 'You international agency people actually support governments whether their powers are separated or totally fused around the barrel of a gun. In either case, the same abuses remain, and the same faces are doing them. So, you achieve your interests as agents of global capitalism by co-opting people into your net. And George Bush is laughing at us, while our teachers are raped by policemen, who get away with it because the judges are unable to act against the executive because they've been so repeatedly compromised by it.'

It's a tough line, one that sets the key elements of Pakistan's internationally backed governance and Poverty Reduction reforms firmly between the rock of War on Terror geopolitical realignments, and the hard place of Pakistani realpolitik: fractious, patrimonial, powerfully ensconced. Pakistan 2004 is a hell of a context to be talking about pro-poor good governance and the core tenets of liberal reformism: the rational 'separation of powers' between executive (local bureaucrats, including the police), elected councillors and politicians, and the judicial system. Or about installing that NIE market-contractualism in social governance: 'inform, compete, enforce' *and* pushing citizen voice into the spaces this creates in local politics. And about getting the three way accountability relations between service providers, local government and citizens right ... but this is exactly what every one engaged in Pakistan's radical devolution process has to do now, as these reforms roll out. The alternative, a major failed reform, here, is unthinkable.

He sits. You stand. 'Thanks for your comments. As you arrived, I was explaining the results of our study of devolution in Pakistan. Let's see.' You look at your notes, then at the 20-odd people gathered for the meeting. There's a chill wind in the trees around the compound, blowing a bit of dust, and after Zafar, people are talking distractedly among themselves. Begin. 'Improved service delivery in decentralization doctrine depends on getting the three way relations between policy-makers, that is, elected councillors like many of you, and service providers, lined up with what citizens want. What will get policy-makers to listen to what citizens want? How will NGOs help articulate citizen voices?' you ask, pointing to reps of Social Participation Organization. 'And how will they get service providers, the staff of local governments and others to respond?' The wind whips again, blowing a few of your bullet point pages off the table. 'Pakistan has pretty much adopted a text book approach to devolution and service delivery for Poverty Reduction,' you say, holding up the latest 2004 World Development Report. 'You've probably seen the Poverty Reduction Strategy Paper, released in December last year?' You hold that up, in the other hand.

'So what Pakistan has done is really a three step process.'

'First, new *laws* are introduced to restructure government – ensuring among other things the separation of powers, and their reassignment to different levels of government. These detail roles and responsibilities of different organs of government, and what federal, province and local governments are supposed to do.'

'Second, the idea is that these new structural arrangements will change the *incentives* that make key policy-makers like local councillors and the service providers want to use resources more efficiently, make their staff perform better, all because they're much closer to and therefore better informed by the people who vote for them. The key is this accountability.'

'If they do this, then the third step, *better service delivery* should follow, and this will be good for the poor.'

'And it's all tied at the top, as we've seen, by the Poverty Reduction Strategy which will get the federal government to allocate public finance according to pro-poor outcomes, health, education, better policing, and so on. We've got sector reforms going on in these areas too, and you know we're supporting a public resource reform program in this province, and that's going to introduce medium-term budget and expenditure management techniques. Pro-poor outcomes, in other words, come at the end of this chain of reforms. So this is the promise the Poverty Reduction Strategy Paper holds out. Of course, we all know this needs debating . . .'

Ali Khan, Zafar sahib's cousin, is on his feet, ready for exactly this point. 'Yes, sir. But you know the Big Problem is that this account-ability thing is **not** on the ground. There's parts of it, sure, but alongside bigger parts from before, and probably parts still there from the Moguls. And I don't know either about this idea of citizen rights filling up the statutory spaces. Yes the rights are clear. Last year our President said (he reads from a battered notebook) "It shall be the endeavour of my government to facilitate the creation of an environment in which every Pakistani citizen will find an opportunity to lead life with dignity and freedom" he also said that "killing in the name of honour is murder and will be treated as such". And just last week we have the Council of Islamic Ideology telling us that women patients can't be seen by male doctors, and that women's clothes can't be sewn by male tailors. I ask you. Our local hospital's only female doctor just ran off after her life was threatened because she wouldn't sign a false death certifi-cate of a women murdered by her brothers. The police, the courts, the administration now seems so powerless and at the same time, completely out of control.'

He stops, fixes you with his index finger. 'In between your laws, your bright new institutional structures, between those lovely incen-tives you talk about, well, there's so many spaces around here already. Here, in Pakistan, a border is a place for crossing. Look at the opium coming over the Afghanistan border. The border continues to harbour Al Qaeda and Taliban elements and the world will hold us respon-sible. We are again threatened by a bumper crop of opium and terrorists. It's all funnelling down this way – and our police are too busy profiting from it to take any notice of your elected councillors moaning on about this. They are running amok. This is Pakistan, Sir, these spaces were filled well before the devolution laws you brought were enacted. All you've done is open up new spaces, in fact, so many damn spaces that its all cracking up and in pieces. And watch who walks in through those cracks.'

Introduction: Liberal governance in a multiply territorialized place

Pakistan's remarkable approach to Poverty Reduction and decentralized governance was certainly produced in fraught conditions: a 1999 military coup, joined on the one hand to a heady combination of local and international (post 9/11) crisis, and on the other, to a reform experiment hard driven by political and technical determination. Against the precipitously rising security stakes of the War on Terror, Pakistan became at once a recipient of enormous international aid and a bellwether of governance reforms that, it was hoped, could provide the antidote to extremes of both poverty and politics. As elsewhere, 'inclusive' Liberalism's reforms were propelled by crisis and could, as in Uganda, reach a compromise, an accommodation with the highly illiberal interests of a military/executive government. The pattern of reforms will be familiar from Uganda. First, having thoroughly accentuated the sense of crisis, the military government sought refuge and legitimacy in the 'common man'; citizens, it was argued, were tired of being dominated by patrimonial political interests, they wanted and were entitled to 'efficient service delivery' and 'access to justice at their doorstep'. At withering pace, a host of executive orders were enacted to create new institutional arrangements for security, policing, fiscal discipline and devolution of responsibility for service delivery to local citizens acting either directly or through councillors elected on a 'no-party' basis. The local state was disaggregated through the separation of powers and by providing a wide range of what will immediately be recognizable as 'quasi-territorial' arrangements to give the clients of devolved services many new points at which they could directly demand greater accountability from local politicians and administrators.

As we will see, on a fast changing and unstable platform of geo-politics, across an extraordinary array of local circumstances, Pakistan was laying bare 150 years of accumulated territorial governance traditions to the full panoply of PRS's sharply formalized, second-generation liberal reforms. In Pakistan basic security and stability issues are being put on the line in ways unimaginable in Vietnam or Uganda to directly assail entrenched territorial patrimonies that have been the basis of a shaky stability. Interlocking patrimonies – military, executive, police and bureaucratic dominance – are prised apart, and apparently replaced by the travelling rationalities of 2004 WDR style arrangements. Authority is being assigned in plural ways, notably to a host of quasi-territorial local bodies responsible to foster 'local-local' forms of accountability that will be sufficiently strong to discipline the state to behave in pro-poor ways.

What this chapter ultimately turns on is a struggle which sees Liberal governance arrangements pitched into the middle of a chaos of accountability, and there asked to perform social and territorial regulatory tasks beyond their capacity. In Pakistan fulfilling basic social regulatory obligations has never been the state's strong point. But under a decentralized liberal governance regime focused on service delivery, ensuring that public assets are protected and well regulated, that public security and order, observance of basic rights and entitlements, accountability and consistency over time becomes tougher yet. Now, the travelled provenance of Liberal governmentality – where something works for social service delivery in the UK or New Zealand is therefore worth applying elsewhere – becomes a potentially lethal flaw.

The first section explains Pakistan's recent historical context to show the kinds of path dependencies of entrenched patrimonial, territorial power that Pakistan's devolution-led PRS hoped to transform; it discusses the legacy of Partition in 1947, in particular, the dominance of territorial, feudal power in government, the steady consolidation of this influence through an alliance of military/executive officials and how territorial and liberal powers were fused in the office of the executive magistracy for the purpose of local governance. The second section reviews recent evidence of how devolution reforms are effecting local governance. It draws explicitly on the 'three-cornered' accountability framework, conceptualized in 2004 WDR that was summarized in Chapter 4. In the third section we speculate about the possible consequences of a shallow embedding of 'inclusive' Liberalism in Pakistan, and in particular about its impact on the capacity of the state to meet its basic social regulation functions, responsibilities that extend way beyond the provision of health, education or other much needed social services and into the protection of public assets and entitlements, including of course, public safety and protection from the abuse of office. All of this takes us back to our book's early descriptions of Lugardian and subsequent development at places at the very edges of Empire. This sets up our discussion, in New Zealand, where we see similar issues arising in a quite different political and historical context.

WHERE POWER GOES, THE LAW FOLLOWS

During March 1946, a British Cabinet Mission arrived in Karachi, just 18 months before the fateful partition of the sub-continent, to assist Indians to settle their differences and set up a constitution making body and a representative Executive Council to see the country through to independence.[2] Would Pakistan be part of a three-tiered federal union, a sub-continental umbrella federation with Hindustan in which different elements would be governed with a degree of decentralized autonomy? Or would Pakistan be a minimally scaled sovereign territory, too poorly resourced and politically fractious to sustain itself? Muhammad Ali Jinnah, Muslim League leader and subsequently founder of Pakistan, favoured the former, but it was politically fraught because the landed powers of the League's conservative provincial governments were resisting all forms of central governance. The mission ultimately had no formula for a sustainable, acceptably territorialized compromise, but they knew that unless the implacably disparate positions were drawn into a settlement the 'thin crust of order', as Ayesha Jalal remarks, maintained by the British and their collaborators for 150 years would break down and release unprecedented disorder.[3] But a compromise eluded all parties; talks sponsored by the mission collapsed after six days.

An outbreak of killings in Calcutta, 16–20 August 1946, unleashed anarchy and broke that 'thin crust of order'; 4,000 people were killed and 15,000 injured, all dreams of compromise were destroyed.[4] Thus began the 'cyclonic revolution' and the horrors of Partition when 8 million people, mostly poor agriculturalists, entered Pakistan in the largest, most rapid and bloody migration in history.[5] Pakistan has struggled with the aftermath, a nation state popularly seen as 'unviable' – what Jinnah called 'the moth-eaten rump' – split into two parts, 1,000 miles apart, its society militarized and pumped up through four bloody and debilitating confrontations with India over Kashmir, and not least, the legacy of speculation, dispossession, nepotism and violence as landed/administrative classes became entrenched and any pretence to social justice was lost.[6] Thus began a history that would track right through to the 'fundamental transformations' that were pinned together by the military government in Pakistan's latest round of neoliberal devolution reforms.

Pakistan has been under military rule four times, for almost half of its existence. No army chief has ever given up power voluntarily, and no elected civilian government has ever completed its term in office.[7] Since inception, state power has been concentrated in an alliance of Pakistan's military-civilian bureaucracy, which has struggled, consistently but unsuccessfully, to durably legitimate its power by reaching variously liberal and territorial accommodations with its citizens. This is not to say that the alliance has remained unchanged or unchallenged, but efforts to install elected civilian democracy have failed to create a countervailing power, rooted in an organized, citizen based party politics. The basic interests of Pakistan's political class have always been guaranteed by an easily patronized civilian-military executive, thus allowing the army, the seventh largest in the world, to consolidate its interests in land, banking, shipping, urban development, manufacturing, insurance, transport and so on and eventually account for 3 per cent of Pakistan's GDP.[8]

The construction of Territorial, executive power

Understanding decentralization, governance and poverty reduction in today's Pakistan requires engaging with accounts of history and identity. Pakistanis frequently attribute growing poverty to the failure of democracy to take root. About this, they ask many questions: is it that our country was born with a profound sense of vulnerability, one that left us prone to the unreliable affections of international alliances? Is it because we lack a territorial identity borne of a common anti-colonial struggle,[9] or that the country is so polarized into regional, ethno-linguistic, religious or other differences that it is the archetypal 'imagined community'?[10] Are we psychologically prone to sycophancy, or to authoritarian or inegalitarian solutions?[11] Was all lost when Jinnah simultaneously took hold of the powers of Governor General, head of state and head of the Muslim League, thus ensuring we never achieved a constitutional tradition, never effectively separated powers? And perhaps all this made it impossible to agree on a progressive social programme and meant that we never invested in ourselves to create the human capital so necessary for informed, stable electoral politics?[12]

Whatever their favoured explanation, Pakistanis agree that as the country emerged from British rule, their political institutions were weak and dominated by a patrimonial, rural-territorial elite: the 'feudals'. And, too, that this undermined the credibility of all political classes. It may be true, as Herbert Feldman says in his famous Omnibus of the country's history, that 'one of Pakistan's misfortunes, from the day of its creation, has been undue participation in public affairs by people to whom the management of the parish pump could not be safely entrusted'.[13] But the fact remains that it was easy for rounds of in-coming military leaders to discredit all politicians as pretentious and self-interested, an unprincipled obstacle to social or political development. By the first coup in October 1958, it seemed plausible that General Ayub Khan should deride politicians for their selfishness, 'ruthless thirst for power, lack of patriotism, and general misconduct'.[14] The military in fact became regarded as the only stable, national institution, and one of the only avenues for upward social mobility for middle- and lower-class families. Over time this deeply conservative institution became, as Tariq Ali remarks, Pakistan's 'spinal cord' binding together political, patrimonial, territorial, and therefore economic interests.[15] And thus, far from an ephemeral influence on politics, intervening directly only to sort out the mess created during brief periods of civilian rule, its eventual return to the barracks after regular bouts of military rule came merely to signal that it was prepared, until its interests were threatened again, to exercise power through executive, bureaucratic rule.[16]

From the time of independence the majority in constituent assemblies, national and provincial, were lawyers, landowners and businessmen all marked by their commitment to preserve the status quo, their dedication to 'the single-minded pursuit of their narrow interests'.[17] Democratic institutions were submerged by their determination that no social reform agenda – for land reform, or even agricultural taxation or other mildly redistributive moves – would ever be taken seriously. This was classic patrimonial territorial governance: and the effects on efforts to create liberal governance institutions should be obvious. Bennett Jones cites the example of

Amir Mohammed Khan, the *nawab* of Kalabagh who, at the age of 14 became the undisputed master of his family's estate in southern Punjab. As with many feudals, he was not initially a supporter of Jinnah's modernist liberal vision, but he joined the Muslim League once it became clear which direction political winds were blowing. After independence, he ingratiated himself with the new establishment, gained secure party seats and was by 1965 governor of West Pakistan. He ruled like a feudal – once his orders were delivered, all civil servants, the courts, the police in Punjab were reminded they owed their loyalties not to the Pakistani state, but to the *nawab* alone.[18]

This dominance had three immediate effects. One was that political parties were used by the military, undermined and sabotaged. In this way, it was necessary only to change politicians, not politics itself. In this, they were entrenching a colonial practice, established in the aftermath of the 1857 Indian revolt, in which elections served only a window-dressing 'legitimacy function', an instrument of authoritarian government to preserve the status quo.[19] A second consequence was that political party alliances became fluid and mercurial. Parties and loyalties could burgeon overnight and vanish the next day. Parties would in the future find it difficult to discipline politicians and each time they failed, the patrimonial and highly territorial bases to power would be reinforced from top to bottom in political culture. This established a maxim much honoured in Pakistan that wherever power goes, then the law must follow and this, rather than any connection with citizen interests, encouraged the politics of expediency.[20] Thus, the third consequence was to reinforce the power landed territorial interests and the politics of factionalism and *biraderi* (a networked 'brotherhood') in which the primary interest is to 'bring in one's own' (tribe, *biraderi*, caste) to the exclusion of the 'other'.[21] As the colloquialism has it 'For friends, everything; for enemies, nothing: and what's left can be distributed by the rule of law'. As will become apparent in the second section, this has profound implications for a governance system that depends on elected leaders responding to 'incentive signals' sent by individual 'citizens'. Few politicians and fewer in the electorate have independent votes – that is, the ability to either 'voice' or 'exit' – indeed, the tradition of 'candidate-based' (rather than 'party-based') local elections favoured by military rulers further consolidated territorial power.[22]

Ever since separate representation for Muslim electorates was granted in 1909, Muslim politicians had little incentive to organize parties as a mechanism to consolidate power over constituencies. Colonial authorities contrived to ensure that real discretionary power rested in the executive. Thus with correspondingly few powers of discretion, it became less likely that political leaders would stand on a social manifesto through which to create ties with constituents. Local territorial *influence* (opportunistic favours and factional or tribal allegiance going down, patrimonial factional or tribal loyalty worked in narrow interests going up) was sufficient to get elected. The poor citizen, largely bound by birth into these territorial containers of power, has never been a source of accountability, only a possibility of rent extraction for territorially ensconced patrimonial powers.

This accommodation, in which Liberal principles of governance and the interests of the majority quail before already consolidated patrimonial, territorial power, was established well before Partition. This is best illustrated by what happened through the world's greatest agricultural land engineering experiment – the colonization through irrigation of Punjab's semi-arid lands.[23] The poor were deliberately locked out from the process. A government press communiqué issued in 1914 remarked that in making grants of land, 'tenants, laborers and other landless men should not as a rule be chosen, as their selection involves the *aggravation of difficulty*, already acutely felt, of obtaining agricultural labor'. The communiqué held it would be undesirable to use Government policy to 'upset the existing social and economic order'.[24] More to the point, the creation of the world's largest contiguous expanse of irrigated lands was deliberately used to create territories of political dominance, an arrangement which engineered, if you will, a political system based on patronage that encouraged those already dominant to use legislative, fiscal and administrative instruments to exclude the poor. Thus was established a post-independence pattern of neglect by politicians of public entitlements for primary education, health, water supply. This in turn stymied local democracy, and further entrenched the notion that entitlements of citizenship were simply matters of 'privilege' which depended on the patronage of the 'good landlord'.[25]

Thus, governing by distributing (territorial) access to land and water, concession over agricultural surplus, official position or opportunity, in fact, any territorial public asset, created a polity in which order was maintained through the privatization of public assets. Prescient colonial administrators worried about its long-term pernicious effects. If a regime did succeed in weaning itself off such practices, how would it govern, especially if people learnt that it had deliberately decided to deprive itself of the means of rewarding political favour? But this connection between abeyance and gratification through favours of public office became deeply embedded in the consciousness of the Punjabi politicians that dominate Pakistani politics to this day.[26] So the colonial practice of curtailing local politics and privileging executive/bureaucratic systems of rule created a perfect system for rent-seeking and valorization of position and privilege – with two consequences. One, as Jalal remarks, was a 'license for an almost anarchical autonomy in local affairs'.[27] Another was that the executive arm of the state increasingly became an expression of territorial elite interests. This is evident throughout Pakistan's history: the colonial revenue collection system for land and water, how the rating system was applied in practice, in repeatedly stymied efforts to introduce more efficient systems for taxing water use, or the assessment of harvest tax liabilities.[28] Entrenched interests sabotaged any major changes even though repeated audits of these systems made apparent the extent to which these govern-ance arrangements in fact 'amounted to an infringement by the strong against the rights of the weak'.[29]

Subsequently, executive power was consolidated by the India Act of 1935 and was later enshrined in Constitutional moves to sidestep political schisms between the central and provincial governments. That few polit-ical leaders were experienced in the art of government also increased their reliance on civil bureaucrats and reinforced a highly centralized, praetor-ian culture of executive dominance.[30] This trend, together with the absence of clear party political agendas, repeated incursions by the military into political affairs, and the rise of the modernist, developmental state of the 1950s through to the early 1970s, meant that power over economic policy and fiscal affairs effectively transferred into the hands of the bureaucracy. The executive bureaucracy was then increasingly well placed to establish patron-client relations with private interests, and in this way the state's primary linkage with the citizenry became dominated by those 'with considerable proficiency in administrative routines' who were 'inclined to regard these as the sum total of government'.[31] As we will see, there are strong historical echoes in the new institutionalism of Pakistan's latest round of executive-driven devolution.

Territorial power: joining the police, magistracy and executive

Thus from independence onwards, the new political class derived power mostly through the semi-feudal aristocracy that had thoroughly woven its interests through the local executive instruments of state rule and, for the largest share of Punjabi (and therefore Pakistani) politics, with a military power firmly rooted in rural culture and politics.[32] Lacking a legitimate base in local, representative politics, factional contests were played out around struggles to control the two principle means of governance: first, the *thana* and *kutchery* – the police and the magistracy – and second, distribution of local development largesse and favours of office. There was little need to foster or control a functioning legislature, or to make links with citizens via political parties resting on larger manifestoes of change. It suited both, the executive bureaucracy and political elite that District Commissioners (DCs) rather than the elected middle class should rule local areas. As we will see, a key focus, at times preoccupation of the architects of Pakistan's current devolution reform has been to break the (territorial) powers accumulated in the office of the DC. It is therefore important to understand how this office consolidated administrative, judicial and policing powers for it is their con-temporary separation that takes inclusive neoliberalism into the heart of Pakistan's contemporary governance reforms.

Before the office was finally abolished in 2001, the DC played a critical role as the hub of territorial management, dispensing justice, maintaining law and order, collecting revenues and coordinating functional departments.[33] The DC was the head of prosecution and therefore supervised pre-trial detention by police of citizens alleged to have broken the law. The DC was the first civilian mechanism for local police accountability – citizens could complain directly to the DC's officers, and by virtue of his powers over the police, could gain an immediate remedy. The DC was the custodian of civil rights and therefore in charge of jails – again, a check on police abuses. The DC was also the chief of the executive magistracy and therefore had powers to prosecute, detain and adjudicate on a range of criminal and civil laws. And, as the head of the district executive, the DC could exercise authority over all provincial departments in the jurisdiction – again, through the DC's office, citizens could complain when denied access to their entitlements in respect of land disputes, administrative grievances with government departments, service providers and so on. And, more than just the sum of these, the DC was the inherent authority of the executive allowing him to direct other departments to carry out their functions in accordance with laws.[34]

The DC's powers were, partly by design, and certainly in practice, largely unassailable, not just for directing local development or removing obstructions, but for extending the 'writ of the state' into every corner of life – in fact, the abuse of these powers prompted the first UK 'torture commission', in 1855.[35] The DC could make all the limbs of the state work to sort out the miscreant citizen. Exercising his powers of summary trial under the Code of Criminal Procedure, the DC's raids or *chhapas* were part of the culture of a state exercising its might against encroachers and other delinquent people. There was certainly huge opportunity to abuse these powers. The imposition of the infamous Section 144 of the Code of Criminal Procedure[36] for instance, placed citizens in double jeopardy – for if the court freed a person that had been wrongly detained by government, then the state could simply rearrest the citizen on any number of other security measures – and this, of course, proved most effective in dealing with politicians or organized elements in civil society. Successive military governments added to the DC's formidable arsenal of powers, such as the Maintenance of Public Order promulgated by General Ayub Khan. Section 144 was used by the DC for a wide variety of purposes; prohibiting tenants from removing grain from the threshing floor before the landlord had received his share; prohibiting demonstrations thought to be a risk to 'public order'; prohibiting the use of firewood for brick-burning; prohibiting the entry of non-students and others within a radius of twenty-five yards of school examination centres; prohibiting 'Eve gazing', so as to ensure the licentious eyes of citizens promenading the streets remained fixed on the horizon.[37] Thus, before the abolition of the DC's office and the separation of powers, daily village life saw entourages of magistrates, the movement of the court lock, stock and barrel, with police and 'complainant department' to the site of alleged offence, mostly having the

desired effect, especially with support of the local press. The DC's officers, as complainant, process server, prosecution, witness, judge all rolled
into one would administer swift justice and thus the 'writ of the state'
would be established. The ends of justice were thus met and later the office
records would be filled out to service the purposes of law.

Pakistan's devolution programme of the early 2000s would finally end
this office so often chastised and romanticized in administrative folklore.
It would do so in familiar ways, not just by attacking the office of the DC,
always considered the paragon of colonial governance. Like earlier devo-
lutions, it would also steer clear of any social reform agenda that required
challenging the territorial basis of landed, irrigated power. Rather, as with
all 'path breaking' reforms, Pakistan's latest devolution was launched with
an attack on politicians and political parties while simultaneously mounting
efforts, as we saw in Uganda, to 'reach over their heads' and appeal to
that classic liberal subject, the 'citizen' and empower them with the service
delivery attentions of a newly efficient local state.[38] In this respect, the
task of a devolved state has always been the same. Although seeming
quaint today, General Ayub Khan in the late 1950s saw the task of his
Basic Democracies as

> the inculcation of ethical and civic values; the development of a char-
> acter-pattern; a raising of the cultural and intellectual level, assisting
> women to overcome the social handicaps that confronted them; encour-
> agement of a healthy national spirit; the elimination of sectarianism,
> regionalism, and provincialism, and the teaching of simplicity,
> frugality, and good taste in living standards.[39]

Khan's 80,000 basic democrats nicely preserved the status quo, for a time
at least, by providing the electoral college for presidential elections and
later for elections to the national and provincial assemblies in 1962.
Although the government of Zulfiqar Ali Bhutto (1972–1977) abolished
this system, it was revived again (although based on direct elections) under
General Zia ul Haq (1977–1988). Sporadic gestures were made by civilian
governments led by Benazir Bhutto and Nawaz Sharif, but shorn of reasons
to make a connection with the people, elected governments did not wish
to create competitors at the local level. Efforts to institute local democracies, embed concepts of citizenship, empowerment or wrap a minimum of
services or socially responsible regulation around local governance were
anyway lost in the political centralization prompted by military conflicts
(with India, with the former East Pakistan, and the local effects of the
Soviet-Afghan conflict and the Gulf wars), the legacy of the 1970s phil-
osophy of executive-led development and the persistence of conflicts
unresolved since Partition between the centre and the provinces. But most
important, the high liberal modernism of all previous 'devolution' efforts
– in 1864, 1882, the early 1920s, and three episodes after independence

in 1947 – ran hard and unsuccessfully against the entrenched power of territorialized patrimony which had stitched up resources, landed and governmental, thereby effectively keeping control over the gates of Opportunity. Each time, liberal governance's indirect route of maintaining order and security through locally empowered, healthy, educated and emboldened citizens was retrenched, largely because governments have seen it as just a very roundabout and uncertain way of doing what can better be achieved by the fused and directly territorialized powers of the local executive officer.[40]

DEVOLVED GOVERNANCE MEETS ENTRENCHED PATRIMONIAL, TERRITORIAL POWER

A most favoured pariah state

It is evident that when General Pervaiz Musharraf seized power on 12 October 1999, he was marching to a long tradition. Musharraf appointed himself Chief Executive, suspended Parliament and the Constitution, established a National Security Council and invited the superior judiciary to take a new oath of allegiance. He then turned his attention to securing four critical prerequisites judged by military regimes as being necessary to legitimize taking power: crisis, the citizen's interests, law and order, and ready cash. When, five days after taking power, he announced that the nation was 'on the brink of the social, political and economic abyss' he was using a well-used clasp to consolidate martial law.[41] His appeal to the citizens of Pakistan chimed with a constituency evidently fed-up with the unstable kleptocracy of elected civilian governments. Since the death of General Zia ul Haq in August 1988, the leadership of civilian governments had changed ten times,[42] as living standards continued to decline and the economy 'failed' in the face of rising inequalities and unbridled corruption. Musharraf immediately promised to stop corruption, stabilize the economy, tax the middle classes, eradicate poverty by ensuring that 'services' and 'justice' were delivered efficiently, and restore real democracy from the 'base up'. The coup-makers' *de jure* legitimacy was secured within one month when the Supreme Court invoked the convenience – well-exercised in Pakistan – of the 'doctrine of necessity' which states that the territorial integrity of the state is of paramount importance such that when a liberally-principled constitutional government is overthrown, the usurping authority must be recognized so as to avoid anarchy. The court's accommodation gave him three years to deliver on his reform package before handing power back to elected representatives by October 2002. Comforted by the grace of the Supreme Court, Musharraf then turned to domestic security, law and order, the deteriorating economic outlook and a sceptical international community. When abandoned by donors following nuclear testing on 28 May 1998, more than

60 per cent of Pakistan's fiscal resources were tied to pay the debt and the military. Foreign exchange reserves were low, and the terms of trade were deteriorating due to rising oil prices.[43] As with many countries, the 'real' level and depth of poverty in Pakistan was being hotly debated, but the public had little doubt poverty had increased alarmingly during the 1990s and this was joined to a palpable sense of insecurity fuelled by sectarianism, local civil conflict and increasing lawlessness due to the visible collapse of many government departments and the flouting of social regulation responsibilities by the police, administration and justice system. At home and abroad, Pakistani's felt their nation sliding into the double opprobrium of a failed and a pariah state.

To deal with poverty, Musharraf faced a double dilemma. People were aware that economic growth had not delivered improved living standards, and much less had it 'robustly' linked up with poverty reduction. In fact, studies had shown how poverty sometimes increased at the same time as economic growth.[44] Equally well known was that *increased public expenditures* tended not to create lasting economic growth nor improved social service delivery, much less, reduce poverty.[45] One of the largest efforts ever mounted to impact on poverty through social service delivery was notorious in the public mind for corruption and plunder, with little to show for the additional debt burden. During the two phases of the Social Action Programme (SAP Project 1, 1993/1994 to 1996/1997, and SAP Project 2, 1997/1998 to 1999/2000), around $9 billion was spent to improve the availability, quality and efficiency of services, especially for the poor and for women, in elementary education, basic health care, family planning and rural water supply and sanitation. Outcomes were disappointing, to say the least, particularly since education had received roughly two-thirds of this allocation but overall net enrolment rates had continued to decline, and, after an initial increase, government resumed established patterns of under spending on social services.[46]

So when work began on the Chief Executive's first *Three Year Poverty Plan* and the *Interim Poverty Reduction Strategy Paper*,[47] the strategy not surprisingly thrust economic growth out as the first of its obligatory three legs: opportunity/economic growth, empowerment/human capital and security/social protection. But the first leg was hardly the most politically robust, implying as it did more of the 1990s liberalization policies that had proven chaotic, hugely unpopular and even then, failed to reduce historically high budget deficits.[48] The problem, then, was seen to lie elsewhere. The highly respected Ishrat Husain, Governor of the State Bank of Pakistan, announced just before the coup that Pakistan's problems were problems of 'governance'; an over-centralized and bloated bureaucracy, elite capture of public privilege and benefit, and a disenfranchised majority.[49] Musharraf thus seized on the popular idea that 'the root causes of all our ills has been the absence of good governance'.[50] This view was backed by the World Bank's 2001 Development Policy Review, *A New Dawn*, that advised 'Not only was the [earlier governments'] reform program incomplete as a program of "first-generation" reforms, but also it neglected what are often called second-generation reforms – the complementary institutional reforms that are needed if a country is to benefit from liberalization'.[51]

Thus began Pakistan's latest round of governance reforms, devolution of power and delivery of 'justice to the doorstep'. Musharraf was surely aware that this was an uncertain route full of grand rhetorical overreach. But from the viewpoint of regime legitimacy and stability, it proved to be a short and highly profitable route too. On the heels of public laments that Pakistan had failed to honour any of its previous IMF structural adjustment agreements, IMF increased assistance by 237 per cent from 1998 to 1999 as the Economic Stabilization Adjustment Facility was relabelled the Poverty Reduction Growth Facility and topped up by $1.3 billion. In December 1999, the holders of Pakistani eurobonds approved a restructuring of their assets, and a further $12 billion in debt was restructured. Total aid receipts, having declined by 70 per cent following the nuclear tests in 1998, jumped 88 per cent from $621 million to just on $1.17 billion in fiscal year 2000–2001, and again, by 196 per cent, to $3.4 billion the following year.[52] Pakistan had managed to become, at one and the same time, a most favoured pariah state.

Yet the rewards of this bold commitment, swept higher by the imperatives of the War on Terror, came with unprecedented challenges. Through 2001 and the summer of 2002 the prospect of nuclear war increased on the Indian border, and US involvement in Afghanistan, Central Asia and the Persian Gulf, India's rapid repositioning with Iran, China, Afghanistan and the US, all impacted Pakistan. US military action in Afghanistan destabilized Pakistan's western border, as refugees moved back and forth with arms, smuggled goods, narcotics and shifting alliances. In between, horrific bombings rocked the provincial capitals Quetta, Karachi and Peshawar, and police and paramilitary violence seemed to increase. Months of uncertainty before and after national and provincial elections in October 2002 saw the balance of power constantly shifting but sliding inexorably towards a shaky alliance between Musharraf's Pakistani Muslim League (Quaid I Azam – otherwise, PML Q) and five pro-Taliban, Islamist parties.

Pakistan's devolution project

Devolution in Pakistan was from 2000 nothing if not bold and ambitious. Three features of Pakistan's devolution are relevant here. First, it involved classic decentralization, that is, where government responsibilities are vertically disaggregated and authority, functions and resources are transferred from the central state to subordinate or quasi-independent local authorities. Devolution was tuned to the edict to 'bring power closer to the people'. This was a colossal undertaking for an extraordinarily diverse country of 150 million people. It required complex reassignments of fiscal, political and administrative powers between the federal government and the four provinces in the face what everyone knew would be hostile provincial legislatures (once they were returned to office in October 2002) and executives. These changes were driven by the creation of a three-tiered collegial system of elected local governments (district, *tehsil/taluka* and union), replacing appointed officials, especially those clustered around the DC.

Second, the local state was horizontally disaggregated to a degree never achieved in earlier episodes of liberal reform. Most prominent was the final step in the local separation of powers, by ending the office of the DC and assigning its coordinative and social regulation responsibilities to local elected representatives and the judiciary. As should be evident from the brief trawl through Pakistan's history in the first section, while the formality was modest, this was no mean feat for, as the architects of Pakistan's devolution saw it, it would require ending once and for all the archaic Lugardian system of local governance which fused legislative, judicial and executive powers in the office of the DC.[53] This local-level horizontal disaggregation occurred *within* many core agencies as well – for instance, the police reforms entailed a separation of its law enforcement function from its investigation function, and the creation of an independent prosecution service.

And third, devolution's new laws – for local government and the police – created a host of new arrangements through which it was intended that citizens would directly and indirectly (through elected representatives) participate and hold public officials accountable not just for delivery of social services, but for key areas of social regulation: the resolution of disputes, monitoring local court performance, checking police abuses, regulating labour and business laws, land administration, overseeing law and order. Achieving this all hinged on creating incentives for a 'three way' mechanism of local accountability involving service providers (the executive), local government (elected legislatures) and citizens. The Local Government Ordinances (2001) and the Police Order (2002) were hailed for the fact that they provided a basis in law for a host of new quasi-territorialized bodies that would allow citizens to discipline local government – Citizen Community Boards, Public Safety Commissions and others we will explain shortly. These lined up almost word for word with the model of local-local dialogue, partnering and shared responsibilities later promoted in the 2004 WDR's, *Making Services Work for Poor People*.

Devolution thus had various objectives; to introduce new blood into politics, give opportunities to previously marginalized citizens to be heard in politics, and not least to introduce a measure of stability and legitimacy to the military government. High on the list of expectations was that social service delivery would improve through 'empowerment'. In an important sense, the history of weak articulation of citizen entitlements and abuse of power by state agencies was largely put aside by the presumption that these issues could be addressed in some select bodies bestowed with quasi-territorial powers.

Give us the facts: but don't open Pandora's Box

By September 2001, these reforms were on the ground. By any measure this was a phenomenal achievement; administrative systems had been overhauled, just on 127,000 local government politicians had been elected to 6,458 union, *tehsil* and district councils and (by June 2002) new fiscal systems were also in place.[54] A year later, as Musharraf had promised, elected federal and province legislatures were returned. But the withering pace at which these reforms had been ushered in by executive orders generated tremendous conflict. The ruling party, PML Q, faced storms of protest. Opposition forces at home and abroad were not prepared to legitimate Musharraf's presidency, and hostile province governments were threatening to amend and roll back his devolution and police reforms by promulgating subordinate administrative regulations that would countermand the intention of the parent laws for local government and the police. Foot-dragging and chanting had reduced the federal and province assemblies to farce. Government, unprepared to relent, faced potential coalition collapse.

Although they had hailed the bold and courageous decisions taken since 9/11, donors were also pressing for quick delivery on their priorities. They said that the regime's credibility would be demonstrated only if it channelled the massive fiscal dividend – a result of improved macroeconomic performance, increased aid and budget support, debt restructuring and rapidly increasing foreign exchange reserves – into the empowerment leg of the PRSP, that is, social services and devolution. But how to do this? Few believed that newly devolved systems were ready to handle the huge volume of resources banking up at the federal level and deliver, quickly, on MDG targets. The global alliance was looking for 'direct route' results: 'President Bush wants to see girls behind school desks, not a long discourse on the benefits of local democracy through devolution. It's about delivery now, not promises for things that may be delivered later' said the USAID Country Director.[55] In an odd confluence, donors' worries about devolution – would it do the job, social service-wise – and the hostility of opposition political elites to devolution looked set to come together and prompt a turnabout. If so, it would see not just the adoption of discredited vertical programme arrangements, like the Social Action Programme, but perhaps a complete ditching of devolution.

In this politically charged atmosphere, advocates of devolution's long haul within key Ministries and the National Reconstruction Bureau (NRB) turned to the World Bank and ADB, along with like-minded bilateral, the UK DFID. 'Don't give us any airy fairy nonsense' said Shaukat Aziz, then Finance Minister:

> we want a dispassionate look, the facts. Be direct, to the point, tell us how we're doing against international practices. But most of all, don't

question the framework: we don't want to open Pandora's Box again; focus at the operational level, tell us what we should do next'.[56]

Here we can only summarize relevant parts of the Study that resulted from this request.[57] It provides a remarkable opportunity to illustrate the key theme of this chapter, namely, frailty of the quasi-territorial governance institutions of inclusive liberalism in practice. More particularly, it reveals how the 'flattened' politics of local-local proximate relations between leaders, bureaucrats and citizens – as designed for in 2004 WDR – stacks up against the vertical, patrimonial power of local political economy and entrenched practices by the executive – introduced in the first section. This will allow us in the third section to step back a bit and reflect on the political consequences of PRS's separations of powers, local dis-aggregations and newly created statutory spaces for citizen-state, local-local dialogue, and what this may mean for the poor.

The three-cornered accountability framework: 2004 WDR in action

To understand the Study's frame of reference, we need to recap on the emerging doctrines of service delivery and poverty outlined in Chapter 4. Recall that the *Making Services Work for Poor People* paradigm in vogue since 2000 concluded that economic growth alone would not lift the poor out of poverty; rather they needed to be targeted through services to compensate for the market failures and social justice/equity issues arising with market-led growth. It also held that a persuasive argument could be made for substantially increased external resources if poor countries could show they were using them well – that is, efficiently and effectively progressing towards the MDG goals, while maintaining 2004 WDR-style governance accountabilities.[58]

In 2004 WDR's manifesto for the second-generation reforms, moral commitment by governments is crucial. Applying the technical instruments – 'combining inputs to produce outputs and outcomes more effectively' – was also important. But neither would work without 'reforming the institutions that produce inefficiencies'.[59] Pakistan's PRSP, also published in December 2003, endorsed this, 'the poverty reduction strategy requires a major transformation of governance structures and systems, as well as of political and organizational culture, especially at the local level'.[60] The 2004 WDR framework on the other hand was perilously simple: it put poor citizens' interests at the centre of the triangle of accountability relationships between clients, service providers and policy-makers. Focusing here, and getting these relationships right, was the key to a poverty reduction through social service delivery. The 2004 WDR framework was convenient for the Study team. It nicely separated the political aspects of poverty and governance from the technical. The 'Accountability Triangle' identified a set of accountability relationships (voice, compact/management, client/power) and actors (citizens, politicians/policy-makers, service providers) that could be readily turned into a study methodology for discussing the potential impacts of devolution on service delivery.

Also attractive was that the 2004 WDR framework allowed the study to focus on 'incentives' for pro-poor service delivery. Again, recalling Chapter 4, the NIE's formalistic mantra of 'inform, enforce, compete' for organizing markets, governance and services are held together by institutional structures (norms, regulations, assignments of function) that create incentives for better accountability and performance of actors at each corner of the 'Accountability Triangle'. Having established the non-political basis of citizen interests – in efficiently delivered services and justice – Pakistan's devolution thereafter had a two-step logic, as Figure 7.1 depicts. First, vertical and horizontal structural changes were introduced through executive laws and regulations that were applied in new organizational alignments and assignments of responsibility. Second, these would engineer

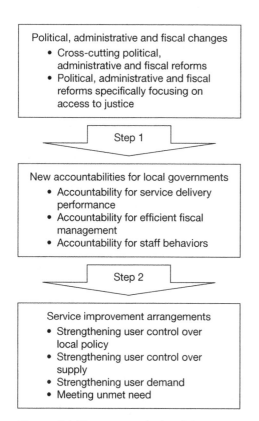

Figure 7.1 The two-step logic of decentralization

new *incentives* that would bear on these actors, thus creating new *account-abilities* between them, in turn disposing each to *perform* better, for instance, to use resources more efficiently for pro-poor service delivery. The result was that the clients of services would have greater voice, more control over policy, and greater ability to direct the executive, the service providers, according to their interests. The Study focused on the role of the new tiers of *elected* politicians – the elected *nazimeen*, (or mayors), and councilors of district, *tehsil* and union governments. In short, it wondered what a well-intentioned *Nazim* or local councillor could and would do when functions, powers and resources had been devolved? Were new incentives and accountabilities being created? And were they resulting in better performance from managers responsible to ensure services were delivered? Both these relations – between citizens and elected leaders, and between the political class and the executive/bureaucracy – while central to the NIE/WDR framework, also lay at the centre of Pakistan's hoped for 'fundamental transformation'.

What did the Study find? 'Achievements and Challenges'

The Study, like 2004 WDR, presumed service delivery improvements would depend on two major factors. The first was political will: were elected councillors, the *nazimeen* or policy-makers really motivated to improve the delivery of services for their citizens, and did they face real incentives to use resources wisely? Political will was contingent on whether elected politicians were really being held accountable by citizen power, (through mechanisms for citizens to have a 'voice', or to 'exit' by choosing to use other services). Beyond this, was the question of whether the fiscal or budgetary incentives (such as predictable fiscal transfers and autonomy in preparing the budget) really motivated good practices. Second, successful reform was seen as depending on enhanced managerial power – were senior officials able to motivate the health workers, teachers and other 'front line workers' to perform well?

The Study, true to the accountability triangle (Figure 7.2), focused on the effects of (1) political incentives and accountabilities, (2) citizen power accountability, and (3) managerial power accountability. In all three cases, the Study found some areas of improvement, along with much more complexity than had previously been imagined, and a good deal of uncertainty about the real effects of the reforms in terms of raising overall accountability. In sum, and not surprisingly given the early stage of the reforms, it showed clearly that for all the apparent comprehensiveness of the accountabilities and incentives, the actual on the ground accountabilities were not in fact well joined-up: rather, there were many limited, overlapping and fragmented accountabilities that were not nearly powerful enough to, yet anyway, produce genuinely pro-poor outcomes.

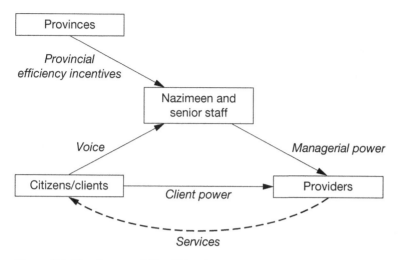

Figure 7.2 The Accountability Triangle

First, the *political accountability* outcomes: were politicians being provided with incentives for fiscal efficiency[61] – for instance, did they have the revenue base, as a result, for example of having been assigned sources of revenue from taxes that were 'buoyant' (increasing in line with the general economy) and 'potent' (could be increased by greater efficiency in collections)? Were transfers being made from province governments according to the new norms of efficiency and timeliness? The answer was yes, and no. While local governments have been given some rights to raise local revenue, the taxes assigned to them are never likely to match their expenditure requirements. And increases in revenue available to province governments tended not to be fairly shared with local governments.[62] Politicians, then, continued to rely as much on political clientelism and 'gamesmanship' to secure resources as they did on transfers guaranteed by regulation. This meant that patron-client relations among local, province and federal politicians and other 'non-transparent' factors continued to shape how much local councils actually received. Accountabilities and controlling relationships were still based on personalized aspects: need or rational entitlement was not the only criteria for budget allocations.

Although most local governments received what they were promised, less than 20 per cent of the transfers from higher to lower levels were provided according to predictable formulae. Local governments also had little autonomy in preparing budgets, and therefore little scope to shift resources according to local and poor people's needs. In fact, they had no effective control over their staff and other administration costs, more than half of their development budgets were under the control of federal and province governments. Development budgets were typically controlled from above through vertical programmes for health, water, education, etc. – similar to the previous chapter's discussion of Uganda – and this undermined local governments' sovereignty; few could plan or act according to their own priorities. In sum, incentives for politicians to be responsive to local or poor people's needs were not significant. Rather, older and more established relations allowed provincial political alignments and coalitions to continue to dominate local councils, and skew resource allocations along established sectoral and patrimonial lines.

Second, *citizen power/voice accountability*. 'Citizen empowerment' is the cornerstone of Pakistan's devolution policy. Were the reforms resulting in local elites in leadership positions responding to citizen demands? There was little doubt that citizens wanted better health clinics staffed by competent workers, water supplies that worked, and more. A lot, however, depended on whether local politicians really needed to, or could respond to constituents' needs. The ability to respond was of course constrained by the vertical controls on budget – noted above.[63] But more striking was that the practical influence of poorly literate, disorganized citizens over local politicians proved frail.[64] Yes, devolution had brought 'new blood' into local politics, but although most new councillors had no previous experience of local politics, the most influential came from 'political families' with long histories of local power.[65]

The very fact that local councils – at three levels, from union, *tehsil* and up to district – were competing for resources to deliver services, and competing with province and federal vertical programmes for control, often, over the same services was weakening the accountability of local leaders to citizens. That multiple agencies, local councils, private sector agents, province departments and special purpose programmes were all busy delivering services meant that there were 'jurisdictional overlaps' – many different agencies with common responsibilities. Family planning services for instance, were being planned and delivered by the federal health ministry, the province department and district health services, not to mention by NGOs and private health workers operating under special, project-specific partnerships. Improving water supplies in one union jurisdiction, for instance, might be hobbled by the failed schemes being abandoned at the fag-end of an IRD project, overlaid by a municipal corporation delivering water supplies under pre-devolution legal arrangements. And overlaid again by the cash-strapped and heavily indebted *tehsil* authority's infrastructure and services department acting as the local agent of the province Civil and Works Department, the Public Health Department or an Area Development Authority under provincial control. This, not surprisingly, was weakening the incentives for councillors to perform better. Local politicians could not easily demonstrate that they were responsible for services. Neither did citizens have much idea who, or which 'government', was providing services. Typically, people remarked that 'in the past, one person had authority (the district commissioner). Now authority has been dispersed'.[66] Where authority is dispersed, where reporting systems are weak and there are jurisdictional overlaps, account-ability is weakened. This encouraged politicians to focus only on core supporters when making service delivery decisions, their neighbourhood, *biraderi* or clan. Councillors focused on local schemes, the classroom, clinic or street lighting, that they could hold up to the neighbourhood they knew directly. While devolution, it was concluded, was not perhaps making the problem of 'private goods' (elite capture) any worse, there was little evidence that things were improving.

Third, were the accountabilities of *managerial power* making service providers work better? In other words, were the *nazimeen* able to get local schools, health clinics and water services to perform better?[67] Again, the results were not encouraging. Small surprise that key local government, health, school and water supply staff saw few reasons to respond to the authority of elected leaders when these had no powers to set terms of employment, that is, to hire or fire staff. In two of Pakistan's four provinces, local councillors had no practical control, because payments for salaries were made under province direction. Staff in different sectors tended to progress through their careers regardless of their performance in previous tenures. And in only exceptional cases were the council's service delivery monitoring committees of elected councillors having any effect.

So what of *client power* and its ability to affect management decisions, accountability and service quality? At the same time as Pakistan's new laws for police and local governments, true to WDRs since 1997, radically dis-aggregated its historically ensconced territorial governing arrangements – not just, but notably the fused powers of the DC – it also provided a legal basis for an unprecedented number of ways for citizens to participate in bodies intended to replace the old, discredited 'check and balance' systems. Table 7.1 lists these quasi-territorial arrangements. This was perhaps the most ambitious feature of Pakistan's devolution experiment. We cannot detail what the Study found for each of these bodies. In general, it was to be expected that given the strength of the political and administrative arrangements that locked local decisions into higher level power, and given the diffused accountabilities resulting from competing, disaggregated service delivery arrangements, both points noted above, these quasi-terri-torial arrangements were, the Study concluded, yet to 'achieve their full potential'. Fairly, the Study's 'overall conclusion (was that) it is too early to tell whether or not devolution has had any impact on client power for the better or the worse'.

It cannot, now, be said whether this will change, or in what direction, but in the next section of this chapter we will dwell on the implications of this situation. Devolution's proponents cite evidence of progress. In some cases, School Management Committees are bringing down the cost of school construction; doctors and health workers appear to be more often at their duty station; some municipal authorities are innovating with Customer

Table 7.1 Citizen voice and power: quasi-territorial arrangements

	Territorial body	Features
External oversight	Accounts Committees	Elected councillors providing representative voice in budget making, financial accountability
	Monitoring Committees	Elected councillors, monitoring and supervision of executives responsible for health, education, public order, etc. performance
	Public Safety Commissions	Monitor, supervise local policing plan, provide recourse for citizens against police excesses or acts of omission
	Insaaf/Justice Committees	Elected councillors, dispute resolution, redressal of complaints against judiciary and creation of community dispute resolution bodies
Citizen dispute resolution	Departmental Grievances Redressal and Complaint Cells	Redress citizen grievances about executive conduct/performance
	District Ombudsman	District formal adjudicative body for grievances against local state.
	Citizen Courts Liaison Committees	Public information, redressal of grievances against judiciary.
	Citizen Police Liaison Committees	Citizen-police accountability, dispute resolution, access to information
	Anjuman I Musalihats	Voluntary, registered bodies for community dispute resolution
Community management of facilities	Citizen Community Boards	Voluntary, registered bodies, entitled to 25% of local council development budget to be allocated to CCB priorities.
	School Management Committees	Voluntary, some budget responsibility at school level, check on accountability of education staff

Service Centres for grievance redressal and public access to information; progressive councils are developing women's resource centres; NGOs on occasion bring police excesses to the attention of councillors and together lobby through Public Safety Commissions for redress. Despite the jurisdictional overlaps, the frequent hostility of province powerbrokers and the shortage of funds under councillors' control, an unprecedented array of local action is occurring.

But more evident to the Study team, overall, was how thin and formalistic the WDR expectations were, and how little purchase such bodies had over realities of both local politics and central control over key services (health, water, education, judiciary, public safety and policing). Rather than a three cornered accountability being clasped together by these quasi-territorial arrangements around particular services, the 'key variable' in all this was politics, or as the Study put it, 'the strength of the provincial incentives to intervene'. Provincial intervention was by no means a predictable factor, either positively or negatively. And even where it could work in favour of better performance, this was largely because local and province political leadership, council, and assembly members shared common fractional or, at times, party alliance. Or where, in remarkably few cases, the local *nazim* had sufficient independent basis of power to resist the predations of unfavourable coalitions at higher levels. But, for the time being at least, this seldom applied: the overwhelming majority of councils in Balochistan, Sindh and North West Frontier Province and large parts of Punjab were controlled by leaders quite out of favour with the leading powerbrokers in province governments. In short, the chain of joined-up local political, citizen and client incentives and accountabilities that features in the triangle was ruptured at several crucial points. Simply, accountabilities and incentives weren't joined-up, and seemed likely to have no predictable impact on either the quality of services, or, even further down the string, on the position of the poor. In 2004 WDR terms, this was not at least so far providing reliable incentives which would *Make Services Work for Poor People*, much less reduce poverty.

MARKETIZED SERVICE DELIVERY VS TERRITORIAL SOCIAL REGULATION

On New Year's Day 2004, Pervaiz Musharraf was finally confirmed as President of Pakistan and obtained a vote of confidence from the National Assembly. The NRB and Pakistan's local government *nazimeen* celebrated a day of thanksgiving. According to the NRB's Chairman, the stage was 'set for a quantum leap in the history of Pakistan (that will) will definitely result in the entrenchment of a new political culture based on transparency, ownership and accountability'.[68] Donors immediately committed to further expediting financial support to Pakistan.[69]

Meanwhile, the Study team was busy crafting evidence and influence to encourage Pakistan's leaders – and their own organizations, the two banks and bilateral donors – to push on with devolution. In sum, the team's 7 goals and 22 recommendations all aimed to achieve one thing, namely, the insulation of the local Accountability Triangle – local policy-makers/councillors, the executive/service providers, and citizens/clients – from hostile political interference by higher level political authorities and to encourage citizens to occupy the political spaces created by new quasi-territorial bodies. A 'disconnect' of high level patrimonial power from its local extension would free local devolved arrangements to work independently of power entrenched in land or other historical privilege: the nodes of the accountability triangle would be able to work in their separately empowered and locally accountable ways, and all three could create effective partnerships and allow the 'nodes' to move appropriately into and out of shared responsibilities that would result in improved service delivery. The urgent need they said was to break the hold of province and federal politicians and bureaucracies over local politics, by removing opportunities for political negotiation and bargaining between higher governments and local governments and replacing these with a range of fiscal and executive measures to achieve this; for instance, specific purpose grants to increase the incentives to raise local taxes, thus reducing dependence on transfers from higher levels.

Stronger incentives to deal with Territorial power

The Study also noted that pushing forward with these local reforms in the face of opposition from province and federal elites, now elected representatives in assemblies at these levels, 'leadership', a 'strong, central coordinative capacity' at the centre – what the 1997 WDR referred to as 'the brains of the system'.[70] However, the reality is that such efforts could flail only weakly against the exigencies of an unstable central politics where Musharraf's government and the 17th Constitutional Amendment that legitimated his own modernist, secular form of military democracy, continued to struggle with a coalition of fundamentalist religious parties known as the Muttahida Majlis-i-Amal and shifting alliances between mainstream political parties.[71] And, in a theme we will develop in our concluding chapter, it was becoming quite evident that the macroeconomic, ready-cash imperatives that had encouraged Musharraf to turn, through adoption of the Poverty Reduction/devolution approach, to the international donor community were quickly weakening. Donor leverage was quickly declining, as by January 2004 by a judicious finessing of the numbers, the State Bank's economic growth projections were up, the stock market was buoyant, and foreign exchange reserves were higher than ever before.[72] Critics lamented that poverty was still growing alarmingly,[73] and pointed to new figures showing that while the richest 10 per cent of households had boosted their incomes by 33 per cent between 1988 and 2002, the share of the poorest had suffered a nearly 10 per cent erosion in purchasing power.[74] But undeniably, government was beginning to argue that the immediate post-coup fiscal gap had all but closed.[75] New imperatives were in the making in the name of infrastructure, and a resolute commitment to a capital accumulation-led economic growth strategy. Thus, Islamabad's wide streets were festooned for the 2004 annual donor consultation, the Pakistan Development Forum,[76] with green banners reminding donors of the new focus: 'Infrastructure for Growth', 'Take off needs Runways',

'Mega-projects bridge to the future', etc. Politics seemed to have lurched back, in Polyanian style, to a Rostovian state developmentalism that would have plenty of conservative and populist appeal, but might leave devolution, investing in pro-poor human services and Poverty Reduction's empowerment leg in distant third place.

As noted earlier, 2004 WDR aimed to redirect attention from the failure of neoliberal market integration to deliver poverty reduction, by the over-reaching assertion that social services, if delivered in the right way, could do the job. But this assertion rests on the dual precepts of NIE's local governance; on the one hand, 'inform, compete, enforce' (the legacy of which we will shortly examine in New Zealand) and on the other, the incentives and sanctions for adopting the new, marketized institutional norms in local politics, as reflected in the Accountability Triangle.

Now, it is the case that citizens do appreciate that Pakistan's devolution has increased the number of opportunities for citizens to meet with elected representatives – after all, there are 127,000 new councillors in their neighbourhoods. But there is a larger presumption that closer proximity, together with institutional arrangements for local-local, horizontal relations between citizens and elected leaders will create in this flattened political space partnership arrangements – a classic 'win-win' compact between weaker and more powerful sections of 'community'. Again, we will shortly observe how even in the apparently best of situations, this is a challenging task. The worry for the Study team was not just that the predominance of vertical political relations was causing local officials and citizens to neglect or discredit the opportunities provided by the new local institutions that depended on community-state partnership arrangements for justice, service delivery, dispute resolution etc. Neither was it simply recognition that they would anyway be frail in the face of entrenched, vertical, patrimonial power, especially when backed by the inflated, post-9/11 fiscal power of Uganda-style vertical programmes for just about every service likely to be valued by citizens. Rather, it became apparent that these new opportunities through which 'community' could exercise discretion over important resource allocation and accountability decisions were being occupied by the personalized power of the executive and local elites. In other words, it seemed likely that these quasi-territorial arrangements might just provide merely a plethora of new opportunities for the exercise of influence by the very 'power-authority' nexus that devolution had sought to prise apart.

A larger worry ran through this realization. It was increasingly evident that the fallout from the local separations of powers that had followed the abolition of the DC's office appeared to be making it *less* likely that anything could be done to avert this capture. Here was not just a worry that local elites and executives would conspire to re-direct or plunder resources intended for health clinics, primary schools and so on into private goods, nor that local corruption was diverting large shares of resources away from such purposes. Rather, there was a greater concern: regardless of their impact on *social service delivery*, the new, three-cornered and political market-promoting arrangements featured in the local government and police laws seemed to be undermining the state's primary, *social regulation* obligations. In some respects *devolved control had meant less control* with regard to the primary public assets and public entitlements. Although hotly contested, there was considerable evidence that basic public regulation was collapsing, that local electoral politics did not create incentives for leaders to invest in these obligations and as we will see, this further indicated that local notables, official or 'informal', might be turning into rogue laws unto themselves. This is an important conclusion; one we want to elaborate, exemplify and relate to the book's core arguments in the paragraphs to follow.

The Study found evidence that devolution had detrimentally affected the maintenance of property rights (public and private), crime control, and criminal justice. There was also rampant disregard for regulations on use of public spaces, water, roads and municipal codes and a decline in the rate of prosecution under local and special laws governing such matters as adulteration of food, petrol products, price controls, hygiene and public health, etc. All this was especially worrying because administrative disputes – for instance, complaints against water and power utilities, the police, revenue and land officers – were already the most common form of dispute, before devolution. And most cases, around 70 per cent, whether civil cases, criminal cases, or administrative grievances, involve land.[77] Thus if what the Study found was generalizable across the country, the collapse in social regulation effected domains of public rights which were already most pressing.

Alarming for local citizens, it seemed that police abuses and illegal detentions were also on the rise. The new Police Order sought to curb this abuse by empowering the *nazim* to inspect police stations, and by making it a criminal offence not to produce arrested persons before magistrates. In addition, it gave new powers to the local judges to inspect police diaries and issue habeus corpus writs. All this was welcomed by the public. But effective actions by the newly empowered *nazimeen*, public safety commissions, the judiciary and citizen police liaison committees to supervise police were almost non-existent. Again, before the separation of powers, the DC had been responsible, in essence, for all local regulation, including crucially checking police abuses. In place, District Public Safety Commissions (DPSC) – comprised of elected and publicly appointed members – were supposed to provide external accountability on the police.[78] But the Study found that 'effective and fully functional DPSCs that enjoy public confidence do not exist in any of the six districts studied'. Perhaps these would begin to function in time, but in Bahawalpur district, a representative of The Great Pakistan Lover's Association, a local NGO said: 'One of the DPSC is a landlord, another is a police tout, a police assistant, another's a lawyer. They're all compromised, none have credibility.'[79] It seemed that the space for client power was quickly being occupied, both by the police, freed now from the controls of the DC, and by well connected elites occupying the new accountability bodies.

Separations, power and social regulation

This outcome was especially significant in that the beneficial effects of the separation of powers lies at the core of 2004 WDR's promotion of wider liberal governance doctrine.[80] It is central to liberal and market principles that legislative, judicial and executive powers should be separate, just as in a similar way it is essential at local level that the powers of client, contractor and adjudicator must be separated for contracts to be made and to be enforceable. Separation, then, is core to a mode of Liberal governance that responds to individual needs within the confines of the law. Service delivery needs may typically be subject to great local variation and change over time, and so are believed best delivered by the private sector. To improve social service delivery devolution had to prise apart the powers held by the DC, in order to put in place the new Accountability Triangle that would be more responsive to local variations. Executive powers – like control over local administrators responsible to collect revenue and to deliver health, education and other services – were assigned to District Coordination Officers, the head of the local government bureaucracy, in turn responsible to the elected mayor/*nazim*. Political responsibilities once held by the DC – as government's representative at the local level – were handed to the local government mayor/*nazim*. And, to adjudicate when this system failed, judicial powers – such as powers to order arrest and detention and to prosecute offenders under the local and special laws mentioned above – were returned to the local judiciary. And all three, as the 2004 WDR had it, were now intended to operate as a unified system, responding to new accountability incentives created by the new laws. But the new three-cornered incentives where not strong enough to ensure these three, the judiciary, the executive/police, and elected leaders, would join up in the intended way. In particular, laws that were essential for local governments to function – and necessary to ensure that citizens were not abused by government officials and local elites – were not being applied. It was not just that elected leaders were not proving effective in directing administrations responsible for social services, as noted above, but moving beyond the realm where the citizen is treated as a client for marketized services, and into the domain of public entitlements like public safety, order and enjoyment of citizen rights to land, irrigated water, or fair labour practices, it seemed devolution's new disaggregated arrangements and their clasping together through these new quasi-territorial bodies were not proving effective.[81]

The larger difficulty, then, was not simply that devolved governance arrangements were not yet significantly impacting on how clients received social services (the primary MDG-donor concern); it was perhaps that framing the entire system of local governance in these terms was disabling the state's already weak ability to deliver on its social regulation obligations, that is, to meet its wider societal accountabilities around security (both of assets like land and irrigated, and person), empowerment (including the protection of basic rights) and opportunity (including such things as business registration, regulation of markets, etc.). What this underlines, we think, is the dual nature of social governance: while in a Liberal sense access to services is a market providable commodity, one that can be regulated by liberal democratic and contractual means, in a territorial sense these elements depend heavily on actually existing power relations affecting services, the law, and political representation. In such a context, assuming that Liberal governance can and will regulate services or provide security is very risky indeed. Such illusions place the poor at risk of both service capture and security perversion, while obscuring and removing from purview their immediate sources. It is these themes we want to elaborate in concluding.

Conclusions

Table 7.2 contrasts the modes of governance suited to individualized social service delivery with the features of a system that must be more firmly territorialized, for the purposes of social regulation.

The central theme of the PRS – 'listening to the voices of the poor' – is found in each of the PRSPs produced by province and federal governments in Pakistan. In this way, the poor citizen, that much referenced 'common man', is pushed forward to lend his legitimating voice to any of the blizzard of assignments given to newly created local government, public-private, and community partnership bodies created by devolution's laws for the purposes of efficient service delivery. The critical point is not that these new institutions are weakly embedded in local political economies. Nor is it that it will take time to embed them, and until then that they will continue to be frail and ineffectual in the face of historically powerful patron-client networks backed by the largesse of vertical programmes for service delivery. Neither is it that it will be difficult to embed the recommendations of the Study and bring about inter-governmental fiscal/grant arrangements that are sufficiently powerful to displace these highly territorialized politics and ensure that decentralization's technical norms win and create spaces of 'local-local' dialogue. In our view, these matters do deserve close attention, for refinements in system design are doubtless possible; but they are not our central concern here.[82]

Pakistan seems to have embarked in March 2000 on a textbook path to 1997 WDR's 'capable state', and, perhaps unconsciously, elaborated this in ways consistent with 2004 WDR's *Making Services Work for Poor People*. In this, it looks set to repeat two worrying trends. First, by radically

Table 7.2 Liberal and Territorial governance: service delivery and social regulation[83]

Marketized service delivery	Territorialized social regulation
Responds to individual needs for particular services	Responds to societal needs to security, wide and non-corrupt access to entitlements
Respects the unequal power of consumers	Seeks to provide on a population-wide basis, according to entitlements.
Requirements defined locally, by consumers and providers seeking market-efficient solutions	Requirements defined society wide: adequacy, access, accountability, removal of systemic abuse
Responds to law, and promotes voice and exit within the confines of law	Regulation by discretionary application of principles to socially interpreted situations
Priorities manifest in budget allocations made by local elected authorities	Choices manifest in legal entitlements, established in law and regulation and applied in practice
Ephemeral, preferences may change regularly, especially in terms of scale and reach of delivery	Durable, at least over medium-term
Amorphous, in that norms are set to the individual	Generalizable, but applied locally as widely recognized adaptations to local norms
Efficiency can be achieved by locally democratic institutions	Equity requires operational autonomy of institutions outside of local partisan control
Can be delivered by private sector	Only possible through substantive accountability regimes of public sector, backed by judicial certainty

pluralizing the assignment of accountability for almost every aspect of public service, it has further diffused accountability in an already chronically corrupted environment. As President Musharraf melodramatically remarked at a UN conference on anti-corruption in April 2004, the executive includes 10 per cent stringently honest and 10 per cent totally corrupt. He didn't elaborate that in between there is a substantial class of grey practices and people who permit their mouths to be sweetened, until they reach the stage where they cannot act without this stimulation.[84] It is not just that misdemeanours, rule breaking, corruption, and benign status quo forces get in the way of a tidy separation of powers. This is a society in which personal safety is constitutionally protected, but in which domestic violence takes place in approximately 80 per cent of the country's households, in which 42 per cent of women accept violence as part of their fate, in which less than 5 per cent take action against it, in which custodial gang rapes by the police are so frequent that 70 per cent of women seeking redress of grievance are subjected to sexual and physical violence in police stations.[85] It is a context where a host of laws and institutional arrangements protect citizen/societal rights but in which child sexual abuse, female child prostitution and trafficking, kidnapping of children, male child prostitution, bonded child labour and hazardous child labour are common-place. Historically, protection has not been available to most citizens, not because of inadequate legal provisions, but predominantly because implementation of the law by local state agencies (administrative and judicial) gives way, as retired Justice Samdani said, 'to the path of least resistance'.[86]

Second, it may prove that Pakistan's NIE- and WDR-charged doctrine of devolution and PRS will make it even less likely the nation will address issues of social justice that have persistently eluded politics since Partition. This is not just witnessed by the fact that barely a murmur was heard from domestic or international politics when the then incoming Prime Minister Jamali in 2003 announced that 'there will be no land reform during my watch'; ironically, about exactly the same time as Pakistan's multi-volume PPA was published and again reinforced the close correlation between poverty and that product of territorial power, landlessness.[87] Indeed, in a liberal governance system shaped up for the delivery of social services to individual clients expressing their changing and different needs through local political institutions, it becomes increasingly difficult to ensure that critical, less changeable, society-wide accountabilities are kept in view. For where the abuse of public entitlements – security of person, tenure and access to usufruct rights for land, water, natural resources, etc. – increasingly goes unchecked, unchallenged by prosecution, there are fewer cases/instances against which citizens, individually or organized through civil society or their elected representatives, can gain purchase on these rights. Over time, it is plausible that there will remain even fewer rallying points around which larger, more ambitious claims might be made by common citizens, or the poor, about historical injustices in the ways these rights have been assigned, institutionalized or frozen into law.

As we intimated in Uganda, it may be that Pakistan too soon discovers the mounting *transaction costs* associated with the system of decentralized planning, budget and expenditure control, audit and compliance for social service delivery that must be run down from global partnerships to local governments and up again in reportage mount in fact to a huge *opportunity cost*. In terms of social outcomes, a large part of this cost is the weakening of more territorial modes of governance that rely on regulatory, disciplinary and redistributive powers assigned to higher levels of authority, acting beyond the market-ized domains of 'local-local' dialogue.

8   New Zealand

Joining up governance after New Institutionalism

'I think we've learned to value ourselves and stand up for ourselves now and say, Hey, we're *key* to the infrastructure of this Community, and this Society. Listen to us. We've got a voice. I think the Community is quite good at it and have got better and better at it. And tapped into the key players a lot more. Local agencies have also been able to see how we have effected change down in Wellington. We spoke up, we didn't sit back, we were just empowered by it. We wrote letters, we had meetings, we got dialogue going. Just a few experiences like that is enough to say, Hey, we *do* know what works and what doesn't work. And the other thing is that I think that Government has got wise, that they now recognize the power of the Community, that we're a source to be tapped into, and that we have a wisdom that they don't have because we're in touch. And so that's the other key thing – that they *know* that the best services come from an empowered Community who are taking ownership of what's happening'.

'. . . It absolutely bores me. I get nothing out of the meetings. It all seems a great big talkfest with no actions. What do we achieve?' 'Waitakere has had some successes around the process stuff, but there's still a feeling that we're fiddling round the edges and not tackling the hard things'.

Community activists, Waitakere City[1]

Waitakere City, well resourced in first world New Zealand, may seem an odd destination for a book on Poverty Reduction and impoverished governance. Obviously, we have reasons. First, New Zealand is a place where unprecedented reign has been given to the governmental reforms seen in previous chapters. Arriving first were both the radical decentralizations and fragmentations of NIE, and the sharpened rational accountabilities of NPM. This radical decentralizing and marketizing phase (1987–1998) became in New Zealand a period that 'history will record as a fundamental revolution in the manner . . . governance functions'.[2] As we will see, NIE and NPM reforms pluralized and undermined territorial governance, while imposing regimes of vertical but market-ready accountability that would long constrain poverty action. Second, New Zealand has since 1999 embraced the Third Way, 'joined-up-inclusive' approach to social governance that has attempted to roll back more extreme effects of Liberal governance, partly by re-embedding marketized services in the quasi-territories of community and local partnerships. In this, Waitakere City's experience is regarded not just in New Zealand as exemplary.

New Zealand also has other, perhaps surprising affinities to poor countries' experience. It has long been a test tube of reform borrowed or travelled from core polities, and plonked down at the planet's opposite pole. New Zealand's lurchy, semi-peripheral ship of state might be classed as reform-prone, repeatedly turned on a historical sixpence, and pushed quickly through thoroughgoing change. As with Chile in the 1970s, neoliberal reforms had travelled to New Zealand in the 1980s in the potent form of uncut theory, taking a rationalist razor to what it called statist excess. Suddenly, reforms only imagined in other OECD countries could be implemented. When in 1993 John Williamson of Washington Consensus notoriety reviewed structural adjustment cases to construct an ideal type of 'technopol' activity – the driving of reforms by technically savvy politicians – New Zealand offered the strongest exemplar.[3] And the travel of technocratic reform didn't stop in New Zealand: rather, as in Chile, Uganda and Vietnam, consultants and policy academics capitalized on the amplification and plausibility that actual reforms lend to abstract doctrine.[4] Soon New Zealand became consultants' golden cachet of 'analogy to elsewhere', so that perhaps more than any country, New Zealand's governance reforms have been a stick for poor countries' backs.

This, particularly in the 1990s, when it became the poster case for NIE and NPM's institutional cheerleaders. Admirers from *The Economist* to visiting US academics lavished attention:

> The reformed [New Zealand] State sector is testament to the power of ideas and the inventiveness of its architects. It is a singular accomplishment in the development of modern public administration, and it will influence the future course of management both in New Zealand and other countries. It is worth briefly reviewing the roll call of some of its pioneering accomplishments. New Zealand has been the first country to fully adopt cost based accounting and budgeting; the first to successfully implement techniques of output budgeting; the first to give managers full discretion in using inputs; the first to introduce strong incentives for the efficient use of capital; the first to require advance specification of the outputs to be purchased; the first to establish a comprehensive accountability regime.[5]

It was the crystal-clear, costed and output-oriented *accountabilities* that attracted most regard, especially for the ways they raised executive managerial power (and, it was hoped, local responsiveness) within a highly marketized and decentralized system. Consistently rated among the world's least corrupt countries, in an age where Good Governance is root and branch of Development, had New Zealand hit on something of global salience?

As often, actual outcomes were not a clear triumph. Even as New Zealand's reforms were internationally trumpeted, in places like Waitakere, a backlash was already underway. In community networks, advocacy coalitions and service-coordination partnerships, all increasingly engaged by local government, NIE's mantra of 'inform-enforce-compete' was resisted and displaced by innovative modes of *collaboration*. By the time an 'inclusive' neoliberal Labour government was elected in 1999, a national groundswell had gathered against NIE's trust-destroying, contractualized accountability regime, and many different efforts were being made to re-embed governance in community, sustainability and deliberative local process.

But this, as we will see, was no return to a communitarian nirvana. While the new government regarded itself as moving beyond neoliberalism, it was committed to preserving core aspects of neoliberal reforms, especially in economic policy. Fairly quickly it also became clear neither the political nor the technical arms of the public service were willing or able to set aside core aspects of NIE-style accountability regimes. They did, however, let a thousand partnerships blossom. The result was a strange new hybrid, already familiar to this book's readers: partnership *and* competitive contracts, inclusion *and* sharp discipline, free markets *and* community. But crucially for us here, these kinds of accommodation served to ensconce NIE practices in ways that would disable territorially accountable collaboration around poverty, and create impossible transaction costs and slippery multilevel accountabilities.

Unexpectedly, perhaps, Waitakere in New Zealand thus offers a cameo of the whole book's argument, an occasion to revisit its central themes, some time after initial implementation. Here we see what happens when travelling reforms are radically implemented, and then partially rejected and replaced by the joined-up approaches we have seen in less mature reforms in Uganda and Pakistan. Now, after six years of trying to rectify excesses of stark NIE reform, we can ask what is there to show and tell in terms of process, service delivery, accountability, and pro-poor outcomes? And, as important, we can ask these questions in a context where there is clean and open democracy; a capable, well-remunerated and highly professional public service at central and local levels; a vociferous and well resourced civil society; and where the security concerns of a single party state do not set the basic parameters and underwrite authoritarian approaches to poverty reduction. The Waitakere–New Zealand focus, in other words, allows us to observe Poverty Reduction-like inclusive Liberalism at work freed from the obvious constraints of a Uganda or Pakistan situation, to see what happens when the Liberal governance project shifts from neoliberal and NIE reform into a Poverty Reduction-like 'inclusive' Liberalism, with its quasi-territorialized participation and accountabilities to 'community'.[6]

The first section of this chapter charts New Zealand's shift from a terri-torialized (though still Liberal) welfare regime to the hyper-Liberalism of the NIE revolution, and the rapid pace at which NIE flaws became evident. Here we will see how governance reforms created both a perversely narrow set of accountabilities to the central government's policies, and profound fragmentation in knowledge and services in local/peripheral settings. The second section presents a more grounded reforms perspective, focussing on Waitakere City, home to 180,000 people in western Auckland, New Zealand's metropolitan centre. We show how Waitakere weathered and responded to the turbulent 1980s, and how innovations in multi-agency collaboration grew in the face of NIE's disaggregating moves to become, as we will explain in the third section, by the late 1990s a key referent for the new Labour Government's more 'inclusive' Liberalism. But we will also show how constrained the inclusive reaction has been against the path dependencies created by NIE reforms: that is, how hard it is to put the Humpty Dumpty of a socially accountable governance back together, especially from a locality basis. We show how attempts to retrofit local collaboration and shared social 'outcome' accountabilities to NPM and NIE frameworks run into powerful obstacles and generate only marginal initiatives. As the transaction costs of such reassembly become apparent, 'collaboration fatigue' rises just as complexity increases and accountability becomes increasingly diffused, technical and localized. Learning from these experiences, the locality seeks sharper engagements with central govern-ment, and much greater alignment of resources, responsibilities and accountabilities.

POVERTY AND GOVERNANCE IN NEW ZEALAND: HISTORY, REVOLUTION AND REACTION

New Zealand was established by the 1840 Treaty of Waitangi between the British crown and a plurality of territorially and genealogically distinguished Maori tribes, each of which claimed authority or *manawhenua*, among other rights, over territory and resources. Rather than establish Pakistan-style or proto-Lugardian indirect rule process, the Treaty sought to incorporate Maori within the British state at large (to create citizens, rather than sub-jects), thereby establishing them as liberal property owners who owned, and therefore could sell their land. Swamped and compelled by migration, mil-itary setback, land sales and confiscations, Maori and their territorial rights were submerged within Westphalian-cum-national governance, until they and the Treaty re-emerged as potent governance shapers (and an unlikely, short lived ally of NIE reform) in the 1980s. As in similar, Anglo post-colonial situations, poverty in New Zealand remains powerfully ensconced among these indigenous people, and more recently, other Polynesian migrant groups, who find asset accumulation especially difficult.[7]

While initially established as a series of provinces of obviously white settler provenance (Auckland, Wellington, Nelson, Christchurch . . .), the country quickly became politically and administratively centralized, partly because of huge loans raised to pay for national infrastructure projects. An activist and land-hungry white population joined to an innovative parliament soon had New Zealand on the way as an international bellwether in 'social Liberal' policy (widened franchise, women getting the vote, old age pension). New Zealand was nationalist in economic development too, struggling with debt dependence, and ensconced trade deficits.

A Labour government elaborated New Zealand's Keynesian welfare state after the 1930s depression and retained bipartisan political support into the 1980s. Yet despite 'cradle to grave' welfare, New Zealand remains a 'liberal welfare-capitalist regime'.[8] Thus social protection transfers ('benefits') while incredibly generous by poor country standards are residual and sharply targeted, compared for example to Scandinavian 'social democratic' regimes. In Liberal regimes, welfare (and wellbeing) primarily revolves around being in work.[9] In New Zealand, this meant a 'wage-earners' Liberal regime, with a national awards-based 'social' wage supported by innovative social policy and education, workers compensation, public health provision, and more. After the Second World War, New Zealanders enjoyed high levels of home ownership, low inequality, and world-leading public health indicators, all hinged on high employment and privileged commodity market access. By the 1950s, through the late 1980s, social governance and services were a hydra-headed outreach of the central state, a command economy of services and direct transfers, with a modicum of regional governance around deconcentrated central departments. District Offices of Social Welfare, and regional elected Boards of Education were branch offices to enact and enforce central standards and programmes and convey central government monies down departmental 'silos'.

At the first oil shock in 1974, New Zealand was a strange epitome: semi-peripheral, but enjoying first world socio-economic conditions on the back of (preferential) primary agricultural commodity trade, Keynes-esque national demand management, and mildly territorial import-substitution industrialization. In this, its most obvious affinities were with Australia, and (somewhat less successful) Latin America. Its position – at the ends of the earth, but not quite at the edge of capitalism – has had a number of implications. The territorialized mass production–mass consumption of Fordism never really happened in New Zealand, except in uneven fits and starts. New Zealanders have felt less than secure in their OECD status, and disposed to desperate economic and social innovations, while prone to agricultural commodity swings and related financial instability. With a uni-cameral parliamentary system providing few political checks and balances on its policy executive, New Zealand from 1950 to 1984 overreached in the national development strategies of import substitution and debt-funded demand management. Successive post-Depression governments were determined to maintain demand, full employment, protection and, after the 1970s oil shocks, energy self-sufficiency. When the mid-1970s crisis came in the form of oil price shocks the government borrowed heavily to sustain a desperate rearguard action against adjustment through to 1984, piling up price controls, export subsidies, and ill begotten 'Think Big' investments in major energy and industrial projects.

When the neoliberal backlash came (under the 1984–1990 Labour government), New Zealand's adjustment was especially doctrinaire. Initial deregulations were an atavistic purge, the policy burn-off deliberately so rapid that critics and opponents were left debating last week's reforms.[10] In retrospect, reform sequencing was profoundly mismanaged: the currency was floated, just as industry protection was wound back. Attracted by soaring interest rates and low sovereign risk hot money poured in, exacerbating febrile exchange rate fallout in productive sectors. Industries given market shock therapy didn't adjust (as they did under Australia's more gradualist regime), but just turned up their toes. Unemployment soared, especially among groups primarily engaged in manufacture, notably Maori and Pacific peoples. Radical tax cuts given to upper decile income earners saw a growth in inequality over the late 1980s and early 1990s with no OECD precedent. Comprador financiers and foreign corporations skimmed enormous privatization profits, many then exiting the country.

In the familiar 'technopols' tradition, technocratic capture of high Treasury office was given political rein by a reformist government entering office in the teeth of fiscal crisis. Rapidly, as we will describe below, social sector management was taken from its traditional silos and shaken about in an NIE/NPM test-tube. As in Thatcher's Britain, the term Poverty was banned from official parlance.

The governmental revolution

Travelling rationalities and bureaucratic capture

As with economic deregulation, governmental reforms proceeded under explicit doctrinal guidance. In the heart of Treasury, institutional economist Peter Gorringe was a personal acquaintance of NIE gurus Ronald Coase and Douglass North. But he remained a theoretical magpie, whose denial of his own economic rationalism, eschewal of rational actor theory and fear of capture of the apparatus of government by a small cartel of policy advisors now seem absurd, in the light of his own cartel's radical, hyper-rationalist capture of New Zealand governance. Re-visiting the Gorringe papers,[11] or Treasury's author-unacknowledged but notorious, theoretically tortured *Government Management: Brief to the 1987 Incoming Government*,[12] it is difficult not to conclude Gorringe was a dangerous crank, given, by dint of historical accident, far too much rope.

Gorringe's policy ruminations were heady with reference and extensive verbatim block-quotation from Williamson, Hayek, North and Coase, ironically especially when they are insisting that change should be informed by localized knowledge. As he recalled in his own review of the *Methodology of Treasury's Policy Advice* (1991):

> Thinking about my own knowledge of economics, much of it comes from the battery of metaphors I can bring to bear on an economic problem, such as the following:
>
> - the concept of supply, demand and equilibrium
> - the concept of general equilibrium
> - the Hayekian view of the distribution of knowledge in society and the price system as a way of communicating this
> - the concept of the implications of rational utility maximizing agents
> - the Marxian view of the world, stressing 'the big picture', including distributional and power concerns, ideology, institutions, technology and class struggle
> - the idea of market failure
> - the transaction cost critique of market failure and the comparative institutional approach
> - the idea of rational expectations and the importance of expectations
> - North's (1990) ideas of the failure of rationality and the importance of people's incorrect subjective models of the world
> - the property rights paradigm
> - the public choice paradigm
> - paradigms which emphasize processes such as Austrian and evolutionary economics
> - agency theory.[13]

Gorringe wasn't kidding around: each of his Treasury papers involves multiply articulated NIE-like rationalisms applied to an extraordinary range of governmental problems, where his working maxim appears to be that many theories (and many accountabilities) will add up to a clearer, more accountable governance. Again, *Government Management* shows best a strained mix of faith and scepticism about market rationality, and the ability of central technocrats like himself to govern via decentralized, market attuned mechanisms.

From this book's Chapter 4 onwards, the decentralization and accountability aspects should now be familiar:

> ... like private individuals, the state will face difficulties dealing with scarcity and interdependencies. The reason of course is that the state is made up of individuals subject to the same limitations as private economic actors.

Whereas in a decentralized setting individuals accept risks and adapt to unexpected occurrences, with those who are caught out suffering the consequences of their mistakes ... With central planning, mistakes tend to be excessively costly and impact on everyone, with few alternatives being available when things go wrong.

The bounded rationality of central planners and the complexity of the world creates strains for the relative efficacy of centrally determined solutions ... the complexity of the problem solving state decision makers are expected to engage in may place even more severe demands on their bounded rationality than is the case for private planning. The only safeguard is conscious and purposive policy review. However given the fact that state decision makers may not bear all the costs of poor decisions and therefore may face weak incentives, this conclusion raises serious concerns.

Centralized decision making faces major information disabilities. The information relevant to a decision may be hard to obtain ... diverse, and may include unavailable information on consumer preferences, or alternative production technologies, or alternative ways of organizing activities. Alternatively information may be possessed by individuals who are difficult to locate, or be of a nature that is difficult to communicate from one agent to another ... The information costs underlying centralized decision making therefore militate against its successful execution. It is likely to be based on incomplete information with consequent adverse effects.[14]

Rolling out NIE in New Zealand was a project of decentring of the state in terms that long anticipated 2002 WDR rubrics: 'inform, enforce, compete', and the familiar three-cornered accountability terms of 2004 WDR: 'voice, compacts (here, contracts), and competition'. Narratives of crisis and grotesque waste in the public service were used to sharpen vertical and managerial accountabilities along NPM lines, and to instigate NIE style service delivery markets. In this decentring and de/reterritorializing process, some basic doctrine-driven 'splits' were crucial: policy-operations, funder-provider, and outcome-output. While these have become commonplace, it's important here we grasp their territorial effects, and their spillover into subsequent 'joined-up-inclusive' reforms.

In the outcome-output split, government ministers would set outcomes, and government and other agencies would deliver outputs that together would create the outcome. Outcome responsibility, presumably involving complex interactions of social, economic, local, territorial forces, was thus removed from social service delivery operations, and restricted to a narrow politicized control. Outputs, on the other hand, could be and were narrowly conceived, costed, and allocated within sharp regimes of downwards managerial accountability, which had the advantage that they could apparently be delivered by almost anyone from anywhere. In the policy-operations spilt, policy in for example Social Welfare became a separate function, controlled centrally by the Ministry of Social Policy, but its operational 'provision' was opened to and marvellously informed by the competing, creative ideas of consultants. Separate from policy, operations (workforce inclusion, income support) were implemented by other agencies, government and/or private. Funder-provider splits similarly separated functions of fund-holding and management from on the ground delivery, enabling a plurality of market competing providers to deliver particular services within any territorial or sectoral jurisdiction. What these provisions especially split was however regional (i.e. territorial) governance of key public sectors. The net effect was a radical decentralization: District Health Boards (DHBs) abolished, and individual hospitals set up as Crown Health Enterprises, competing within much wider regions to provide services to make a profit; Regional Education boards were abolished, and individual schools were bulk funded for all recurrent costs, and made to compete on open education markets. As we will see, local governments too would be reformed along similar lines.

Table 8.1 New Zealand's version of NIE reforms

	Market	*Civil society*	*Disciplinary governance*
2000 WDR *Attacking Poverty*	Opportunity	Empowerment	Security
2002 WDR *Making Institutions Work for Markets*	Compete	Inform	Enforce
2004 WDR *Making Services Work* for *Poor People*	Exit and entry choices	Voice and participation	Stronger compacts between vertically and horizontally disaggregated agencies, and clients
New Zealand experiment 1987–1999	Funder-provider split, contestable service delivery, competing schools, hospitals become individual/competing 'Crown Health Enterprises'	Policy-operations split, client choice and 'exit', consumer charters and rights, individual school boards	(Competitive) contractualism, output based funding and accountability, multiple points of client exit

Within this pluralization and specification, NPM modes of accountability were crucial to ultimate outcomes. An apparently innocent technical move from input to output accounting meant that producing outputs was controlled by managers, who according to doctrine were empowered to act and held accountable to policy outputs. Management meant output and cost, rather than, say, outcomes (e.g. in relation to poverty) experienced in particular places. In theory, competition would now 'incentivize' a knowledge market involving information about price, what competing local providers were doing, and local knowledge of client needs. The bureaucratic capture and 'heavy transaction costs' of 'inefficient centralized decision making' (and other territorialized, social-outcome focused regimens) would decline, and allocative efficiency improve. An even more radical deterritorialization of subject rights was involved. The direct nexus between the *citizen* and the state through elected regional health boards, for instance, would be replaced by privatized service deliverers in which market-oriented *clients'* rights would become a key node for accountability.[15]

The nation state thus became an internally deterritorialized domain, wherein markets would transparently allocate services to whichever providers offered best rates and minimum compliance. Deterritorialization was particularly marked in the health sector, where, as we will see, it generated a potent political backlash against the reforms. With the outcomes focus far removed from localities, overall population characteristics or assets determination health outcomes for people living in a particular jurisdiction became regionally and locally ungovernable. Considerations as to what other 'non-health' sector actions might need to be coordinated to improve wellbeing slipped from view altogether. Rather the Ministry of Health became entangled in a costly, extenuated and ultimately fruitless exercise to define health as a set of 'core services' and turn these into costed 'outputs' that could be micro-contracted and enforced by 'bean counters'. This drastically limited the ability of remaining territorial governance (especially at local levels) to coordinate and achieve any mutual accountability of sectoral or social policy or service providers. Local coordination, much less collaboration, was explicitly *not* funded. What local coordination did happen, as we will see, was voluntaristic, driven by people reacting against local fragmentation, trying to join things up using their own scarce resources. As then, so now. Similarly, within tight vertical NPM output accountabilities, well off the agenda was the prospect of promoting shared accountabilities – either horizontally across higher levels of government, or inter-governmentally, between central and local government – for wider outcomes.

The unravelling and reaction

Fairly quickly, a series of flaws became apparent, and in Liberal and Territorial governance terms alike. The regime actively rewarded contractors for simply 'sticking to their knitting'; delivering services to the letter of the contracted outputs regardless of the complexity of client needs; all received fragmented services from multiple agencies, none of which were accountable for the client's overall wellbeing. A client with drug and alcohol dependency, and perhaps mental health issues would become the subject of diverse agencies' attentions. One would be contracted to deliver a number of counseling sessions, another to oversee living arrangements, another to provide regular mental health assessments. Contractualism destroyed collegial trust and professional cooperation between agencies, meaning none of these people were talking to each other. Morale and job satisfaction imploded, as social workers, now employed by competing agencies, found themselves unable to respond to crosscutting family issues, and got bogged down in compliance reporting. Funding negotiations between government and private or NGO contractors turned into mean, nitpicking exercises, focused on cost cutting and compliance with the detail of service standards. As frontline staff resources were cut and cut again, managerial and compliance costs mounted. Junior doctors would find their numbers cut and responsibilities extended, only to see budget savings spent on management salaries and meeting the high transaction costs of contractualism.[16]

Where the deterritorializing aspects of delivery were most strongly felt, however, was where poverty was most strongly territorialized: among rural (and marginally urban) indigenous populations. Initially Maori groups, both rural and urban, had been approached by a Hayekian Minister of Health desperate to gain legitimacy for the reforms, by enrolling key civil society elements into service provision. Maori, with worse health outcomes than other groups, were offered the incentive of population-enrolled health care provision, with some local autonomy and devolved funding. Here was a political incentive for Maori: if funding for service provision might be controlled by tribal or urban corporate groups, wider self-determination possibilities might also be opened up. In Waitakere, a pan-tribal urban authority stitched together a large number of different contracts, in a bid to provide comprehensive services to its members, and provide a wider corporate capacity base. While multiple contracts meant horrific transaction/compliance costs, they enabled the group to get established. Elsewhere, however, tribal concerns over integrated service delivery were more complicated, as local state hospitals and other services were crucial to health, yet under fragmented governance arrangements could not be coordinated to meet tribal (territorial) population health needs. While, as below, wider political backlash over user fees and privatization fears in health became the first important turning point in rolling back ultra-liberal governance in New Zealand, it was the manifest weakness of the reforms in responding to territorialized Maori health needs that provided some of the most telling examples of coordinative and outcome accountability failure.[17]

At a general level, ill-conceived notions about minimum market size for specialized services and about how clients would travel to points where services could be most efficiently delivered meant that markets for secondary and tertiary health care and education simply failed to eventuate.[105] For example, the mis-scaled health purchasing authorities struggled to organize services across highly differentiated spaces: surgery patients from north of Auckland were by no means keen on travelling to a small town in the middle of the country for elective surgery, just because their operation had been allocated there on cost basis. Similarly deterritorialized schools could compete for the best or richer students from anywhere. But once an elite, or, very quickly, even a mid-range school had filled its roll, it would unilaterally reterritorialize its catchment zone and cherry-pick other elite students. The idea of students voting with their feet and motivating educational improvement became a farce, especially in the poorest of schools. The dominance of central hospitals and elite schools was reinforced while, at the same time, powerful national level NGOs hoovered up contracts across multiple jurisdictions, but with no specific local or wider territorial accountabilities.

Quickly, then, a range of fragmentation, inequity and coordination issues became apparent. Ultimately, when in several sectors local coordination issues approached crisis, frontline staff, strategic brokers, contract managers and case-managers were given the job of joining it all up again. But even here, the coordinating focus was rarely territorial, focused on for example the basic resourcing of clients, or the overall social profile of citizens within their territorial jurisdiction: rather the focus of coordination remained on the managerial and contractual risk to outputs. Worse, this happened alongside unprecedented rises in unemployment and unprecedented increases in inequality.[19]

There remain enormous practical and theoretical ironies around New Zealand's NIE reforms. Beyond the practical farce created by the combination of bean-counting, risk aversion, and knitting-focus, many spectacular failures were theoretical. While Gorringe made frequent reference to the scarcity and perverse, gamed use of knowledge in for example, the prisoner's dilemma, he failed to anticipate how Treasury's reforms would destroy shared knowledge and cooperation; how the policy-operations split would marginalize substantive local knowledge from the central policy-making departments; and how the reforms would generate less knowledge rather than more, by reducing relations between services and people to minimal, formally contracted outputs. Outputs were faithfully enumerated and reported, but in a way devoid of specific qualitative or substantive content. In short, within government and social services, Northesque assumptions that linked a decentralized competitive environment to better information that in turn would result in better services proved hopelessly naive; indeed markets and useful shared information proved inimical in profound ways. In a lopsided power situation where policy departments worked in centralized isolation and local operations units let contracts after rounds of competitive bidding, the information that mattered was minimal compliance requirements around covering narrow objectives. Substantive local information gathering (related, say, to social outcomes) was not just no-one's job; it was systemically disincentivized. Thus 'inform', 'enforce' and 'compete' either reinforced each other in perverse ways (fear based risk avoidance, competition destroying information), or generated a race to the bottom in terms of quality service delivery. Perversely, this often left no-one better off in terms of the cost of services.[20]

A consumer-led political backlash occurred first in the health sector: the government had to back off 'user pays' principles in hospitals, the restructuring of hospitals as commercial enterprises, and the shuffling of surgical patients around regional markets. But while consumers (and citizens) had plenty to say about this new market for social services, their 'voice' was less able to cause improvements in service delivery from either contracted service providers or government policy, funding or regulating agencies. Contractors were bindingly accountable to their funders, far less so to the clients of their narrowly defined services.[21] Similarly, from the clients' viewpoint, accountability for outputs on the government side was so highly diffused through multiple levels and agents responsible for policy, standard setting, contracting, contract monitoring, compliance audits and basic enforcement that substantive accountability for systemic failure was simply disaggregated away. In the end, it was raw political resistance, as measured at the ballot box and in opinion polls that enabled effective (citizen, not client) voice.

Meantime, rapid developments in the hard science of social epidemiology were being picked up in New Zealand. Social epidemiology's 'social determinants' and 'health inequalities' approaches were able to demonstrate above all else that health outcomes were determined not by access to services, but by underlying 'social determinants' including (chiefly) income, and other resources. In New Zealand, researchers linked income, housing, transport and other measures from the census into a 1–10 'deprivation index' score where household scores were combined by small geographical areas ('meshblocks') of some 80 persons, and which (as we will see in Waitakere) could be mapped against both local neighbourhoods and wider territorial jurisdictions such as Waitakere council. Here, health outcomes (mortality, morbidity, violence, exposure to risk factors) were shown to be powerfully related to deprivation in all age groups and for most diseases, across a social gradient from rich to poor. Figure 8.1 is one of dozens in the Ministry of Health's 2000 *Social Inequalities in Health* report.[22] When these meshblocks are colour-coded and mapped, the territorial ensconcement of poverty related health outcomes becomes apparent. While relative deprivation is only one factor, and must not be mistaken for (much more crucial) absolute deprivation, its implications for territorial population health outcomes are substantial. This not least because, as we will see in Waitakere, the ability of post-NIE social governance to respond to peripheralizing shifts in distribution of poorer deciles across territories is particularly low.

Assessments of both the process and outcomes of contractualism ultimately resulted in a series of shifts, including the joined-up-inclusive-partnerships approaches adopted in 1999 by the incoming Labour government. Long before this, however, Waitakere City's activists were already swimming quite consciously against the neoliberal/NIE tide.

REFORM AND REACTION ON THE GROUND: TOWARDS COMMUNITY WELLBEING IN WAITAKERE CITY

Waitakere City comprises Auckland's western quarter, a sprawling suburban hinterland, set against one of the largest areas of temperate rainforest close to a major city anywhere. World famous in New Zealand for its boy racer petrol-heads with bad haircuts, and its distinctive eco-Westie identity, it is famous in governance and urban development circles for entirely other reasons.

All cause mortality, males, by deprivation area of residence
and age group, 1996–1997

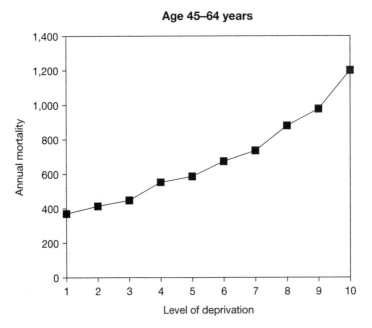

Figure 8.1 Deprivation and wellbeing, New Zealand

Source: Howden, Chapman and Tobias (eds), 2000, 25.

Figure 8.2 Waitakere/Auckland map

In governance terms, Waitakere punches above its weight, as New Zealand's own, consciously branded 'Eco-city', a bellwether for the 1992 Rio summit's *Agenda 21*. At the 2002 *Rio Plus 10* summit in Johannesburg, Waitakere Mayor Bob Harvey co-chaired the parallel Local Government forum, foregrounding the environmental and community partnership successes that have given Waitakere a place alongside other 'star' cities (Curitabo, Oslo). In all this, devolution of powerful resource management governance mandates and tools to local government has strengthened Waitakere's ambit. But so has political incumbency, in a fairly stable succession of left-leaning elected councils. This, in some ways against the odds: in New Zealand, land taxes (known as 'rates') have long formed the overwhelming source of revenue for local government. Local body politics reflect this residence-based franchise and is usually more responsive to conservative homeowners than other residents. However, Waitakere has tended to return Labour Members of Parliament in Wellington *and* a Labour-Green council.

Waitakere's rapid population growth has in recent decades included a relatively homogeneous mix of working class, indigenous and migrant people, with a trim of middle and professional classes occupying rainforest ranges and waterfront subdivisions. New Zealanders may feel precarious about their OECD status, but this remains the first world. Auckland (pop. 1.25 million) is consistently ranked among the ten most liveable cities on the planet. But Auckland is also the New Zealand city where social disparities are most apparent. Waitakere, however, is not yet characterized by the fault line social disparities of regional Auckland's other sub-cities, where areas of poverty sit alongside pockets of real estate affluence and security paranoia. On the Deprivation Index, as seen in Figure 8.3, Waitakere until recently had relatively small populations in the richest (1) and most deprived (9–10) deciles, its population mesh blocks spread fairly evenly across the middle to lower deciles, peaking around deprivation decile 7 in the 1996 Census, but slipping to peak at decile 8 by 2000. Maori make up 15 per cent of the population; 32 per cent are non-European.

The area's explosive growth during the 1960s and 1970s as a greenfields, lower middle/working-class suburbia contributed to its social service makeup. The absence of services motivated a range of innovators and service pioneers in the community, who found themselves drawn rapidly into advocacy and organizational development roles. In their words:

> The growth of community organizations in the West was quite organic ... In terms of working together it was key people that were the critical factor and made things happen, rather than $$$. We didn't really need much funding to get things happening initially.[23]

Community service provision and its politics remains powerfully grounded in local networks, specialized local knowledge and local efficacy:

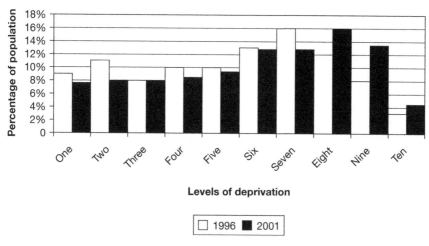

Levels of deprivation

☐ 1996 ■ 2001

Figure 8.3 Deprivation in Waitakere, 1996–2001
Source: NZ Census data 1996, 2001.

We strongly believed in the need to have our own identity, to be in control of our own destiny. People together making decisions for themselves – not being 'done to'. We just did it. We had a strong belief in it and in our ability to do it. It was for the benefit of our children and we grew strong as a community from it.

Local resistance to NIE fragmentation

Following its formation in a late 1980s round of council amalgamations, the Waitakere City Council adopted a range of big picture, headline approaches to social and environmental activism, most notably to the development of Waitakere as an Eco-City. Council amalgamation was both deterritorializing and reterritorializing: an economic deterritorialization, as many local governments privatized their commercial and infrastructure assets, or cunningly placed them into arms-length commercial trusts. Internally, along NIE and competitive contract lines, councils had their responsibility for provision and maintenance of basic infrastructure, key service functions, regulation and enforcement (and sometimes core administrative functions) outsourced or opened to competitive contracting. On the other hand, after council amalgamation, both the territories and resource management functions assigned to local governments increased their policy and administrative scope. Waitakere's Eco-City strategy reflected this, as did council's embrace of a wider 'wellbeing' mandate.

In sum, the reforms left Waitakere City Council in the unenviable state of being both 'policy rich' and relatively 'asset and income poor'. On the one hand, it has a thin base of own-sourced revenue (largely residential land tax income),[24] relatively little industrial development, and a growing population dominated by lower middle income earners and centre-left voters. On the other hand, the council hired some exceptionally able strategic policy leaders, analysts and community focused activists, who would invest the Eco-City and its wellbeing ambit with drive and ambition, sometimes contributing to a fiscal overreach that prompted council to sell-off assets to balance budgets. One consequence of the disjuncture between high policy ambition and tight budgetary reality was the motivation to incrementally accrue programmatic arms and legs (injury, safety, children, the environment) by entrepreneurially and opportunistically engaging in the 'pilot project economy'. Here, over time, Waitakere City earned a reputation as 'a safe pair of hands' for demonstration projects, under the motto 'if it can be done, it can be done in Waitakere'.

The Eco-City vision hybridized local resource management regulatory arrangements together with global, Rio 1992 *Earth Summit Agenda 21* commitments to sustainable, triple bottom line reporting. It grew to include pioneering commitments to the Treaty of Waitangi with Maori, and effectively picked up the 1974 LGA's unfunded and undefined mandate to address 'citizen wellbeing'. This statutory mandate ultimately supported a range of innovative activities in Waitakere that extended from environmental activism to community strategy coordination, facilitation and advocacy, and even to 'strategic real estate acquisition'. When the LGA was redrawn in 2002, Waitakere's experience was influential in entrenching and expanding this wellbeing mandate; although, as we will see, its functional content remained a 'to-be-filled' blank, mired in conflicting jurisdictional overlaps, and badly wanting assignment of matching revenue.

Waitakere also became a nationally conspicuous site for other wider, locally coordinative and collaborative initiatives. In the 1980s, before NIE, a remarkable interagency youth justice programme was initiated by justice agencies in Waitakere, which was eventually scaled up into the 1989 Children Young People and their Families (CYPF) Act. Originally drafted just prior to the outcomes-output split, this Act conceived overall objectives in this area in broad outcome terms,[25] and enabled local agencies to coordinate activities to these ends. Initially exemplary for its collaborative approach, which drew on the broad, local social care concerns of many of its justice and social worker contributors, it soon became exemplary of how NIE could undermine the best of local intentions. The NIE outputs emphasis overlaid (and only rewarded) a narrow specification of process outputs (for example, mere convening and punctual attendance of meetings funded for a limited number and time), and failing to recognize the range of interagency commitments and underpinning coordinative activities (managing relationships, coordinating load-sharing, going the extra mile) necessary to sustain the wider outcomes envisaged by the CYPF Act.

By 2000, both the youth justice system and its primary agency were in profound crisis, with a major critical review sheeting responsibility for its demise emphatically back to ineffective coordination, over-specification (but poor conception) of outputs, and chronic under-funding.[26] This shows how even the kinds of disciplinary inclusion social governance functions (which link surveillance and sanction to security and property rights protection) in which 'inclusive' Liberalism can plausibly claim success are undermined by hegemonic NIE and NPM models.

Such was the recurring pattern through the 1990s, with ramifications continuing into the 2000s. Once the basic NIE revolution was in place, reactions to fallout occurred in a range of social sectors in Waitakere and beyond. While coordination, reaching enforceable agreements, and resolving jurisdictional overlaps and disputes ultimately became a fundable output class of its own, accountability for achieving this was again seen as best achieved at frontline worker or 'community' level. In other words, the problems of diffused accountability, jurisdictional overlaps, discordant policy, planning, budget and expenditure management, etc., by vertically disaggregated agencies were pushed down the system to land in the laps of front-line service providers as 'local problems' requiring 'local solutions'. Again, Waitakere innovators went into the breech, piloting an interagency, 'wraparound' case management modality eventually called *Strengthening Families*. Once again, Waitakere's enthusiastic social sector had found a local solution that tapped the vast reservoir of community activism in Waitakere. These goodwill-heavy solutions were quickly upscaled by a central government grateful for having local coordination to fill in the gaps. But under-resourcing and narrow shared accountabilities undermined this shared case management, with the programme plagued by unallocated cases, and lack of real compulsion for core agencies (police, justice, social development) to accountably contribute.[27] By 2004, poor outcomes here drove management of family and children's crises into a media frenzy, and a long series of highly critical crisis reviews.

More generally, a familiar 'developing country' pattern was established: local coordination was managed as a narrow output (read: a special purpose 'project result area'), or co-production contract (say, between NGO and government and project agency) coordinated by a management unit operating in parallel to the on-going business of separately managed departments. Wider coordination (or, in caseworker's own terms, *collaboration*) had to be done on the cheap, or over and above peoples' and organizations' standard, contracted outputs. The contractual regime failed to deal with coordination or collaboration imperatives requiring a restructuring of the fragmented arrangements in the mainstream government system for policy setting, planning, budgeting, setting performance standards, etc. This meant that high level strategic decisions about service provision were never accountable to on-going delivery experiences, as perceived either by service providers themselves or their clients. On the ground, where substantive joining up was needed, vulnerable 'clients' simply fell through the cracks.

A localized, operations-focused coordination around outputs was all that NIE realities permitted. Strategic co-ordination and effective trust remained narrow, with interagency goodwill vulnerable when contracts were up for competitive renewal. Nonetheless, the competitive urge was widely recognized as destructive, and at a number of levels organizations and frontline staff resisted and, over time, managed to build non-competitive information swapping, networking, and informal collaborative arrangements. Here, Waitakere City Council was an active partnerships innovator. Perhaps the best success came in developing a strategic alliance initiative in community safety, called *Safe Waitakere*. Here, *Safe Waitakere* has come to combine injury prevention, crime prevention, road safety and alcohol initiatives in a joined-up programme explicitly based on partnership principles, strategic planning and shared, devolved funding. As government community safety policy (and most recently child safety and domestic violence) have moved towards devolving funding to locally collaborating groups, the Waitakere model, difficult and organizationally complex, has again been exported, and in this upscaling Waitakere has again attracted substantial initial pilot project funds.

The Waitakere way and the 'muffin economy'

Another kind of local-led, reterritorialized coordination arose in the late 1990s, as Council instigated forums that gave local activists a deliberative platform for lobbying and information sharing in support of their 'wellbeing' mandate. Internationally, 'wellbeing' has provided the high tent pole under which the complex territoriality and accountability issues of locally 're-joining up' social governance have played out. Wellbeing is certainly the sort of polysemous term that can cast an inclusive, normative ring around the whole spatial domain of the city, offering a call to high consensus, while glossing structural fragmentation. In Waitakere, it has made it possible to get the collaboration ball rolling, as the council interpreted unfunded wellbeing mandate in enabling terms. 'Wellbeing', as the 1996 *Towards Wellbeing in Waitakere* report noted, 'is a far reaching concept . . . The possible factors that affect individual, community and city wellbeing are divergent, wide ranging yet inextricably linked: housing, income, employment, mental health, crime, safety, leisure, recreation, the environment and family relationships are but a few'.[28]

But how to effectively corral, coordinate and fund these areas from a territorial perspective? As everywhere, each of these mandates and their money remained assigned, that is, siloed, in central government agencies and their local branches. Whereas functional responsibilities for funding, planning, policy, regulation, the actual delivery of services, then monitoring standards, enforcing contracts, etc. were spread across multiple levels of government and contracted commercial and NGO service providers. More, the central factors that determined social outcomes (returns to labour, income security and inequality, basic wellbeing-related resourcing of families, and housing affordability) were emphatically not within the strategic control of the local government or the ambit of their community partnerships. Nor could progress towards them be achieved only by delivering on outputs. What, then, could the wellbeing mandate expect to achieve in terms of sufficient marshalling of budget resources and services planning to high wellbeing ends?

In true Waitakere fashion, activists led these collaborative initiatives and drew on the extensive goodwill and expertise council and community networks had gained through this process. Initially, they also drew on awareness of the need to lobby politicians from a locality (i.e. territorial) basis about local outcomes. Results so far have been mixed, and by and large marginal against wider forces shaping social outcomes. Overall, quasi-territorialized approaches (like local partnerships) and mandates (like the council's wellbeing one), where possibilities for concerted local action are raised but not funded or statutorily enforced, have both raised and eroded hopes. But this is to jump ahead. What we need to understand now are the kinds of process (and transaction cost) investments made by council to promote coordinated efforts around wellbeing, over the period 1996–2004.

From 1996, the iterative consultation, dialogue and strategic processes around the *Wellbeing Report* became emblematic of what was later branded the 'Waitakere Way', a three-way (community, council, central government) partnership. Sustaining the Waitakere Way has meant endless engaging, networking, collaborating, promoting dialogue, and exercising both political and technical sensibility. In the words of one:

> There's no way you could do this sort of stuff on your own. It's about understanding how politics works, understanding how to actually get things done from within a bureaucratic organization. It's understanding your own community in the sense of the dynamics of that community – who can get things to happen, community wise. That's knowing all about leadership and knowing how to exercise leadership; how to tap into leadership and develop it – all those sorts of things. Process, process, process – absolutely critical.

No accounting was ever made of the 'people-process' costs incurred in networking, getting a strategic and tactical sense of local and national developments in other sectors, getting to know significant local others, and not least by providing good muffins, orange juice and coffee at meetings. But these processes enabled some participants to recover from NIE's fragmentation and to reaffirm their sense of agency:

> There's been a big investment of everyone's time upfront, over the years. One of the gains is that once you've been through the rounds

a few times sitting round a table, in Wellbeing Network or inter-sector things, you really know who people are and where they're coming from. Sometimes it gets to the point where it's just a matter of one phone call, you can get straight to the person you know can make a decision. People who've worked in other places notice that as a difference out here. I think community organizations feel they can do it too, and often government agencies, though it's always a matter of new people finding their feet. [29]

Just as clearly, there were limitations, and doubters; many wondered whether all these meetings 'really achieved anything', or simply 'wasted a lot of people's time'.[30] But the doubters rarely outnumbered the pushers, at least in public: overwhelmingly, local activists want to collaborate to high ends. Over time, the limits to the kinds of progress and accountability to be expected of this kind of community wellbeing process have become clearer. Ironically, the intensity of the process (and the limited resources on the table) seems to crowd out wider focus on (and accountability to) regional or even city-wide poverty issues. And because participation in the wellbeing process is voluntary, getting solid accountability tied to substantive budget around such territorial issues is severely constrained, whether in or out of meetings. In sector and interagency forums, both political and technical accountability issues have had to be raised in non-threatening ways and, *pace* 2004 WDR, by separating political from technical accountabilities. Putting elected leaders, council staff and bureaucrats in the same room was soon recognized as a recipe for political grandstanding and crowding out the practical: here, political accountability was too pointed and, besides, it was not leveraged to anything politicians could actually control. Directing accountability questions to senior bureaucrats merely ensured their withdrawal from forums for similar reasons. Here, collaboration niceties sometimes blunted advocacy and lobbying. In general, the accountability generated was largely voluntaristic, based in peer esteem, reputation, a desire to demonstrate capability; and hedged about with a need to let everyone save face in public forums.

The limits of voluntarism were also the limits of bottom-up strategy: accountability to outcomes was unlikely without links between the local collaborative activities with substantive central government planning, budgeting and operations. While professionals enjoyed and benefited from the networking, there were neither substantive institutional nor fiscal incentives to engage with for example restructuring service delivery and other social governance arrangements to address poverty. Participation in local forums was sometimes made explicit in individual employment and in NGO/private sector service delivery contracts. But neither departmental budgets nor special purpose allocations were directed towards collaborative ventures. This meant innovation was largely restricted to the edges: to rhetorics of wellbeing strategy, or to micro-level partnerships around subcontractual service delivery, pilot projects or information sharing. This 'inclusive', bottom-up participatory process could not be monitored against social outcome indicators.

Nevertheless, important lessons were learnt, and new modes of coordination experimented with. Further iterations of the *Wellbeing Strategy* moved away from general, bottom-up coordination to focus on specific areas of project sponsored actions. In 2002–2003, a committee of central and local government and community leaders prepared seven headline, outcome oriented 'Calls to Action', e.g. 'Families give their children a great start', 'Every child has access to a quality early childhood education', or 'Violence against children and women is reduced'. Voluntary groups of government and community agencies met regularly to instigate a range of small to medium projects[31] to pilot national initiatives and build networks. Popular and in some circles nationally recognized, and able to 'umbrella' and legitimate funding for a range of pilot initiatives and new sectoral gatherings,[32] this 'Collaboration Strategy' iteration of the community wellbeing process has been more activity focused than previous ones, and drawn more directly on bottom-up sectoral and interagency activism. But it has also suffered from the same difficulties as previous iterations: intensive on process and transaction costs, its resourcing and accountabilities (to both collaborate and deliver) voluntary and largely (outside of particular projects) unenforceable, and with little or no claim on core/higher budgets, its ambit has been largely at the innovative and coordinative margins of social services in Waitakere.

In wellbeing collaboration, by running well ahead of the central government and its resourcing, local agencies and activists have been able to experiment. Some have been able to capture money for pilots. Core strategic leaders members became aware of the need to progress to what they called the 'hard stuff' of aligning budgets and plans in strategic areas, in a context where operational budgets contained 'not one red cent' for this type of coordination activity, and where other key outcomes and outputs dominated priorities. By 2004, it became clearer that what was needed was both more substantive, core-budget linked local coordination, and more local discretion, incentivized and enabled for example via an LDF-style devolved fund. But in the absence of better shared accountabilities, and in consideration of the impact of such on other budgets, a devolved fund too has been resisted by potential (government agency) funders.[33] Meanwhile, 'collaboration-fatigued' community activists, who have born many of the transaction costs of collaboration themselves, were by 2004 beginning to suspect the consensual, collaborative approach to community wellbeing, and wanted repoliticization of the community wellbeing forum into a 'community wellbeing caucus', with a primary policy/advocacy focus.

This fatigue has been especially apparent, as departments of all ilks have been pressed to be more consultative and collaborative around local strategy. At one point in Waitakere in 2002, five different agencies were involved in consultation and services mapping around child and youth services and strategy. Here, the usual 'community' people were being consulted yet again, and, not just privately, they made some pretty straight complaints. Often, very little of the material from previous consultations appeared to have been read. Here again, contracted consultants and fresh agency staff were working from a blank page were wanting to fit it all into their particular project or territorial boundaries, within their own short timeframes. Here were more strategies and plans, not linked to any guarantees of substantive money. So whose interests, activists asked out loud, was all this process in?

THE THIRD WAY, 'INCLUSIVE' LIBERAL TURN: PARTNERSHIPS, JOINED-UP INCLUSION, MANAGING FOR OUTCOMES

As Waitakere shows, community-reactionary aspects of New Zealand's turn to an 'inclusive' Liberal orientation were well underway by the mid-1990s. Politically, the backlash had to wait until the country's fifth Labour government was elected in 1999, on a Third Way markets-*and*-communities platform. A re-embedding of a kind had begun: in policy documents market-led growth was recast as a means to higher ends, re-framed around the national economic interest, sustainability, and social development. While core neoliberal legislation was retained (the Reserve Bank Act which makes inflation targeting the aim of monetary and fiscal policy, the Public Sector Finance Act which frames outcome and output accountabilities), this government saw reforms that blurred NIE separations with the 'soft' institutions of partnership, inclusion, and joined-up governance.

The incoming government made much of the inequalities and fragmented services it inherited, and of its intention to undo resulting social fragmentation, especially through partnerships with communities. Their basic policy statement, *Key Government Goals to Guide Public Sector Policy and Performance*, required all departments to re-configure programmes around cross cutting goals.[34] Most breathtaking (and shortest lived) was Goal Five, which sought to attack poverty through (better coordinated) services. Or, in its own terms, to gain better outcomes for poor (particularly Maori and Pacific peoples) communities, 'through education, better health, housing and employment, and better coordination of policy across sectors, so that we may reduce the gaps that currently divide our society and offer a good future for all'. This goal, pushing social change on the string of joined-up services, was however soon dropped, as it became obvious the gaps weren't going to close anytime soon, coordination or none.[35] Nonetheless, projects of putting the Humpty Dumpty of socially accountable governance back together were begun. While some of the policy-operations splits were reversed (merging, for example, the Ministry of Social Policy, and Work and Income New Zealand to create the Ministry of Social Development), more ambitious possible realignments (e.g. combining health and social development) fell away. But the broad orientation to inclusive governance continued to be fleshed out in strategic policy, and this prompted some experimentation in partnerships, local strategy and regional coordination.[36] This 'soft' institutionalism was joined to harder NIE types.

Waitakere's leadership was underscored as the government's *Social Development Strategy* was launched in the city, at an event attended by the Prime Minister and the Chief Executive Officer of the newly merged Ministry of Social Development. This strategy was a familiar Third Way cobbling together of 'market inclusiveness' (training for work), social investment (child welfare, etc.), and a mild workfare of disciplinary obligation: 'Those who fail to take up suitable jobs that are offered to them will not receive taxpayer support'.[37] Here, partnerships 'with the voluntary sector, with local government, and with business' would provide 'backing' so that communities could 'find local solutions to local issues'.[38] As a subsequent high level policy statement noted:

> The partnership approach that government has taken means open relationships based on trust and understanding ... This commitment to partnership also means that government agencies will need to be better co-ordinated in their dealings with others ... the government expects that others will recognize the partnership approach as our normal way of doing business.[39]

Underpinning these orientations were some core governance reforms, which need mention here. These included *Review of the Centre* report, which addressed overall state sector integration; a shift in the accountability regimen from outputs to outcomes; a reterritorialization of the health sector; and a review of the LGA.

Local government's collaborative mandate was underlined in the 2002 LGA. With Waitakere's dextrous deployment of the 1974 wellbeing mandate explicitly in mind, the new Act encouraged councils to do whatever might plausibly impact favourably on their citizens wellbeing, introducing a statutory obligation to consider all existing and proposed activities against the rubric of the 'four wellbeings': social, cultural, environmental and economic. But this generously defined, apparently territorial mandate was entirely unfunded and, given the diverse political, technical and economic capacity of local governments, will be very differently applied across the country. The Act also demanded a potentially coordinative (and highly consultative) planning process from all councils, dubbed the Long-Term Council Community Plan (LTCCP), which may result in some further council, community and central government co-production of plans and services. But again, expenditure responsibilities for the 'community outcomes' mandate are unfunded and there is little statutory requirement on

Table 8.2 Reform and beyond, 1987–2004

2002 WDR	Compete	Inform	Enforce
New Zealand experiment 1987–1999	Funder-provider split, contestable service delivery, competing schools, hospitals become individual and competing 'Crown Health Enterprises'	Policy-operations split, client choice and 'exit', consumer charters and rights, individual school boards	(Competitive) contractualism, output based funding and accountability, multiple points of client exit
Outcomes 1990–1996	Vertical and horizontal disaggregation; uneven service provision; inequalities between elite and other schools grow; low morale, lack of trust and fractious relations between NGOs, and NGOs/government	Destruction of incentives to produce shared information: 'sticking to your knitting' client needs 'falling through the cracks'. Policy-operations split means central policy uninformed by grounded experience/issues in service delivery	'Triumph of the beancounters', narrow output classes, little purchase on outcomes. Avoidance of 'difficult' social/big picture outcomes in contracting process. Bureaucratic risk averting. Depoliticized services
Reaction, and re-embedding 1999–2004	Spontaneous local partnerships, regional coordination bodies for integrated service delivery. Funder-provider split retained but watered down; health 'a service not a business': devolving of power/funding/function to DHBs, removal of school competition; rise and official embrace of partnerships and collaboration, joined-up and wraparound service delivery	Policy-operations split removed; partnerships and collaboration welcomed as building trust; appointment of regional policy advisors and information brokers, consultation obligations ramped up	Rejection of output basis, move to outcomes based funding, 'managing for outcomes' 'Partnership agreements', 'Wellbeing processes'; reducing numbers of departments and contracts

silo-sector budget holders to align plans and strategy to LTCCP outcome indications. Both wellbeing and community outcome mandates, then, are for all practical purposes, only quasi-territorial in their governance effects.

More substantive was the establishment of DHBs with devolved funding regimes, and population health mandates. Here, on the surface, is real territorial accountability, via elected boards, and with social determinants of health issues surely clamouring in these territorially, population based wellbeing regimes. Yet again, siloed vertical accountabilities and funding regimens have dominated these devolved entities, with most funding, old and new, devoted to expensive, centralized clinical facilities, and governance focused on getting comparable service standards across New Zealand's 23 Health Boards. Substantive address to regional or citywide health issues through strategic partnerships will require intergovernmental alignment and assignment of responsibilities for plans, budgets and accountabilities, and so, remains a hope for somewhere down the track. Even if the DHB devolution has for the time being at least stabilized messy and slippery scale assignments in health,[40] it hasn't removed the basic hegemonic parameters of health funding, which are overwhelmingly about vertical underfunding and cost restriction. Nor has it resolved any subsidiarity issues about the levels of scale to address a whole range of vital wellbeing and poverty issues (housing, transport, ghettoizing of low income families, etc.), because in all potential partner departments, subsidiarity and regional capacity issues remain fraught, and pilot-restricted. Again the net effect of an apparently territorial devolution is a merely quasi-territorial effective wellbeing ambit.

But what has made joined-up traction and shared accountabilities hardest to achieve has in fact had little to do with either mildly enhanced wellbeing or population health mandates at council or DHB level. Rather, it is the legacy of Peter Gorringe's Liberal market-oriented governance reforms, wherein it is clear that while fragmentation was quickly ensconced under NIE/NPM, undoing it requires radical re-engineering the whole system.[41] To their credit, the fifth Labour government did attempt reform. Recognizing the ongoing effects of fragmentation, but wanting to maintain core NIE/NPM accountability frames, the government's *Review of the Centre* process recommended 'alignment improving innovations', including:

- establishing networks of related agencies to better integrate policy, delivery and capability building;
- an *accountability and reporting* system that puts more emphasis on *outcomes* and high level priorities, as well as output specification;
- changes to vote structures to facilitate a *greater outcome focus and better prioritization across agencies*; and
- gradual *structural consolidation* targeting: small agencies; Crown entities required to give effect to Government policy; policy/operations

splits, and sectors where there are Ministerial concerns about agency performance or alignment.[42]

The intentions were high: but the on-ground reality highlights the difficulties of rebuilding after the policy and operational rash of NIE. The NIE/NPM accountability system has proved more reform-resistant than imagined, and re-building (especially shared/joined-up) social governance *outcomes* has proven tough. At this writing, the *Review of the Centre* process was moribund.[43] The difficulties of reconfiguring governance around social outcomes after the Poverty Reduction-style transformations are crucial for this book's argument, and why this is so needs some closer technical detailing before closing the chapter.

Managing for outcomes

The shift from an 'output' to an 'outcomes' based accountability regime – or rather, retrieving outcomes from being something only ministers were concerned with, to something to be 'managed for' on a day-to-day basis – was supposed to support reduced fragmentation, and create an ethos where interagency action could prosper. Certainly, such high rhetorical goals are often framed within New Zealand government departments' annual, outcome-oriented 'statements of intent'.[44] However, practically, outcomes are still something for which, under the retained NPM framework, managers in all contracting parties are accountable for only within departmental silos and/or their particular contracted output classes. In current practice, high reaching outcomes statements are seen to be operationally realized through cascades of lower level outcomes and outputs. For example, a key outcome for the Ministry of Social development is that 'Families . . . are strong and richly interconnected with their communities. They are able to support their members' wellbeing, identity, participation in society and interdependence'.[45] Intermediate outcomes are envisaged as supporting this:

- Families . . . are strong with the resources and capabilities to play full, vibrant and functional roles for their members and communities.
- Families . . . have a strong voice in decision making, and are valued as key institutions in both current society and as trustees for future generations.

Below that several outputs make up the bulk of practical support (for example, income support, support for getting into work). Thus the whole edifice of outcomes depend on settings determined by the wider economy (getting into work) and/or the actual settings and generosity welfare transfer system (income support); that is, on political economic and political decisions beyond the Ministry's actual management (and accountability) ambit.

Here, the same Ministry shows in its 2004 Social Report, that the percentage of population living in family units with incomes below 60 per cent of median incomes had increased slightly to 22 per cent; while the percentage of households with no income from paid employment had decreased only very slightly at 16 per cent.[46] Changing such real world outcomes would rely on the unstated political commitment to income support, which as we will see has been less than generous.[47] Practically now, where outcomes are managed for, common practice has involved specifying outcomes so narrow as to represent 'hairy outputs' (e.g. 'all students will leave school with a plan'). Hence, in this frame, it is possible to merely (technically) 'manage for outcomes', and not (politically, or territorially) campaign and structurally plan for them.

But the key issue is more basic than this. For all the high rhetoric and goals, the outcomes orientation is not in fact primarily about producing social change outcomes, but about attempting (and largely failing) to widen narrow public service accountabilities. Were outcomes in fact primarily concerned with social change, they would need from inception to operationalize some kind of theory or model of how social change actually happens, rather than trying to retrofit existing output accountabilities to high-level rhetorical goals. The result is that under NPM/NIE, and even under 'joined-up-inclusive' versions of public service, accountability remains primarily oriented not to producing social change outcomes, but to ensuring narrow process accountability, managing risk,[48] and ultimately cost containment. Thus New Zealand's experience should raise some issues for poorer countries hoping to get to better poverty outcomes through NIE/NPM reforms linked to decentralization.

Conclusions

> The new system brought accountability at the expense of responsibility, contestability was more ideal than reality, strategic capacity was underdeveloped, managers had a narrow view of their work, transactions costs were high, and most contracts lacked means of enforcement. The model worked, but to what end?
>
> Alan Schick, *Reflections on the New Zealand Model*, 2001[49]

Its early days yet, but New Zealand's experience is a sharp reminder that the fragmentation legacy of NIE reforms runs deep. Even in Waitakere, compared with our other case stories, a rich, highly literate citizenry, a formidably strong civil society and an activist, high capability local government operating in a comparatively stable political environment can leverage little social outcome purchase around health, housing, poverty, and wellbeing. In fact, if anything, strengths in community and local government activism and coordination might be seen as working against substantive reform: if so much can be mobilized voluntarily, the long-term structural reforms needed to enable better regionally shared accountabilities can perhaps be deferred for a while longer. Meanwhile political representatives, the executive and citizens all face a complex and overwhelming mess of regional and local assignments and disaggregations: 'herding the cats' of contractualized service delivery NGOs via endless voluntarily attended meetings, into a blizzard of process and strategy making, punctuated by more incoherent consultation and pilot projects looming on the horizon. In the absence of higher level coordinative planning, and budgeting and enforcement arrangements to support them, 'strategic brokers' and front-line workers struggle to improve coordination around services, let alone create territorial strategy. But while conveying an air of voice and accountability, this process makes no substantive claims on budgets, nor results in the reassignment of any substantive governance to local levels. Overall, more accountabilities therefore add up to less (effective) accountability either at the level of outputs – for clients – or wider social change outcomes – for citizens.

In this difficult context, Waitakere must struggle to push beyond 'soft' institutions and parnerships to harder shared accountability for outcomes, if it wants more than marginal gains on collaboration. So far, however, for reasons beyond its control, it has not been able to achieve more than soft, quasi-territorial dimensions of local joining up: general consensus and innovation without substantive claim on budget or sustained, accountable reference to outcomes. While many thoroughly territorial outcomes will remain beyond most of its ambit, it must seek smart territorial leverage in a number of areas (such as housing, transport, health outcomes), if it is not to be a mere receiver of wider market force outcomes. This means making it clear in Wellington that '*laissez faire* partnership' and 'inclusion-delusion' around hairy outputs is not enough, and that in specific areas, substantive commitments to shared accountability, funding and clearer alignments and assignments are needed.[50]

However, in terms of real wellbeing outcomes in Waitakere, the news is better. Chief here has been the sustained high levels of growth in the commodity price dependent national economy and steep falls in unemployment.[51] Here, in an ideal environment, and with an OECD-low minimum wage, 'inclusive' Liberalism's strong suit of disciplinary inclusion and mandatory market integration has ultimately found its feet, with fierce case-management of the unemployed resulting in rapid workforce re-insertions, most recently under a programme called the 'jobs jolt'. In Waitakere, this successful inclusion, abetted by some of the 'Call to Action' skills and employment collaborations, has seen unemployment fall by double figure annual reductions to 20 year lows. In mid-2005, New Zealand had the lowest unemployment in the OECD, at 3.6 per cent.

But New Zealand still languishes well below most OECD economic averages and remains profoundly structurally semi-peripheral. Foreign ownership of New Zealand assets is at globally extreme levels and current account deficits are entrenched. New Zealanders have the lowest rate of savings in the OECD, and unprecedented household debt. With the housing boom, housing affordability for the poor remains as low as ever, and real estate markets are more active than ever in shifting the poor into urban peripheries. It is now clear that in Waitakere, as Figure 8.3 showed, all through the period of wellbeing strategy, local coordinators and service providers have been fighting a rising tide of representation of relatively poor people among their clients and citizens, and the implications of this shift are only now beginning to come home. The key factors are enormously potent: underlying economic inequality manifests on the ground through the 'spatial sorting' effects of Auckland's real estate markets, which have continued to push poorer families into less affordable housing on the city's periphery. Such factors are well beyond the ambit of current joined-up inclusive wellbeing approaches. Addressing such mobile, but locally concentrating poverty would require either considerable central level redistribution or a major governmental shift, beyond mere inclusive coordination. But currently, even service delivery funding is not directly referenced to local social deprivation levels.[52] Small wonder local activist NGOs are now trying to build coalitions to repoliticise poverty related issues.

Even where government has moved politically (and substantively) towards redistribution, a disciplinary inclusion ambit continues to dominate. Many hoped for substantive redress to inequality from the government's 2004 budget, hailed as the largest pro-poor income re-distribution in thirty years. Certainly for working poor, the new family tax credits, the 'in work payment' and the wider 'making work pay' ambit will leave them considerably better off, while underlining the historical 'inclusion means the workforce' premise of New Zealand's Liberal, wage earners' welfare regime. But the budget expressly did not extend similar support to benefit dependant families, who will now face even greater relative deprivation. More of these people and their children, it is certain, will be moving to Waitakere in the next decade. Yet even these limited, 'inclusive' Liberal redistributions have occurred in the context of a wider political backlash against Maori and the alleged special treatment they get.

Meantime, governmentally, the partnerships approach is in Waitakere at least being increasingly recognized for what it is: an invitation to work together for shared ends, but at the same time, a shifting of coordinative and outcome responsibilities down, but usually without the requisite alignment of funding, function and mandates that would enable local success. At worst, such 'partnership' has been a classic Third Way triumph of form over content and of Tiny Symbolic Gesture over vague good intentions. But even at best, it remains a feebly reterritorialized mode of governance, undercut by risk aversion, high transaction costs, and a lack of substantive realignments and reassignments from the centre. In the absence of such, what remains central and formative is the dry bones framework of NIE market-voluntaristic, individual agent orientation. What's built on it has the oozy accountability structures of a multilayer pudding. Also oozing outwards, however, is the realization that the existing reforms are highly resistant to pragmatic realignment. As Alan Schick notes:

> In contrast to other countries in which reform meant adding peripheral elements to the pre-existing managerial system, in New Zealand, the reforms are the system. There is no other managerial system. This means that dismantling the reforms would require the government to divest itself of the ways in which it prepares and administers the budget, runs departments, links ministers and managers, and decides what to do. In other countries, an unsuccessful reform can be stripped away, leaving the core system intact. This would be more challenging in New Zealand . . . It remains to be seen, however, what will be left if critical elements are stripped away.[53]

9 Conclusions

Accountability and Development beyond neoliberalism?

Six years after General Musharraf's coup and buoyed by economic growth, the Pakistani Prime Minister, Shaukat Aziz, said to donors gathered in Islamabad for the country's April 2005 Pakistan Development Forum[1]: 'We have been emboldened by our successes. Liberalization, privatization, decentralization and deregulation, all this we've done. But the top-down, one size fits all, straight-jacket approach of PRSP has ended.' He then outlined a 'customized approach' he said was tuned to Pakistan's special needs. 'Investments in infrastructure, rapid economic growth and human capital.' These renewed 'second-generation' poverty reduction reforms 'demand' he said 'strong institutions, for markets and financial security, and this demands institutional integrity'. Or, at least, a strong, perhaps authoritarian state counterpart.

Aziz, for all his nationalist, IFI-resisting stance, is certainly not alone in taking these kinds of positions. Elsewhere too, second-generation Poverty Reduction looks set to reaffirm Development's commitment to institutions and services, while deepening its engagement with infrastructure, and pursuing nationalist capital accumulation strategies in which, given the flush state of global capital markets, the IFIs are just one potential investment source.[2] For some strong, security-significant countries, such approaches will enable IFI-regarded success in Poverty Reduction, 'Trollope-ploy' accommodation within wider security and rising sphere of influence contests, and the kinds of nationalist territorial Development successful governments love to claim credit for. But wider outcomes for the poor in the periphery will be fearsomely uneven; as we have seen in Asia and Africa, some areas will integrate rapidly and successfully, carving out niches or swathes of production and outsourced services (and the social classes producing them) from other economies. Other places, barring much greater international commitment, will remain nightmares of exclusion, failed growth, failed infrastructure, failed security.

Under current conditions, getting more substantive shared accountability around poverty outcomes is going to be an enormous struggle. Markets may provide cheap capital and generate growth, but they are poor guarantors of equitable outcomes, and can actually make it harder to achieve outcome accountabilities for the poor in their places. They can also make it harder for territories to distribute crumbs from rich tables, including their own. Currently, territorially redistributable resources are fought over and divided between tax relief for corporations and the rich, service and public good subsidies for rich country middle classes (including farmers), and billions of poor people. Lacking political power in core countries, the really peripheral poor should not perhaps expect too much from this contest. The corporate drivers of mobile capital will devote enormous energy to avoiding social or territorial traps both at home and abroad, to reducing costs (labour, again tax), and avoiding liability for externalities (environmental account-abilities, social disruption).[3] Rich governments will face increasing pressure to meet their citizens' rising aspirations for health, education, and aged care, all politically powerful rivals for aid budgets. Flat taxers are increas-ingly at their door, both as competing territories and as internal lobbyists. Beyond this, any country that breeches the narrow fiscal responsibility codes of finance markets risks precipitous punishment; unless it is fortu-itously aligned with security imperatives. At the same time, rich countries will face rising pressures to shore up their boundaries against the move-ment of poor people desperate to crash the party. In fact, Poverty Reduction may well be largely correct in its prognosis: what the poor will get will be overwhelmingly determined by their ability to grasp opportunities in internationally integrated markets. That, for many, will be nowhere near enough.

In this chapter, after summing the basic parameters of Development's current, shifting situation, we in the first section sketch two different scenarios for poor and peripheral countries. We conclude in the second section with a consideration of alternatives, each exploring parameters for accountability that go beyond the limits of current neoliberal institution-alism and inclusive neoliberal Development. In particular, we argue for a reconsideration of the balance between Liberal and territorialized approaches and accountabilities, and argue that here primary accountability lies not with poor countries themselves, but with wider agents of the inter-national Liberal and security order.

Looking back, looking forward: our argument in this book

Despite numerous embedding and reterritorializing moves, Liberal ambits have continued to dominate Development's lead agendas. But now as ever, Liberalism has been a slippery, contingent project. Since the late 1980s, a raw economic neoliberal agenda has been augmented by neoliberal insti-tutionalist elaborations and, through the 1990s, has been wrapped around in the consensual politics of 'inclusive' neoliberalism's Poverty Reduction. All this elaboration was driven by real Polanyian concerns, over security, failed states, and failed institutions, impacting capital security, economic growth and the poor themselves. Yet what happened was not Polanyian in the sense of a bottom up, broad movement of enlightened reactionaries, concerned at the social disruptions of volatile market integration. Rather, this was in large a reaction led by major Development institutions (and their political backers), concerned about how failure undermined the legitimacy of their institutions and the wider Liberal order.

Hence, while this phase sought deeper 'ownership', accommodation and contextualization with poor countries' governments, the reaction was characterized by the elaboration of a disciplinary, juridical, and 'enabling' service delivery orientation, focused on improving institutions. Development agencies were abetted in this elaboration by two groups: (1) recipient, especially authoritarian, often conflicted governments themselves in lead PRSP countries, and (2) the undiminished gamut of private entrepreneurs, NGOs and plausibly 'civil society' groups eager to 'partner' in these 'inclusive' arrangements and gain a new prominence around service delivery. The travel and embedding of Development's new orthodoxies, then, was not simply one-way traffic. Rather, recipient governments and donors became involved in complex relationships, wherein governments could at once be genuinely reformist (often in the wake of alarming crises), and equally concerned to find ways to be seen to be complying with Good Governance frames to ensure an uninterrupted flow of global largesse (while widely known 'other' practices persisted).

For these kinds of recipient government, the crisis or HIPC-driven windfall redistribution of largesse through Social Funds was a tolerable affair, capable of moving considerable funds from central to local levels. They offered ways to reach over the heads of mid-level governors, greedy elites and patrons, and urban middle-class political cliques, and distribute largesse and directly patronize rural and urban poor in their peripheral places. But the more elaborate, depoliticized framings of PRSP and decentralized governance – from PRA to LDF to whole-of-budget framings and intergovernmental re-assignments of expenditure – were especially attractive. They offered both Liberal governance respectability and legitimacy, and the opportunity to institutionalize patronage transfers down directly into poor places, engaging local governance structures and, sometimes, providing enough discipline to ensure that peripheral cadres made the transfer system work.

In all, the results of neoliberal institutionalism have been mixed. The last twenty years have shown extraordinary growth in strong, though somewhat illiberally governed states (China, Vietnam): and, in many other contexts to date, meagre returns on investments in Liberal institutionalist governance approaches. But as noted, we are arguably too early in the latest round to judge much more than the early effects of implementation and commitment. In this book, we have seen variety in ownership of Liberal governance reforms. In Vietnam, Liberal local governance modes disappear like water off a strongly territorialized local state's back. But we have also seen in Uganda a government charging ahead of the pack, eager to experiment with innovative, internationally funded governance in order to restabilize its political situation through Liberal patronizing of the local. We saw in Pakistan Liberal governance techniques and practices deployed in the most illiberal of places, used to deliberately break up local patrimonies by a military government concerned to ensconce its own legitimacy. In all cases, at this early stage service delivery outcomes were generally at least no worse; but what we have referred to as social regulatory outcomes, however, seem so far, much less favourable. Similarly in New Zealand's very Liberal contexts, we saw quasi-territorialized players struggle to strategize or deliver around real social outcomes, especially around poverty, while more powerful market forces called the tune.

This book has argued that the now dominant combination of neoliberal institutionalist governance (NIE, PEM, normative decentralization) with 'inclusive' neoliberal politics and participatory democracy (PRA, PPA, PRSP) is sorely inadequate. The overwhelming focus on institutions, especially its NIE version of 'inform, enforce, compete' provides little more than a narrow formalist disciplinary regimen; one that narrows local politics around service delivery mandates, while realpolitik interests continue to operate with impunity; one that coordinates only weakly and voluntaristically, around SWAps, sectoral and local 'strategies' or contractual partnerships; and that disaggregates governance in ways that sends outcome accountability to the four winds. Despite general donor consensus, a high level of competition, parallel and overlapping reforms, and fragmentation prevails in practice. With few exceptions, actual leverage and accountability remains weak in areas involving the basic political economy of poor governance and public sector financing: revenue bases for paying public servants, corruption, democratic accountability through representative politics, social regulation and impunity of the powerful. Alongside this, the formally elaborated neoliberal institutionalist frame tends to under-engage substantive aspects of the territorial economy (productivity, competitiveness, labour relations, offsetting particular geographical and sectoral conditions). Rather, when laid down across existing political and economic realities, the frailty of these kinds of institutional solution at best offer an 'inclusion delusion', a sense of something multifaceted, involving plural partners, including civil society, responding to the voices of the poor. But accountable, in the end, to no-one, unless it is the individual donors insistent on moving the money in their own current budgetary timeframes.

As Part I of the book suggested, there is a wider familiarity in this story. Thus has capitalism always spread, as powerful external trading interests have reshaped and aligned with powerful internal political and economic interests to de- and reterritorialize places, in order to open them up and secure them for trade. But it would, as we have argued throughout, be wrong to describe the outcome of current reterritorializing rounds as simply a neoliberalizing or marketizing of peripheral space. Rather, obviously, what are emerging are (increasingly complex) hybrids, wherein both market and local territorial powers – and approaches to governance and poverty reduction – co-exist. As we have shown, the ways they co-exist, especially currently in their multiple forms (social funds, Local Development Funds, community scaled transfers to inventive mixes of NGOs, local governments and private entrepreneurs) and in continuously shifting alliances, often has perverse implications for accountabilities around poverty outcomes. These modes of governance, deployed into local spaces, do indeed open them up, but the hybrid frames they establish – community, partnership and the like – are quasi, weak, ephemeral, and easily re-colonized not by citizens, but by NGOs, patrimonial powers, rogues and masqueraders. Here, when programmes telescope, shuffle assignments of responsibility, leave vast gaps or jurisdictional overlaps, fail to confront or mobilize raw political economic power, who, ultimately, is responsible?

It must be acknowledged that in many semi-peripheral places, these quasi-territories and the multiple governance modes deployed in them have been enough to quell local conflicts, and create a hybrid security order that keeps a lid on things. Thus, it might be thought, have the enlightened reactionaries of Development succeeding in inoculating against a real Polanyian social movement. They, together with various more obviously disciplinary, policing operations, have done just enough by way of embedding, co-opting, legitimation, and ideological snowing to keep both the stroppiest and weakest peripheral states and regions in line. In this, the War on Terror has certainly raised the stakes (and, dramatically, several countries' aid budgets), making top down Polanyian re-embedding more significant than ever. But so has an even graver spectre: the possibility of the re-emergence of sphere-of-influence conflicts, across Asia, Africa and the Pacific. Rising imperial rivalry ultimately undid the Liberal order of the nineteenth century. Now, the agents of multilateral Liberal governance may well find themselves drawn to one side (the neo-conservative one) of a potentially catastrophic sphere of influence (and basic resource) contest with more nationalist territorial interests represented by China and to a lesser extent Russia. In this context, as in previous rivalry phases from Imperialism to the Cold War, the territorial allegiance of powerful authoritarian, especially oil rich states becomes a potent element in their Trollope ploy. In all this, the current shallow and disciplinary Polanyian double movement may be displaced by much more powerful kinds of double movement reactions.

WHERE TO NOW: DEVELOPMENT BEYOND NEOLIBERALISM?

If such a larger Polanyian double movement is actually underway, we might expect a range of security and geopolitical issues to re-emerge in

Development. IFI-led Poverty Reduction's institutional governance ambits won't disappear, but they will be linked to more powerful tools of patronage: politically conditional funds and infrastructure loans. Here, though no doubt still arm in arm with neoliberal institutionalist programmes, a more security-led Development will emerge 'beyond neoliberalism'. Here, IFIs and like-minded harmonizers will find their roles under serious review. The IFIs, which have had their Washington Consensus convincingly shattered by US Treasury fiscal profligacy, may now find that their major shareholders are increasingly assertive, if not unilateralist in their approach to what, already, are barely multilateral institutions. If so, the IFIs will struggle to maintain even the appearance of technical, depoliticized neutrality that neoliberal institutionalism helped to sustain, even, some say, as it has built Liberal empire.[4] For the poor, it is clear that in such changed arrangements some countries will benefit a great deal. More particularly, we might postulate two divergent futures for poorer countries: the strong/strategic states, for whom accommodation with Liberal modes will bring powerful rewards; and the truly peripheral state, which can expect certain kinds of transfers, and, if it is lucky, plenty of policing.

Scenario 1: Stronger strategic states making territorial (or market) tradeoffs

For both strong and weak poor states, what is coming from Development will likely appear as a variant of Poverty Reduction, with more or less disciplinary requirements depending on the sponsor's perception of the country: this is the essence of the performance aid allocation systems noted in Chapter 4: the Millennium Challenge Account,[5] the PRSP. But, from an IFI/G7 perspective, underlying security will rely on the strength of accommodations between IFIs' Liberal governance modes and national regimes. This accommodation, its incentives ramped up through heavy infrastructure lending, and pushed down to the local via Good Governance oriented decentralized funds, will be reached with peripheral states using the best security clients as leading cases.

But here things may get very interesting indeed: there are signs that some of the more powerful states are finding their bargaining ability vis-á-vis IFIs is greatly enhanced. This is not because, like DFID and other donors, they realize that 'security and development are linked (and that) insecurity, lawlessness, crime and conflict are among the biggest obstacles to achievement of the MDGs'.[6] Rather, in the face of successful capital accumulation strategies, soaring stock markets and readily available capital, usual IFI concessionary loan schemes have a great deal less policy leverage. In Pakistan, the immediate post-coup fiscal and legitimacy crisis in 1999–2001 saw IFIs leverage accommodations around an unprecedented range of PRSP-governance reforms and lending. Six years later, free of the disciplinary framework of the IMF's Poverty Reduction Growth Facility, and aware of their strategic role in the War on Terror, the government is putting aside what it sees as IFI policy straight-jackets. By 2005, World Bank staffers lamented that policy conditionality had been all but evacuated from the latest rounds of negotiation for their second PRSC for Pakistan. Thus stronger poor country governments will try to take advantage of this reduced leverage, and re-colonize and redefine Liberal governance arrangements. In emerging new accommodations, countries are able to push for top-end budget support – ironically, quite consistent with calls by the High Level Donor Forums for harmonization with governments' own systems – while resisting demands that they deliver on the usual governance and poverty reduction conditionalities attached to concessional financing.

In a highly competitive lending market across the developing world poor countries are once again also seeking major loans for mega-infrastructure investments, which now perform a number of important tasks for them. They aid their capital accumulation strategies by feeding directly into GDP growth multipliers. But this is not just economic Keynesianism resurgent. As in the first Development Decades, conspicuous infrastructure development consolidates nationalist parties' image as national developers. It also relieves the typical bottlenecks emerging in urban peripheries where the newly urbanized poor cluster, and maintains wider legitimacy of investing in basic services (water, transport, electricity). Infrastructure investment is also a notorious vehicle for politicians bestowing patronage and skimming commissions. In this, China's burgeoning development assistance to many poor countries has been especially significant. From electricity and roads throughout Africa to subsidized factories in Cambodia, Chinese soft loans and other concessions fulfil explicit political functions, including creating deliberate rivalry with G7 backed IFIs. Smart strong peripheral states are now in a position to play off donors against each other, reaping the benefits of both Poverty Reduction and another approach to development (and politics) which, as many point out, has certainly been no less successful in reducing poverty. IFI's response to this will be interesting to watch, especially under current leadership.

Scenario 2: Peripheralization

Strong, strategic and poor, then, may not be a bad place to be, especially if you are adept at Trollope ploys. Which indeed governments from Addis Ababa and Nukualofa to Khartoum and Phnom Penh are learning to be. But there is another, darker side to current 'inclusive' neoliberal reterritorializations: the nasty territorial consequences of pushing poverty responsibilities down into weaker, more peripheral contexts, especially via quasi-territorialized domains of joined-up service delivery, decentralized and disaggregated accountabilities.

It is strange testimony to the duplicity of Liberal development that the same story of Liberal empowerment and *inclusion*, and their techniques can also be told in terms of security-driven walls and *exclusion*. Liberalism's duplicity here is essential: liberty, in Liberal terms, can only exist within the law, and expanding markets require the defence of existing and exclusive property rights. Clearly, this is a recipe for the strong prevailing, whether their power is state or market-oriented. But it is also a recipe for the weak failing, and becoming criminalized in so doing. And just as clearly, *inclusion* in domains of law-bound liberty requires at least the spectre of its other: *exclusion* in domains of lawless abjection. Here, political and economic 'otherness' is currently increasing fears and causing the ramping up of security. As the gaps between rich and poor countries grow, the sense of poverty as being a (risky, threatening) other is compounded. At one level, much was made of the economic impact of the Bali bombings (reducing Indonesia's GDP by 0.5 per cent) and World Bank estimates that some 10 million people were pushed into poverty as a result of the 9/11 attacks. A more far-reaching concern is raised by research showing that lower levels of GDP, that is, more poverty, is clearly associated with violent conflict.[7] Poverty also ramps up the incentives for poor people to try to get to rich countries, along with the resentment and despair when they can't.

On simple rich country poor country comparisons, the poor are now more 'other' than ever. In terms of relative incomes and inequalities, the qualifiers and defenders are unequivocal: the poor won't catch up anytime soon.[8] The perversity of Liberal doctrines is most apparent in the hyper-Liberal Bhalla's invitation to 'imagine there are no countries'.[9] As it should be clear, this is a signature Liberal imagination: the stark reality is that there *are* national and all kinds of other boundaries, and that poverty is profoundly ensconced in territorial aspects, micro and macro. Ultimately, it is these domains that Liberal governance via Development is most concerned to 'include', depoliticize, and assuage guilt, fear, and its own frail legitimacy in relation to.

Yet as we have seen, treating the poor as universally equal gets Liberal developmentalism only half the distance: it has, at the end of the day, to be prepared to defend the inequalities, and support the Liberal market and governance mechanisms that in all likelihood will mean the inequalities remain. Thus the longer history of Liberal developmentalism reveals a much more elaborate making of walls that *both include and exclude*. Inclusion was achieved by creating sentimental, minor territories and which territorialized in minor, sentimental ways. Throughout history, variants on 'the local community' made it possible to elaborate the boundaries between inside and outside in ways that were legible, could be represented ('the poor's voice') and could be disciplined. At the same time, from Lugard's dotted frontiers on maps, to the real walls of economic-order exclusion (the West Bank, closed Immigration counters), boundaries were created beyond which other, less than Liberal rules and standards were applied.

Now again, 'inclusive' neoliberalism masks exclusion, and this will happen more strongly if, as seems likely, it begins to take a more conservative security turn. This involves not the creation of governance techniques of *equality*, but a governmentalizing in ways that underscore *difference* (conceived in Liberal/market terms), and which detail out mechanisms for further peripheralizing should the threat of instability appear too marked. This disciplinary approach requires both the credible threat of exclusion and creation of liminal thresholds – phased steps, and performance benchmarks – that have the effect of *steepening the slope* between exclusion and inclusion. To work in practice, the differences and the steps up from abject exclusion to acceptable inclusion need to be made clear, and the consequences of default equally elaborated in sanctions. All this extenuates the spaces on the slope where empowerment and security mesh. The edges of the zone of inclusion are extended, incentivized by cheap loans, lubricated by promises of technical assistance. Yet the morality is fiercely Liberal: universal, and ahistorical. These stepwise inclusion rubrics, from the selective scoring of PRSP to the performance based allocation systems and fiduciary risk assessments for ODA, are applied globally, notwithstanding the fact that most peripheral countries are there because their colonial and then Development experience profoundly malterritorialized them, and left them prey to future failings.

Here we conclude that a wider territoriality of Development lurks in the shadows. In the longer run of Development, the poor have repeatedly been subject to projects whose unspoken ambit is to keep them in their place, where they are subject both to imposed security order and to the kinds of localized enablements (especially services) hoping to persuade them that a responsible subject can rise from poverty *in situ*. Here, a long history of uneven, illiberal policy around migration and movement again rears its head. Colonial rights of passage were usually one way, and post-colonial rights in a global system, while substantive, have always been demanding and highly selective. The 1970s and early 1980s were marked by feeble efforts to keep the poor where they were, through IRD and last-first Basic Needs approaches. In a bigger picture, this stabilizing of the poor was typically achieved through governance devices (from Lugardian Indirect rule, to Pakistan's 'tribal areas', to South Africa's bantustans, to variants of decentralization) that both legitimized socially unjust governing regimes and created local containers for culture and power. Within these power containers, the poor can be addressed in the depoliticized terms of need and vulnerability, or of participation and partnership. In these kinds of zones special policing occurs, rights are denied, and labour is made to function within quarry economies or globally deterritorialized ghettos of Export Processing Zones in which unequal commodity exchange is the norm. Soft walls, in other words, are often just as useful as hard ones. Surely, the soft walls of Poverty Reduction's governance mechanisms are more elaborated, but so too are the rubrics for disciplinary inclusion more sharply inclined.

Its unprecedented transaction, compliance and opportunity costs are justified on the basis that these devices will offer countries a passport to the globally integrated world of FDI attraction. Here is the ultimate in 'stay at home' poverty governance, where problems of 'poverty as exclusion' will ultimately be resolved by capital coming to you. Many poor, in this context, would be foolish to wait.

VEXED ACCOUNTABILITIES, AND WHAT TO DO ABOUT THEM

Inclusion/exclusion duplicity not withstanding, Poverty Reduction and Good Governance together raise the prospect of some heightened, shared accountabilities around what happens to poor people in their places. This, in several different ways: better service delivery by government and others; more actors, more choice and thus more power for consumers; closer monitoring of poverty status via the MDGs; donor harmonization and coordination; more locally responsive policy and operations. Some, we see as plausible, and necessary: others, tricks of Liberal perspective, which need disabusing and even debunking if anything progressive is to come of them.

As we made clear in opening this chapter, getting serious accountability to the poor in these complex contexts is extraordinarily fraught. Nonetheless, there are now both critical experiences and some small prospect of new technical and political ways forward. Here, we want to chart both. In particular, we want to move the debate about Poverty Reduction away from its current 'liberal markets, allocative efficiency and client-oriented service delivery' modes, towards some more adequately territorialized modes, within which both technical and political accountabilities might be more effectively pursued. Focussing on questions of social and territorial outcomes, social regulation and accountability we think provides a necessary balance to unrealistic Liberal, market-oriented hopes around allocative efficiency, service delivery and consumer voice. It also, we think, refocuses attention to accountability questions out of local community and frontline contexts, and moves it up the political and institutional scale. In doing so, it will shine some accountability light on not just local governance issues, but on Development's own institutions and their geo-political sponsors.

In particular, we hope these emerging, wider accountability foci will continue to question three of 'inclusive' neoliberalism's central 'inclusion delusions', one for each leg of Poverty Reduction:

1   Opportunity: market inclusion with good institutions and social services is enough to address poverty, at least in the long-term;
2   Empowerment: governments can pull themselves up by their own bootstraps and by better internal discipline (or with help from civil society), and

3 Security: more services delivered by whoever, wherever will do much of the job of poverty reduction, and do it more efficiently.

In response we argue:

1 Opportunity. It is vital we understand some of the limits of neoliberal political-economic understanding, and of the empowerment and inclusion rhetorics of positive Liberalism. And, therefore, that we understand opportunities and constraints in real historical, placed, and political terms. It is also crucial that we continue to point out as loud and clear as possible where Liberal agencies and powerful governments are being two faced about openness, for example, in trade, and migration, while systematically neglecting more heterodox policy solutions to the problem of economic growth.

2 Empowerment. Governance reforms have often involved fragmenting local governance arrangements, and then mounting efforts to tie accountability together through quasi-territorial bodies that mix representatives of local executive, elected leaders and civil society (such as Pakistan's DPSC). But, as Waitakere's experience shows, this seldom provides the strong accountability relationships necessary for empowerment, even in the 'best' of circumstances. Strengthening accountability means not just disciplinary institutionalist surveillance of the state (via MTEFs, PEM and SWAps), but also investing substantively in strengthening government by professionalizing the public sector and institutionalizing substantial budget support. It also means assessing and facing up to the transaction and opportunity costs involved in current fragmented arrangements: costs often submerged in internal donor/NGO/private sector cost structures.

3 Security: It's a perverse irony that neoliberal and NIE doctrines, which often assume the basic ungovernability of markets and even social situations, should have in fact contributed directly to the be-fuddling of even the most necessary social governance, and so undermined social security. But while governance remains ungovernable, poverty will remain beyond anyone's accountability. To get to sharper accountabilities beyond ungovernable neoliberal mess, we need to move towards much smarter reterritorializations of poverty, outcomes and accountabilities.

4 Finally, we add a fourth, more explicitly political call, something we think is glaringly missing from Poverty Reduction's current technical and institutional ambits.

1. Beyond narrow neoliberal political economies of opportunity

Here, it's encouraging to note some important progress. In their core strategy documents, Development agencies are at last moving beyond the naive abstractions of marketizing neoliberalism ('open up and hope', 'balance the books, get the prices right and go home early'; 'liberalize, rationalize, privatize . . .') and are taking account of what we have called territorial constraints and opportunities. Both the UN's *Millennium Report* and the 2006 WDR *Equity and Development* devote extensive coverage to issues of geographical, historical, demographic constraints, gender and equity issues, and poverty traps that no amount of simple market integration will overcome any time soon. These substantive causes of poverty, it should be clear, will require more than neoliberal institutionalist change (despite the fact that the 2006 WDR still leans overwhelmingly in such directions). Helping these countries and regions out of poverty (and poverty-shaped poor governance) will need substantial resource transfers, and even once unthinkable approaches like fostering these countries industrial competitiveness and productivity. For many, framing Poverty Reduction only in MDG terms is restrictive and unimaginative. Yet, even in these terms PRSP targets are constrained by short sighted economic policies, too often obsessed with price stability and pay too little attention to the need to boost domestic savings or public investment strategies. The Chinese government's approach is interesting here; for expressly political reasons it offers a range of concessions to Chinese businesses willing to set up in parts of Asia, Africa and China itself, it considers strategic or risky. Could not other agencies offer similar concessions and incentives, for more poverty-reducing reasons?[10] Clearly, it is possible, technically and strategically, that more heterodox solutions to poverty reduction could be considered as part of a second-generation look at Poverty Reduction's Opportunity.[11] These, surely, must include another look at domestic national and local taxation systems, in which even the IMF has acknowledged not enough has been done to deal with the vertical equity issues of orthodox approaches, or to confront powerful vested interests that block more inventive, pro-poor use of tax and other market-regulatory instruments;[12] or to deal with chronic problems of capital formation through investment-led (rather than PRSP's predominantly consumption-led) public expenditure.[13]

Of course there are less contentious, more consistently Liberal approaches. Writing in his 2004 book *Trade Policy and Global Poverty,*[14] William Cline argues that poverty will be heavily impacted by changes to industrial-country trade policy and laws.[15] Cline calculates that a regime of global free trade would confer income gains of 'at least \$90 billion annually', and, in combination with what are called 'dynamic effects', \$200 billion each year. As Cline notes, '[e]liminating industrial-country protection alone would provide long-term gains to developing countries of about \$100 billion annually – about twice as much as annual aid. The overall income gains would reduce the number of people in poverty globally by about 500 million by 2015, or by about one-fourth'. Here, progress would seem to be a matter of Liberalism being true to its free trading rhetoric: becoming more than a one-sided yet duplicitous doctrine of convenience for rich countries, and developing the political strength of its technical convictions. But it's more than this. In imagining a trading regime in which trade rules are configured to maximize development potential, rather than asking 'how do we maximize trade and market access?', Dani Rodrik argues that savvy negotiators should be asking 'how do we enable countries to grow out of poverty'.[16] By focussing on the agricultural-liberalization-centred agenda of Doha's 'Development' Round, he says, developing countries have been tied into a debate that at best will have ambiguous results for the poor and ensure that the real winners will be taxpayers and consumers in the North and traders and intermediaries in the South. The issues are complex. But given the potential impact of trade on Poverty, when can we look forward to, for example, a WDR on trade issues? Such a WDR would, it now seems inescapable, direct attention to three critical issues: space for heterodox policy approaches, migration and creating more articulated, territorially specific economic strategies.

Liberal Development would also do better by being internally coherent about migration. While developing country negotiators may have been bamboozled into concentrating on winning better market access rules for agriculture and other commodities not yet fully deterritorialized by rich countries, are they neglecting potential benefits from liberalization of labour flows? Global markets may be increasingly liberal, deterritorialized and commodified, but as Polanyi recognized, Labour ultimately is not. Rather, in stark contrast to the last great period of Liberal free trade, fierce immigration regimes mean labour is perhaps the most territorialized factor in the 'global' economy.[17] If capital and commodities are free traded, with the removal of barriers to their movement subject to active political badgering by powerful global actors, then why not labour? Here, Liberal duplicity about territorial aspects gets palpable: the poor's labour in particular remains locally territorialized, as poor people are simply stuck in low wage settings. Meanwhile, rich countries benefit enormously from territorialized low wage regimes, enjoying cheap manufactures while ramping up migration barriers.

Here, of course, we are deep in sensitive politics. But we are also deep in the real structures of opportunity for the poor. However, some places and situations seem to offer fewer demographic and political challenges than others. Consider the sparsely populated South Pacific where, in response to public order collapse and anarchic civil strife in the islands of Melanesia during the 1990s, governments in Australia and New Zealand at once maintained restricted labour migration (and tracked down overstayers at home), while sending ever larger contingents of armed security forces, police and aid workers (for water, sanitation and community development-driven stabilization measures). Now, across the Pacific, budget subsidy is offered along with increasingly strident demands for compliance with disciplinary good governance standards. Australia, which demographically could absorb most migrants with least impact, has been the shrillest of discipliners while playing to its audience as a dubious deputy sheriff for the US. This, while maintaining a spectacularly hard line against illegal migrants, and using small Pacific states as detention centres for imprisoned queue jumpers. We should expect such containment activities and costs to continue to burgeon, unless many more opportunities are given for at least short-term migration for young, stranded populations. It seems politically feasible: leaders in the US and OECD seem to have begun to open the policy door to resuscitating temporary migrant worker schemes of the past. Rodrik has roughly calculated that the benefits of a temporary labour migration scheme, amounting to no more than 3 per cent of rich countries' labour, could easily yield $200 billion annually for the citizens of poor countries, twice the hoped for $100 billion a year being trumped for global aid.[18] Again, what about for starters committing the resources of a WDR on migration, highlighting the mutual benefits of such a global redistribution of Opportunity?

2. Empowerment: enabling the state to do its job by making sure it has the resources

It should be clear now that we think the notion that local voice, marketized, social service solutions and locally owned, decentralized governance will deliver an adequate counterweight to locational and competitive disadvantage in a global economy is perverse. We think it is time to review what 'civil society' organizations, and the client power mobilized through them really add to accountability in either voice or service delivery terms. This, especially where what calls itself 'civil society' is in reality a plethora of professional international and local elites occupying NGO positions, and aiding the fragmenting of services and accountabilities. In such a context, to have service accountabilities pushed down to hybrid NGO/local authorities, moral accountabilities pushed down to clients, and coordinative responsibilities pushed down to strategic brokers and caseworkers is no substitute for empowerment based on potent, central to local, accountable transfers of assets that have been decided through durable forums of elected political leadership. What is needed are forms of regulation of those assets that remain free of capture by parochial (including both government and NGO) interests, and channel these assets into productive and substantive investments which directly impact on poor people's income and vulnerabilities.

Simply, as we have shown in this book, once things have been so thoroughly disaggregated and localized, and/or never adequately aggregated and embedded in professionalized public institutions under the control of leaders accountable to citizens, more *accountabilities* do not make for better *accountability,* especially around social equity, justice and security outcomes. But within NIE-style 'inform, enforce, compete' regulatory and

accountability frames, basic issues of fragmentation and associated institutional weakness in scope and regulatory power are routinely set aside.
Instead, the power of clients' voice is routinely overstated, as is the power
of law, information and competition to create incentive structures to drive
political economy in pro-poor ways.

Market and civil society agencies do have important roles to play in delivering services and raising voices. But even in the latter they often operate
at some remove from political accountabilities. And when given a privileged position as the prime means of articulating citizen voices, can have
the effect of diluting the accountability of state agencies to their citizens.
When actively incorporated in the political domain of important social regulatory issues – such as access to land, labour rights, public safety and
environmental assets – they can undermine and depoliticize accountability
by enacting weak proxies (participatory community processes, as opposed
to political activism and party manifestos, perfunctory community budget
conferences as opposed to hard-edged public accounts committee debates),
and soaking up activist energies in narrow project contexts. In areas of service delivery, the assignment of key functions to complex mixes of civil
society agencies often further atomizes accountabilities, and thus exacer-
bates the governance problems faced by local authorities. Whether in the
domain of service delivery or social regulation, where governance is structured around marketized ideas about a 'level playing field' and consensual
'local-local' dialogue, this is often a recipe not just for elite capture, but for
ensuring that local political choices become preoccupied with responding
to the changeable client politics and delivery of private goods. Substantive
investments that require medium-term, region-wide or inter-generational
planning and financing (population-wide public health [e.g. HIV/AIDS[19]],
or investments to sustain the environmental or productive assets of local
economies) typically lose out or become parochialized. True public goods
and entitlements, especially those of keen interest to the poor, require a more
durable political process, and often the assignment of responsibility and
power to far less 'local' and often far less 'populist' institutions. As we have
shown, pushing responsibilities for joining up governance down to local
levels, in the apparent interests of the local poor, usually means coordinative agencies (again, like Pakistan's police complaints and accountability
bodies) are occupied by the politically powerful who can exploit their slippery accountabilities. The coordinated agencies themselves have limited
resources, limited scope and capacity, and are required to operate in narrow,
compliance focused ways that are socially and territorially impotent.

And, as has become painfully clear, none of this necessarily costs any
less than more powerfully institutionalized systems. Here, compliance,
opportunity and transaction costs multiply, despite the ability of government, donors and NGOs to disguise these costs in recurrent programme
overhead, monitoring and evaluation, or, as commonly happens in governance focused development, to finance such process costs by taxing the actual

development/investment budget allocation available to the poor. Thus it is important to sheet accountability issues right back to the donors themselves. When these transaction costs – both indirect, as a cost of marshalling local governance systems in this way, and direct, though hidden in donor/service deliverer overheads – consume so much of monies raised in the name of the poor, questions need to be asked about who benefits from all this programming. Put it another way: the combined programming costs of all the NGOs or stand alone projects operating in say Cambodia, if aggregated and made visible, could more than pay adequate professional salaries to a much more effective public service than currently exists. But while donor countries persist, in the name of 'country absorptive capacity', in allocating substantial aid shares to free-ranging NGO/private service delivery programmes, such accounting and wider accountability is systemically undermined. These wider costs must now be more comprehensively reconsidered against NGO claims of cheaper, more responsive service delivery, or of representing civil society. This is not, we should stress, and argument for the full re-nationalization of soft NGO budgets: rather they could be up-scaled, with NGOs and donor-privatized Project Management Units competing in-country for larger contracts to deliver SWAp-directed services, using donor money channelled through government budget, expenditure and audit systems. But in large, we agree unequivocally with the Millennium report's findings:[20] ultimately, public servants must be paid properly if you are to expect any kind of accountability from them. If regaining public control over privatized aid flows helps this, we think that should be a priority over the proliferation of transaction intensive and market responsive NGO/private contractors.

Until this happens, donor rivalries and paradigmatic contest will continue to be exacerbated by the plurality of agencies and techniques on the ground. As programmes of similar but separate provenance are rolled out across multiple levels of governance, mess and chaos ensue. In Pakistan for instance, in the field of reproductive health services, a local government mandate, in many districts it is possible to find two or three national and provincial vertical programmes directly involved, alongside special purpose donor financed district projects, the local government's own staff and these in various competitive and partnering arrangements with contractualized NGOs. Social funds and other direct route transfers under the control of vertical programmes (as seen in Uganda and Pakistan, for example) can undermine local political and other accountabilities, while devolved funds remain focused on marginal, incremental infrastructure developments, and stipulate expensive, unsustainable participation, monitoring and compliance mechanisms. The collision and telescoping effects of pluralization of accountability, and how this wrenches and distorts accountabilities has been well illustrated. If things get really disaggregated, with NGOs and pilots dominating core aspects of governance, the unevenness of poverty outcomes across local jurisdictions can become quite stark. Technically capable and politically aligned localities benefit, while weaker jurisdictions either miss out, or lack competence to deliver. Poor people's access to moneys for services becomes not a matter of population based entitlement, but an accident of geography, or history, of being in the right place by chance, all matters over which they have no control and puny voice. The poor, in other words, are pushed further into the periphery by pluralized, localized and otherwise disaggregated regimes. The effects are a long way from empowerment.

3. (Social) security with scarce resources: smart reterritorializing, better joined-up interventions, common accountability platforms

Even if a reassessment of such transaction and fragmentation-fallout costs led to reallocation of sizeable aid monies towards substantive governance, the resources released would still be scarce, and would need sharpened, smartly reterritorialized accountabilities attached if they were to be optimally spent. Enhancing donor government political confidence about committing growing levels of public money to Development too will need much more technically savvy, locally attuned kinds of harmonization. What is happening in branches of social epidemiology and elsewhere, does indicate that some smarter approaches are in the offing, around which devolved funding is beginning to be aligned with territorial public health mandates and a more outcomes and rights based approach to accountability. Again, much of this is embryonic: realizing their promise will need substantive investment, and a commitment to enabling 'working together' that goes well beyond voluntaristic or weakly contractualized partnerships.

Poverty is currently subject in the main to surveillance regimes that focus attention on shifts in national statistics, or on sectoral and partnership-based allocations of services to regions or communities. But as we saw in Waitakere, the movement of poor people into sub-regional spaces has important implications for services there, and as evidence from everywhere makes clear, the clustering of the poor into marginal spaces compounds their difficulties by heaping them together. Here, urbanization into impoverished mega-cities will constitute much of Poverty Reduction's geographical future. Here, there is a need for approaches to monitoring poverty which make territorial dimensions much clearer, and which problematize fragmented or merely siloed approaches to service delivery in terms of their impact on peripheral and impoverished locations. There's a need to map the constraints to investment – informational, legal, regulatory, political – against opportunities to alleviate these by local action.

In our experience, social mapping of this kind faces technical difficulties and caveats. But it can have a potent effect by generating new kinds of responses and accountability from both officials and politicians.[21] In this regard, the work now being done to map indicators of a locality's endowments in MDG terms is a useful start: but it has quickly become clear that, as we saw in Uganda, focusing on immediate MDG effects alone can generate a range of perversities and unnecessarily limit the scope, again, to matching social deficits to preferential transfers only for service delivery. What's needed are more fine-grained analyses which further *contextualize* indicators, and map, for example, the diseases of poverty against a suite of known causal factors, including income, resources and access to services, and these at the particular periods in the lifecycle where the causal basis of later disease is developed. In public health, for instance, what might be called the technologies for causative analysis and intervention are now available. Here, the poor demographic groups affected by late onset diseases including diabetes and heart disease, and immediate problems including childhood nutrition, trauma and injury can now be geographically identified. We can know where they live; the causal factors are known and can be targeted (income poverty and environmental factors); we know when they do their main damage (*in utero* and in early childhood); we know what they need in terms of basic resourcing and service access; and we know by and large how to get these things to them, (either via central or locally governed agencies). What we need to do now, and increasingly as this kind of preventive intervention sophistication increases medically and epidemiologically, is make such crucial information the basis of smarter territorialized governance and intervention around poverty.

This sort of development is in fact happening in many OECD countries around areas of community safety, injury prevention, health, housing, domestic violence, and so on.[22] Yes, OECD country experience is different, not least because they often have the budget, and the kinds of cultural constraints, for example, that block technically savvy interventions around poor-country maternal-child health are often formidable. But nonetheless, what is relevant for poor countries are the techniques whereby scarce resources can be more reliably targeted. Here, an emerging range of skilled but politically savvy technical activists have emerged, and joined together to address what are clearly multi-sector issues. Increasingly, their work has become a sustained, budgeted part of local government's capacity, strengthening local government's overall ambit against social problems. While responses thus far are often locality and infrastructure focused, the understandings generated by such programmes about the wider social causes of health, violence and similar problems can be politically potent. They have reinforced, for example, the UK government's more systemic response to poverty-related health inequalities, and generated reliable, better contextualized sets of shared indicators around which sectors can collaborate and plan individually and together, and some of what are being called 'open method' coordinations and accountabilities can be built.[23] There, community and local interventions sit alongside a potent tax credit transfer based response to child poverty.[24] Need we say that it is the latter that seems to be having the major impact: but at least here local, joined-up inclusive approaches are themselves joined-up to (and no doubt greatly enabled by) a more substantive and clearly accountable approach to poverty.

Overall, what will help here are *smart reterritorializations*: clever strategic realigning of mandates, accountabilities and budgets to get the best out of them locally, and enabling them to be locally responsive in dealing with multiple dimensions of poverty. This is increasingly linked in the field of population health into an understanding of particular poverty impacts across the lifecycle. All this should create challenging technical (and technocratic) roles, and broaden the ambits of technical activists based in local government and local agencies. What these kinds of smarter reterritorialized accountability frames potentially offer is a more efficient and effective mode of community and interagency engagement, joined to an understanding of how to affect outcomes that seeks to do so more by direct route interventions than by longer route institutionalism. Both will cost; Good Governance always does. But the primary emphasis needs to shift from generic institutional capacity buildings to delivering direct support to the right places. The message here is that its possible to force the allocation of substantive resources, by mapping commitments into resourcing needs, by tracking expenditures against outcomes, and calling the Development establishment to account when the limits of existing policy interventions becomes apparent. Capacity building in these kinds of environments literally comes with the territory: territorial resource allocations can here drive good governance mechanisms, rather than be crowded out by them.

Here too, the creation of evidence-based *common accountability platform* arrangements,[25] whether local or national, might generate tougher accountability regimes that could resonate as an entry point for a more substantive approach to reducing poverty: increase the transfers of income from rich to poor. Before readers collapse in a heap, consider the evidence. The social determinants in health literature are unequivocal: income is the most important single determinant of health and wellbeing outcomes. Beyond that, mother's education is clearly crucial; but this too depends on adequate time and other resources being made available so girls can go to school. In OECD experience, the most spectacular impacts on child poverty in recent years have come through increasing parents' income, whether by sending them back to work with subsidized childcare, or by the more direct route of a universal child tax credit, which turns up weekly in workers' (and beneficiaries') pay packets. In the UK, this approach should halve child poverty by 2010.[26] Thus the vulnerable child, that classic Liberal subject combining marginality with potential, has become the political Trojan horse through which in a neoliberal age, real transfers of income become politically possible. In Development contexts, this is not an argument for child sponsorship or even for tax and income based transfers.

Realistically, it is an argument for finding every plausible way to reduce household expenditures on basic services: by, for example, finding smarter ways to subsidize water and sanitation, medical treatment, housing for the poor, public transport, and of course providing free or better education. It is, in other words, an argument for a more direct route, beyond the elaborate and costly framings currently being rolled out.

These are just some areas where technically savvy activists can push policy: by developing sharper territorial conceptions of political and political economic causation; by designing smart, genuinely joined-up interventions around these and by showing how poor outcomes reflect real material lacks, which need real material attention. It is surely the case that in many places where gender inequality, poor education, or even a basic acceptance that these problems deserve urgent action struggle to find any political footing alongside imperatives that push military or security budgets with impunity. In many cases, to be sure, we are not even at the starting line for these kinds of technically savvy interventions. But until now, their use for surveillance has outstripped the smart interventions. A striking feature of the past decade, as noted above, and surveyed throughout this book, has been development of a wide range of instruments for tracing in fine detail the relationships between different levels of business and government, and through this promoting better accountability to policy commitments, the law and local preferences. The depressing consequence of these developments of course is that the scale of theft of public assets and other resources is now clear. Websites inform us that the average urban Kenyan pays 16 bribes a month, costing one third of average incomes; that 92 per cent of Pakistani's pay bribes for public education, averaging $86 (this, against a poverty line of $170 per adult equivalent); that corruption in Indonesia in 2001–2002 cost almost as much as the amount received in development assistance; that during the Abacha years, Nigeria received $1.1 billion in aid, whilst he and his family looted up to $6 billion, and so it goes on.[27]

But this kind of information also points to the phenomenal positive potential for informing policy with far more sophisticated sources of information, about local needs and endowments, production possibilities, assets and their distribution – and on the other side, exposing the practice of policy to that evidence. This will be difficult, and costly. And can't (and often won't) be done by governments alone. But in Uganda, some well-renowned examples of 'evidence-based' policy has much improved the match between budget outlays and actual spending in schools, hospitals and public utilities, just as they have shown that the apparatus of the top-level MTEF was not yet well articulated down through the sector budgets to the Local Government Budget Framework. But here, in one of the most difficult settings, it has been possible to use evidence to build common platforms for better accountability; of donors to their commitments, of central to local governments to the law; and that country's mass of uncoordinated and jealously independent NGOs and the army of international consultants and domestic technical activists to constitutional commitments hard won through civil war. And this, in turn, has promoted far better appreciation of the poverty of much of the development doctrine visited upon the country in the past decade and not least, the complicity of the international community in aiding and abetting the continuation of the hugely costly military adventures in the northern Acholi lands and the Congo.

Developments in this area will need to be pragmatic, building in all likelihood on existing, flawed arrangements, and trying to generate better accountability in both voluntaristic and mandated ways, in what will remain post-NIE, fragmented local domains. Even in rich contexts, like New Zealand's Waitakere, it means building on existing unfunded mandates, such as the wellbeing and long-term planning processes 'community outcomes' frame. It means using these policy devices to expose the continuing failure of central government to align plans and budgets accountably to local process. Here, the failure of local and national siloed departments to align around social outcomes – instead merely around narrow outputs classes in service delivery contracts – will be exposed, and perhaps force changes to the Public Finance Act 1989 to hold the state accountable to social governance commitments.

But the message for poor countries is again clear. The resources they have in these areas are often tiny. And available resources are scattered among larger NGOs without substantive territorial accountabilities, and shared for a variety of reasons among the many new quasi-territorial hybrids now emerging and being promoted, even more lacking in coordination capacity and incentives. Clearly, the kinds of smart technical intervention framed around common accountabilities we've mentioned need more coordination, not less. Governments and lead donors will have to be both technically smarter and more territorially assertive about common accountabilities than they have been. Certainly, in all this, political aspects still matter: as we saw in New Zealand, inclusive neoliberal elected governments will still actively select against the poorest of the poor even while promoting apparently social democratic redistribution policies. Here, then, smart approaches which ensconce poor people's plights in discourses such as health can bring social security out of the political and into not the market, but the necessarily expanding (and socially/territorially deepening) the 'inclusive' and 'positive' Liberal realm of rights based approaches. We think, even the slipperiest of Third Wayers might be brought to some account by these approaches.

4. Smart re-politicizing? Negotiating the realities of territorial politics and its accommodations

We support the calls for some mild re-politicizing of Development, to make political space for heterodox policy approaches to economic development and poverty reduction of the kind advocated by Dani Rodrik, Robert Wade and Ha-Joon Chang. That said, we are sure that getting the right kinds of political and technical balance is going to be awfully tough, given the nature of current authoritarian/IFI accommodations, wider imperial geopolitics, local patrimonialism and local horizontal (poor on poor) violence.

In its current consensual Good Governance guise, Development routinely proposes technical and consensual solutions for what needs to be addressed politically. This is not just the IFIs' tendency to be cautious: it is potently built into the very universal Liberal governance Poverty Reduction is based on, as well as into the localized, consensual and sentimental communitarianism of partnership and participation in which Poverty Reduction likes to dress up its Liberalism. This is nothing new in periods of international Liberal consensus: it was true too in, for example, late Victorian and early twentieth century Britain. There, however, the poor were able to considerably advance their causes and win a number of citizen rights (the franchise), welfare and workplace concessions (the eight hour day, paid holidays): but only when, crucially, they or their (social-Liberal) advocates organized politically to get them.

What is needed, then, is a much more frank and determined look at where political action is needed to push Poverty Reduction into what it really ought to be doing. And, perhaps surprisingly, this is actually happening in an increasing list of countries. In Pakistan, while it is true that the imperatives of the War on Terror have led donors to accommodate General/President Musharraf's form of military-democracy, they and other subscribers to the PRSP process have listed 'political economy' as the crucial area for their attention. This involves moving out from their focus on executive short-cut solutions, to work with newly elected federal and sub-national legislatures; it involves assisting political parties prepare party manifestoes to include commitments on employment and industry, women in the public sector or gender violence, and budgets to back new commitments, supporting voter education and political literacy, supporting small to medium enterprises in ways that put the productive economy back on the local political agenda, and so on; it involves variously nuanced schemes to strengthen local bar associations' work with local judiciaries, funds to support NGOs to recover a social justice agenda, sponsoring action to create case law to activate environmental tribunals, to challenge local governments to fulfil their obligations to regulate child labour and marriage, tenancy law and the regulation of public irrigation assets.

The enabling of the political, and even strategic repoliticizing, is of course no guarantee of pro-poor outcomes. Representational politics typically orient closely around the left and right of central – which is to say middle income, political power bases. Here, even small pretensions to pro-poor policies can be followed electorally by reactionary governments determined to tip the balance in the interests of the rich and powerful (and, it must be emphasized, of men in general). Nevertheless, repoliticizing strategies can and should be put on the table and this not just at local or national levels. In fact, the interests of poor countries may well prove best served by a repoliticizing of global forums, such as the WTO, and a change of current political hegemony within the IFIs, which are, corporate style, subject to uneven shareholder rather than truly multilateral accountabilities. Such re-politicizing could enable emerging economies to, for example, fight the bunk of the level playing field of the WTO, and demand equal rights in the Banks, the IMF or WTO. They could demand more international accountability to MDGs and related indicators, and apply pressure on countries and IFIs to deliver aid in more accountable, truly locally owned ways.

It is possible that this would change the nature of aid delivery, loan enrolment, and technical assistance, shifting its cost base and changing the kinds of experts hired both to consult and to staff IFIs. At this point in our concluding argument, we're generally reminded that global development doctrine, and what governments can be made to do, is constrained by the hegemonic control of the IFIs, and that they in turn are constrained to go this far because of the strictures of their Charters or because of the countervailing imperatives of their shareholders. As the record of the 1990s shows, the institutional culture of the IFIs is far more elastic than this; when needs be, their Charters and operations can be interpreted to range into just about any domain of social life. There is, we are convinced, a wide terrain of practical possibility for action, especially if one takes the view that the new techniques, data and other resources may themselves be used as sources of power and influence.[28] In the case of the IMF, it is even plausible that a democratization of IMF policy could lead to a shift in the balance from debtor adjustment, to involving more creditor accountability for the nature and conditionalities around loan transfers and rescue packages, much as Keynes imagined. Because of the stakes involved, and exactly whose interests would be affected, we should not expect further IFI democratization any time soon, at least not before the populations of the G7 states are seriously affected by crisis.

Meantime, the greatest precursor of poverty remains conflict: the conflict of a mal-territorialized, failed and excluded state (or family), ripping itself apart. Or, the conflict of a relatively powerful bully state taking advantage of failures in neighbouring and peripheral states to engage in ideological or more forceful interventions. Or internal conflict, in which the protracted conflicts parlayed by governments in Uganda, the Philippines, Central Asia or Pakistan against religious, political or ethnic minorities go unresolved while they declare their commitment to Poverty Reduction. Not for nothing, we note, are many such conflicts described as 'domestic': for they also reference a deep seated structural violence routinely meted out on women. Here too, advocates of Poverty Reduction need to show greater conviction around basic principles. The increased scrutiny by NGOs of IFIs, the exposure of their policy practices to evidence, and their, albeit begrudging, responses hold promise.[29] But it must go further, for while poverty remains ensconced in peripheries, and relative incomes gaps between rich and poor grow, the social psychology of horizontal and gendered violence tells us we should expect more conflicts within and between peripheral states and in peripheral families. In this context, genuine poverty reduction is itself a potent form of peacemaking, and, conversely, its absence will mean increased policing responsibilities for an increasingly security obsessed and vulnerable core.

Thus the likelihood of peripheral nations, regions and communities continuing to war with each other seems extremely high. In recent times, two-fifths of countries that emerged from conflict relapsed into conflict within five years. In this context, deploying Poverty Reduction as a kind of 'defensive modernization' and peace building strategy to create sticks and carrots seems sensible: tying, for example, aid transfers to cessation of conflict, transitional justice and active peacemaking to resolve the underlying cause of conflicts. Using the smart territorializations of decentralized governance to create arrangements whereby fractious regions can have substantive autonomy, and creating long-term incentives for autonomous regions to find ways to be better neighbours. This would require, of course, the mandating of IFIs and the UN to act as more than the moneybags of security driven political alignment, and to engage in substantive multilateral interventions supporting political (and if need be militarily imposed) settlements, in places like Somalia, the Congo, Sudan, etc. This embedding of international military policing within a serious and substantive transfer regime would have its risks. But so does the current alternative, the deployment of territorially frail and perverse travelling rationalities and communitarian sentiment, while ramping up the walls of disciplinary selectivity and exclusion, and creating peripheries of fear and abjection populated only by the desperate, demonized poor and those who would govern them. This again, need we re-emphasize, will have the most profound effects on the most vulnerable, in age, poverty, gender and ethnic terms.

The other rising risk, as we have noted, is the prospect of rising imperial rivalry, and its sponsoring or prolonging of such conflicts. This, we note, is already the case in Sudan. Here, a repoliticizing of Poverty Reduction in rival imperial terms – neo-conservatism on one side, authoritarian loan and business concessions on the other – may suit some stronger states, and benefit their poor. But it is a fraught and dangerous path, subject to all the mutual escalations of paranoias and prejudices that have sustained both global and local wars for centuries. Making policy choices a matter of wider geopolitical alignment is the thin end of the wedge here. Yes, as we've shown, such has always been the case. But its escalation, whether through sharply punishing policy departures, or, by perversely incentivizing a slavish alignment with policy orthodoxy, is bad news for everyone.

Conclusions

Clearly we don't think this situation can be redeemed easily. And, yes, this more structural pessimism perhaps marks us off from the progressive Liberals we have cited approvingly throughout this book.

We are sure that development will be uneven, that its most peripheral forms will be monstered hybrids of the authoritarian (policing and political accommodations) and the travelling Liberal, and that the benefits of all this on the ground, where such exist, will be overwhelmingly weighted towards the more powerful political economic forces involved. We think that the worst place to be is these 'local' places that are policed or selectively worked against, that receive, at best, services in place of substantive equity or social justice, while many of their real resource issues are snowed, or, blind-spotted by permutations of consensual neoliberal hybrids.

Rather, however, than ending with dominant structural machinations and their tragedies, we want to turn the spotlight around again, and shine it on those involved in doing Development on the ground. As Deleuze and Guatarri noted, reterritorializations typically rely on artificially resuscitating archaisms, national or regional identity, sentiment and nostalgia for community and locality.[30] In inclusive neoliberal, Poverty Reduction Development contexts, this is clearly true: true too, is that for many agencies, given the fact that their own processes consume much of the budget, the Third Way penchant for Tiny Symbolic Gestures is all too prevalent. We note too that Development's practitioners are just as prone to a naive faith in technique and the next grand yet intricate travelling rationality. Who hasn't seen Development agents promoting in foreign places various forms of utopian communities or technical decentralizations or extravaganzas of disaggregation in ways they would never dare to at home? It's hard to justify such practice by an appeal to fashionable consensus ('Everyone's doing it') or institutional policy ('I was only following the WDR 04').

In the current consensual world of Poverty Reduction, need we remind anyone close to it, there are enormous incentives to go with the flow: to let spin triumph again over substance and in practice, too, let self-deception and easy self-congratulation prevail. This spin celebrates the little participation, partnership or local success, when the ground is falling out from under the poor, or stealing the poor's assets. It insists that all is on track, that only another round of capacity building, and more donor financing is needed to get people to recognize their 'institutional roles and responsibilities'. There are strong incentives to pretend that the minor power of voice or some participatory event offers some poor person or community the prospect of 'ownership' of their Poverty Reduction programme, or that it will lead to meaningful accountability, a decent claim on budget or a necessary redistribution of assets; to deceive ourselves that both the minor devolved funds and the bigger SAPs and PAFs create sustainable, accountable governance either centrally or locally; or to pretend that the latest round of pilot projects we know will never grow up are doing much more than shoring up ghettos for the poor, transferring responsibility for their own poverty downwards, and creating alibis for powerful international or domestic interests.

Meanwhile then, we will need to learn how to challenge Liberal and quasi-territorialized governance's technical and ideological overreach, to deal with the personal costs of whistle-blowing their sentiment and their formalisms, blind spots and travelling bullshits. We will need to learn not to depoliticize what is political: not to make cheap and toothless appropriations from social democratic and more territorially grounded modes of governance; not to put deceptively soft packaging around that which is designed to conservatively discipline and ensconce existing patterns of property and privilege; not to conceal the stresses and ravages of markets under the stardust of Opportunity. We will need to endlessly shift the debates away from formal principles of doctrine, and into the domain of substantive politics and geographies. This, not just for critical positioning or for ideological and political opposition, though this is important too. If nothing else, seeing beyond the blind spots and soft sentiment is a matter of simple integrity, of bottom line ethics, of mere good practice.

The poor themselves are hardly in a position to fight even the most perverse inclusion delusions of neoliberal institutionalism and 'inclusive' neoliberal Poverty Reduction. At worst, these have weakened their political hand, and undermined their basic security. It beholds, then, those many clever people actively involved in Development to be frank and blunt about the limited political and economic outcomes of their own institutionalized doctrine and practice. And to be as tough and imaginative as they can be about the accountability of their own programmes and their shared accountabilities across Development's sharded programme fields. In this, the real scope of the problem of peripheral insecurity and poverty may become clearer to everybody, along with the scope of the commitments needed, beyond the pretentious charity that trickles from the institutional maw of current Development beyond neoliberalism.

Abbreviations

ADB	Asian Development Bank
CBO	Community Based Organizations
CDB	Commune Development Boards (Vietnam)
CDF	Comprehensive Development Framework
CERPAD	Centre for Rural Planning and Development (Vietnam)
CFAR	Country Financial Accountability Reviews
CPAR	Country Procurement Assessment Reviews
CPIA	Country Policy and Institutional Assessment
CPRGS	Comprehensive Poverty Reduction and Growth Strategy (Vietnam)
CYPF	Children Young People and their Families (New Zealand)
DAC	Development Assistance Committee (of the OECD)
DC	District Commissioner (Pakistan)
DDP	District Development Project (Uganda)
DFID	Department for International Development
DHB	District Health Board (New Zealand)
DPSC	District Public Safety Commissions (Pakistan)
ESAF	Enhanced Structural Adjustment Facility (of the IMF)
FDI	Foreign Direct Investment
FSA	Financial Sector Assessments
FY	Financial Year
GATT	General Agreement on Trade and Tariffs
GDP	Gross Domestic Product
GNP	Gross National Product
G7 (G3, G8)	Group of 7 (major industrial economies)
HIPC	Highly Indebted Poor Countries
IDA	International Development Association (World Bank)
IFI	International Financial Institution (e.g. Asian Development Bank, World Bank)
IMF	International Monetary Fund
IRD	Integrated Rural Development
LDF	Local Development Funds (Vietnam)
LDG	Local Development Grant (Uganda)

LGA	Local Government Act (Uganda, New Zealand)
LGDP	Local Government Development Program (Uganda)
LTCCP	Long Term Council Community Plan (New Zealand)
MDB	Multilateral Development Bank
MDG	Millennium Development Goal
MTEF	Medium Term Expenditure Framework
NGO	Non-Government Organization
NIE	New Institutional Economics
NPE	New Political Economy
NPM	New Public Management
NRB	National Reconstruction Bureau (Pakistan)
NRM	National Resistance Movement (Uganda)
ODA	Overseas Development Assistance
OECD	Organisation for Economic Cooperation and Development
PAF	Poverty Action Fund (Uganda)
PEAP	Poverty Eradication Action Plan
PEM	Public Expenditure Management
PER	Public Expenditure Review
PERC	Public Expenditure Reform Credit
PIN	Press Information Notices
PIU	Project Implementation Units
PMA	Plan for Modernization of Agriculture
PML Q	Pakistani Muslim League – Quaid-e-Azam
PPA	Participatory Poverty Assessment
PRA	Participatory Rapid Appraisal
PRGF	Poverty Reduction and Growth Facility (World Bank)
PRS	Poverty Reduction Strategy
PRSC	Poverty Reduction Strategy Credit
PRSP	Poverty Reduction Strategy Paper
PSIA	Poverty Social Impact Analysis
QNDN	Quang Nam Da Nang Province (Vietnam)
RIDEF	Rural Infrastructure Development Fund (Vietnam)
ROSC	Reports on the Observance of Standards and Codes
SAP	Structural Adjustment Programme
SIP	Sector Investment Programmes
SWAp	Sector Wide Approach
UN	United Nations
UNCDF	United Nations Capital Development Fund
UNDP	United Nations Development Programme
UPE	Universal Primary Education
US	United States of America
WDR	World Development Report
WTO	World Trade Organization

Notes

1   Governing poverty: development beyond neoliberalism?

1   World Bank, *States and Markets*, 2002.
2   See UN Millennium Project 2005.
3   See Sachs 2000; Robinson 2002; Pieterse 2000; Escobar 1995; Sachs 1992.
4   As ODA from DAC countries dropped a real 20 per cent from $60.8 billion (1992), to $48.3 billion (1997); between 1988 and 1997 private financial flows to developing countries increased from $36 billion to $252 billion, although only ten countries received 78 per cent of FDI. UNDP, *Human Development Report* 1999; Therien and Lloyd 2000.
5   See Stiglitz 2002; Rodrik 2001.
6   Polanyi 1944, 3; Stiglitz 2001, 2002, 2003; Gray 1998; Pieper and Taylor 1998.
7   'While monetary interests are voiced solely by the persons to whom they pertain, other interests have a wider constituency. They affect individuals in innumerable ways as neighbours, professional persons, consumers, pedestrians, commuters, sportsmen, hikers, gardeners, patients, mothers or lovers – and are accordingly capable of representation by almost any type of territorial or functional association such as churches, townships, fraternal lodges, clubs, trade unions, or, most commonly, political parties based on broad principles of adherence' (Polanyi 1944: 155).
8   Peck and Tickell 2002.
9   These provide concessional assistance for balance of payments difficulties; the ESAF followed the Structural Adjustment Facility in 1987, and was replaced by the PRGF in 1999; IMF, 'IMF Concessional Financing through the ESAF'. Available online: www.imf.org/external/np/exr/facts/esaf.htm (accessed 8 September 2005).
10   See Porter and Craig 2004; Stewart and Wang 2003; Bretton Woods Project website, www.brettonwoodsproject.org/ (accessed 8 September 2005); Poverty Reduction Strategy Paper Watch www.prsp-watch.de (accessed 8 September 2005); Malaluan and Guttal 2002; Cammack 2004; Weber 2004; and Hanmer *et al.* 1999.
11   Hellinger *et al.* 2001.
12   See IMF and IDA 1999.
13   World Bank 2000a; Government of Cambodia, 2002; Clark 2002; Bush 2001.
14   Government of Pakistan 2004a, 27.
15   UN Millennium Project 2005; The Commission for Africa 2005.
16   E.g. Z. Brzezinski, *Out of Control: Global Turmoil, on the Eve of the 21st Century*, New York: Collier Books, 1993, III–V.
17   W. Lutz W. Sanderson and S. Scherbov, 'Doubling of World Population Likely', *Nature* 387, 1997, 19; and Global Urban Observatory, *Slums of the*

World: the Face of Urban Poverty in the New Millennium? New York, cited in M. Davis, 'Planet of Slums: Urban Involution and the Informal Proletariat', *New Left Review* 26, 2004, March–April, 5–34, p. 6.

18   Davis, op. cit.
19   This, rather than more 'positive liberal' approaches to enabling markets and their participants. See I. Berlin, *Four Essays on Liberty*, London: Oxford University Press, 1969, and later discussion in this chapter.
20   J. Williamson, 'What Should the World Bank Think about the Washington Consensus'? Paper prepared as a background to the World Bank's World Development Report 2000, *Institute for International Economics Papers,* 1999, p. 2. Online. Available http://207.238.152.36/papers/williamson0799.htm
21   World Bank, 'The Challenge of Development', *World Development Report 1991*, Washington DC: World Bank and Oxford University Press, 1991.
22   A basic set of NIE references is provided in Chapters 3, 4 and 8.
23   World Bank, *The State in a Changing World*, World Bank Development Report 1997, Washington DC and New York: Oxford University Press for the World Bank, 1997a, p. 152.
24   See P. Cammack, 2004.
25   See H.-J. Chang, *Kicking Away the Ladder? Policies and Institutions for Economic Development in Historical Perspective*, London: Anthem Press, 2002.
26   B. Jessop, 'Liberalism, Neo-liberalism and Urban Governance: a state-theoretical perspective', *Antipode* 34(3), 2002a, 455.
27   Readers should certainly see the triangle itself, presented in Chapter 7.
28   We acknowledge a range of important influences here: see Arrighi 1994; Arrighi and Silver 1999; Deleuze and Guatarri 1977; Hardt and Negri 2000; Hopkins and Wallerstein 1996; Jessop 2002b; Palat 2003.
29   See e.g. Burchell *et al.* 1991; Rose 1999b.
30   Williamson 1994.
31   Hirschman 1967.
32   Bauman 2002; Deleuze and Guattari 1977.
33   We simplify territoriality horrifically. See: Lefebvre 1991; Soja 1989.

2   Historical hybrids of Liberal and other Development, *c.*1600–1990: markets, territory and security in Development retrospect

1   See Cain and Hopkins 2002; Havinden and Meredith 1993; Cowen and Shenton 1996.
2   Smith 1976 (1759).
3   Bakan 2004.
4   Apter 1999. See also Mason 1985.
5   Cain and Hopkins 2002.
6   Havinden and Meredith 1993.
7   Havinden and Meredith 1993, 88.
8   Cited in Owens 1981, 41.
9   Ingham 1956, 6.
10   See Havinden and Meredith 1993, 51; Mason 1985, 44.
11   Davis 2001.
12   Davis 2001.
13   Stanley, citied in Davis 2001, 313.
14   Davis 2001.
15   Eliot 1966 (1905), 309–10.
16   Mamdani 1996.
17   Lugard 1922.
18   Mamdani 1996, 21.
19   Ranger 1983, 215. See further Meyer 2004, 14–15.

20 See Tinker 1954.
21 Rose 1999b, 4; Mamdani 1996, 49, and Meyer 2004, 27.
22 Perham 1960, 12.
23 Perham 1960, 26.
24 Lugard 1922, 48.
25 Mamdani 1996, 37.
26 Lugard 1922 617–18.
27 Lugard 1922, 617.
28 See Mason 1985, 156.
29 Lugard 1922, xlv.
30 Lugard 1922, 194.
31 Mamdani 1996, 39.
32 Cf. Mason 1985, 55–56, 66.
33 Lugard 1922, 539.
34 Lugard 1922, 95.
35 Lugard 1922, 96.
36 Lugard 1922, 97.
37 Lugard 1922, xxviii.
38 Chanock 1991, 70.
39 Lugard 1922, xlvi.
40 Lugard 1922, 618.
41 Polanyi 1944, 3.
42 Polanyi 1944, 209.
43 Ruggie 1982.
44 Acheson 1969, 3.
45 Sterling 1980, 76.
46 See Skidelsky 2000.
47 Eichengreen 1997.
48 Isaacson and Thomas 1986; Acheson 1969.
49 Acheson 1969, 375.
50 In van Nieuwenhuijze 1969, 15–16.
51 See Schurmann 1974; Reifer *et al.* 1996, 13–37.
52 Acheson 1969.
53 Marshall 1947.
54 Isaacson and Thomas 1986, 429.
55 Isaacson and Thomas 1986, 429.
56 Truman 1949.
57 This mood pervaded the immediate post-war years, see Hoselitz 1952, 8–21;
    Wohl 1952, 1; Lewis 1955; Morawetz 1977, 10–13.
58 Rostow 1960.
59 Rostow 1960, 1.
60 Rostow 1960, 9.
61 Rostow 1960, 6.
62 Easterly 2001.
63 Lewis 1955; Rostow 1960.
64 Easterly 2001.
65 See Thacker 1999, 37–38; Killick 1995; Momani 2004.
66 McNamara 1996, 314.
67 McNamara 1996, 333 (emphasis in original).
68 McNamara 1968, 144.
69 See Ayres 1983, 227.
70 Ayres 1983, 233.
71 McMaster 1997, 161; see further Porter 1995b.
72 Ayres 1983, 45.
73 Moss 1984, 94.

74   Ayres 1983, 43.
75   Reich 1992.
76   Williamson 1994.
77   Kuczynski and Williamson 2003.
78   Friedman 2000.
79   Easterly 2001.
80   See Dezalay and Garth 2002.
81   Kapur *et al.* 1997, 606.
82   See Skidelsky 2000; Galbraith 2003; Berger 2004, 122.
83   Kapur *et al.* 1997, 607.
84   Mukand and Rodrik 2005, 374–83.
85   Williamson 1990, 18–23.
86   Williamson 1990, 7.
87   Williamson 1999, 2.
88   See Stiglitz 2002.
89   Williamson 2002.
90   Naim 1999.
91   Kuczynski and Williamson 2003.
92   Kanbur 1999.
93   See SAPRI 2001.
94   Chang 2002.
95   Toye 1987.
96   See Jessop 2002a.
97   Porter 1996, 26; Porter 1991, 54–75.
98   See Spiro 1996.
99   See World Bank, *Accelerated Development*, 1981.
100   Bates 1981; Commins 1988.
101   See Chapter 3, Toye 1987 and World Bank, *The State in a Changing World*, 1997a.
102   Ghai and de Alcantara 1991.
103   Fowler 1992; Rugendyke 1991.
104   Porter 1996, 64.
105   OECD 1988.
106   Van der Heijden 1989.
107   World Bank 1989.
108   Bratton 1989; Brodhead 1987; Edwards and Hulme 1997; Hellinger *et al.* 1988. See also Hearn 2001; Jessop 1997; Sen 1999, 329.
109   Fergusson 1990.

3   The rise of governance since 1990: the capable state, poverty reduction and 'inclusive' neoliberalism

1   Brunner and Marmot 1999.
2   Frank 2000.
3   Reregulation refers to the range of instruments by which governing institutions sought to reimpose obligations or constraints on private sector behaviour. E.g. see OECD 1995.
4   Amin 2004.
5   Ahrens 2001, 54–90.
6   See O'Brien *et al.* 2000.
7   Porter 1993a, 53–67.
8   Porter 1996, 19.
9   Bayart *et al.* 1999; Wolfensohn 2001.
10   Brzezinski 1993, iii–v.
11   Camilleri and Falk 1992; Kaplan 1994, 24; Berman 2003, 117–19.

12 Slater and Bell 2002.
13 Therien and Lloyd 2000, 27.
14 UNDP, *Human Development*, 1999. This trend continued through the decade. The share of developing countries in world FDI fell from the peak in 1994 of 40 per cent to less than 20 per cent in 2000. See Wade 2003a, 8–14.
15 World Bank, *The State in a Changing World*, 1997a, 16.
16 World Bank, *Accelerated Development in Sub-Saharan Africa*, 1981.
17 Mosley *et al.* 1991, 4.
18 Helleiner 1983, 22.
19 Hutchful 1995, 399.
20 World Bank, *Long Term Perspective Study,* 1990.
21 World Bank, *The Challenge of Development*, 1991, 131.
22 Krippner 2001.
23 World Bank, *The Challenge of Development*, 1991, 10.
24 World Bank, *Long Term Perspective Study,* 1990.
25 Naim 1994.
26 World Bank, 'Managing Development', 1991, 10.
27 World Bank, *Helping Countries Combat Corruption*, 1997b.
28 Wolfensohn 1999.
29 World Bank, 'Managing Development', 1991, 1.
30 World Bank, 'Managing Development', 1991, 1.
31 Walton 1993, 41.
32 Camdessus 1992.
33 World Bank, *Sub-Saharan Africa*, 1981, 5.
34 World Bank, *Governance and Development*, 1992, 58.
35 World Bank, *Governance and Development*, 1992, 3.
36 World Bank, *Governance and Development*, 1992, 5.
37 Girishanker 2001.
38 See *International Monetary Fund Survey*, September 1997, 9.
39 Gill 1999, 190.
40 Quoted in IMF 2002.
41 World Bank, *Governance and Development*, 1992.
42 Girishanker 2001, 14.
43 Collier and Gunning 1999; Goldstein 2001; Kapur and Webb 2000.
44 Girishanker 2001.
45 Naim 2000; Collier and Gunning 1999, 319; Caiden 1994, 111.
46 See Wapenhans 1993.
47 World Bank, *Governance: the World Bank's Experience*, 1994.
48 ADB 1995; ADB 1997.
49 See Harriss 2001.
50 Carlsson and Ramphal 1995, 335; see also Douglas 2000.
51 Wade 1990; Berger 2004. See further, Wade, 'What Strategies are Viable for Developing Countries Today', 2003; Went 2004.
52 See Palat 2003.
53 See further Blustein 2001.
54 See Singh 1999.
55 Singh 1999, 9.
56 Stiglitz 2000.
57 See especially Cammack 2004; Stiglitz 1999, 2002.
58 Singh 1999; Berger 2004; Blustein 2001.
59 See Palat 2003; Wade 1998; Wade 2001d.
60 'There are two superpowers in the world today in my opinion. There's the United States and there's Moody's Bond Rating Service. The United States can destroy you by dropping bombs, and Moody's can destroy you by downgrading your bonds. And believe me, it's not clear sometimes who's more

powerful.' Interview comments from *The News Hour with Jim Lehrer: Interview with Thomas L. Friedman* (PBS television broadcast, February 13 1996). Available online: www.pbs.org/newshour/gergen/friedman.html (accessed 7 September 2005).

61   See Stiglitz 1998.
62   International Monetary Fund 2003, 1.
63   Soederberg 2001.
64   Fisher 2001.
65   Dollar and Svensson 1998; Hoogvelt 1997, 170.
66   From International Monetary Fund/International Development Association 1999, 6; World Bank, *Governance*, 1994; Bolt and Fujimura 2002, 152. See also Wolfensohn 1998.
67   See Booth 2003, 155; Killick 1998, 86–8; Johnson and Wasty 1993; and Haggard and Kaufman 1992.
68   Malaluan and Guttal 2002, 3.
69   Blair 1998, 4.
70   Roelvink and Craig 2005.
71   Cited in Levitas 1998, 123.
72   Short 1997, 5; Blair 2000, 6.
73   World Bank, *Governance and Development*, 1992, 4.
74   World Bank, *Reviewing Poverty Reduction Strategy Program*, 2001.
75   See e.g. Chambers 1994a, b, c, 1997.
76   Narayan *et al.* 2000.
77   Norton, with others 2001; Oxford Policy Management 2001; Whitehead and Lockwood 1999; Hanmer *et al.* 1999.
78   Thus Pakistan's PPA was 'not just a new type of study of poverty and its causes'; it equally aimed to build a constituency for the poor, increase accountability of government institutions, and define better policies and interventions' (Government of Pakistan 2003b).
79   Here, it is relevant to note the strong language of the International Monetary Fund/World Bank's review of PRSP that 'The macro-economic policy and structural reform agenda – for example, trade liberalization and privatisation – are, however, sometimes not even on the table for discussion' (International Monetary Fund/World Bank 2002, 24); compare International Monetary Fund/IDA 1999.
80   See Wade, 'Showdown at the World Bank', 2001d.
81   Sen 1999.
82   Zizek 1989.
83   Porter and Craig 2005; Stewart and Wang 2003.
84   Malaluan and Guttal 2002; McKinley 2005, 7–8.
85   World Bank 2003b, 11. The average now is 26 months.
86   United Nations Conference on Trade and Development 2002.
87   Ohno 2004; Wilks and Lefrancois 2002.
88   Oxfam 2003, 1.
89   Integrated Regional Information Networks 2001.
90   World Bank, *Poverty Reduction and the World Bank*, 2002d, 23.
91   Government of Pakistan 2003a, 5.
92   See www.un.org/millenniumgoals/ (accessed 9 September 2005).
93   Devarajan *et al.* 2002. More recently, the Millennium Project report estimates requires a quintupling of present ODA levels. See UN Millennium Project 2005.
94   World Bank and International Monetary Fund 1999a; World Bank, *PRSPs – Operational Issues*, 2004d; and World Bank, 'The Poverty Reduction Strategy Initiative', 2004c.
95   World Bank and International Monetary Fund 1999, 2002.

96 World Bank, *Toward Country-led Development*, 2003b. In fact, country 'ownership' was arguably not a developing country objective – cf. McKinley 2005, 7–9.

97 These perverse effects were fully noted later. See World Bank 2004c, 6ff.

98 As reflected in Kakwani and Pernia 2000.

99 World Bank 2004c, op. cit., 23. See also International Monetary Fund, *Report on the Evaluation of Poverty Reduction Strategy Papers*, 2004.

100 Again, this is acknowledged in World Bank/International Monetary Fund reflections on PRSP. See World Bank 2004c, 15, 45. The new interest in expanded public investment in productive sectors is discussed in Chapter 9. See also McKinley 2005.

101 World Bank 2004c, op. cit., 15, Table 2.2; and World Bank/International Monetary Fund, 1999b, 6.

102 Influential here has been Holzmann and Jorgensen 2002; and Sinha and Lipton 2000; World Bank, *Poverty Reduction and the World Bank*, 2002.

103 World Bank 2004c, op. cit., 40. Total IDA lending to the 35 countries with Poverty Reduction Strategy Papers at the end of 2003 rose from $13.1 billion in FY 1996–1999 to $15.9 billion in FY 2000–CY2003. By comparison, total IDA lending to 31 other active IDA-eligible countries fell from $13.1 billion in FY 1996–1999 to $11 billion in FY 2000–CY2003.

104 World Bank 2004c, 18.

105 World Bank/International Monetary Fund, op. cit., 8.

106 Based on World Bank 2004c, op. cit., 40, Figure 4.2.

107 International Monetary Fund/World Bank, *Poverty Reduction Strategy Papers*, 2001.

108 See Porter and Craig 2005.

109 The roots of 'positive' Liberalism we elaborate a little in Chapter 2: they go back to J. S. Mill and forward to Sen 1992; 1999 and de Soto 2000. See also Berlin 1969.

110 New Zealand Treasury 2001, 15.

111 Sen, op. cit.

112 Esping-Andersen 1990.

113 Bush 2001.

114 Bush 2002.

4   Local institutions for poverty reduction? 1997–2005: re-imagining a joined-up, decentralized governance

1 World Bank, *Governance and Development*, 1992, 11.

2 Pascal 1670, as quoted by Santiso 2002, 3.

3 Government of Pakistan, *Accelerating Economic Growth and Reducing Poverty*, 2003a, 53.

4 Branson and Hanna 2000, 4; Cammack 2004. See also Hoff and Stiglitz 2001.

5 See World Bank, *The State in a Changing World*, 1997a, iv.

6 Cornia 1998, 32–38.

7 World Bank, *The State in a Changing World*, 1997a; *Governance and Development*, 1992, 1, 14–31; Porter, 'Scenes from Childhood', 1995b, 63–86.

8 Larry Summers quoted in Blanchard *et al.* 1994, vol. 1, 252.

9 World Bank, *The State in a Changing World*, 1997a; *Governance and Development*, 1992, 1.

10 Soederberg 2001, 852; Sachs 2001, 187–98.

11 Camdessus, M., *International Monetary Fund Survey*, 14 December 1992.

12 Adapted from World Bank World Bank, *The State in a Changing World*, 1997a; *Governance and Development*, 1992, 152.

13 Bates 1981, see also Hyden.

14 Moore 1998, 40; Leys 1996.
15 Hutchful 1995, 404.
16 Burki and Perry 1998, 11; North 1981, 1990.
17 Burki and Perry 1998, 3.
18 Williamson, op. cit., p. 599.
19 World Bank, *The State in a Changing World*, 1997a, 117.
20 World Bank, *The State in a Changing World*, 1997a, 8; North 1981, 1990; Haggard 1990.
21 Stiglitz 2003, 111–39. See also Berger 2004, 163ff.
22 World Bank, *Building Institutions for Markets*, 2002a 25.
23 Harriss 2001; Chandhoke 2002; and Mouffe 2000, 108–27.
24 Pritchett and Woolcock 2004, 191–212.
25 Olson 1965. The large literature on decentralization will be referenced at various points in this book. Manor 1999 is a useful survey, whereas the late 1970s–early 1980s round of decentralization is exemplified by Mawhood 1983. The links with NIE, as suggested, are strong. cf. North *et al.* 1990; Wunsch 2000; Wunsch and D. Olowu 1990; Olowu 2001.
26 World Bank, *Governance and Development,* 1992, 21ff.
27 World Bank, *Uganda: Growing Out of Poverty*, 1993, 143.
28 World Bank 1996a, 10.
29 World Bank, *Community Driven Development in Africa*, 2001.
30 International Monetary Fund, *Aligning the Poverty Reduction Growth Facility*, 2003, 1. See also Foster and Mackintosh-Walker 2001.
31 Government of Pakistan 2003a, 34.
32 The transition from SIPs to SWAps is examined in Hill 2002. See also Cassels 1997.
33 Foster and Mackintosh-Walker 2001, 1; Schacter 2001; Foster *et al.* 2000; Harrold 1995; Land and Hauck 2003.
34 Parker and Serrano 2000, 13.
35 World Bank, *Community Driven Development in Africa*, 2001.
36 The concept of LDFs is examined in Romeo 1996.
37 Tendler and Serrano 1999.
38 World Bank, *The State in a Changing World*, 1997a, 131.
39 G7 Finance Ministers 2000, 31.
40 See World Bank, *Getting Serious About Meeting the Millennium Development Goals*, 2004a; Development Assistance Committee 1999; Wolfensohn and Fischer 2000; cf. Pender 2001.
41 Cf Petesch 1996; World Bank, *Assessing Aid*, 1998.
42 IDA, for the World Bank, and ADF, for the ADB.
43 Asian Development Bank 2000, 1; World Bank, *Country Assessments and IDA Allocations*, 2000d, 1.
44 Asian Development Bank 2000, 1.
45 World Bank, *Attacking Poverty*, 2000a, 1.
46 World Bank 2000a, 6.
47 US and France 'show up far down in both policy and poverty selectivity indexes, indicating that much of the assistance of these big countries goes to recipients that are not particularly poor and not particularly well governed' (Dollar and Levin 2004, 5).
48 Kaufmann *et al.* 1999.
49 See the papers by Kaufmann, Kraay and Zoido-Lobaton on World Bank governance website.
50 Mosley *et al.* 1991, found no clear association between the intensity of conditionality and success in implementation of promised reforms. See also Collier and Gunning 1999; Standing 2000.

51   World Bank, *The State in a Changing World*, 1997a, 124.
52   DFID 2004, detailed the circumstances under which aid could be reduced.
53   World Bank, *The State in a Changing World*, 1997a, 101. See also Development Assistance Committee 2003; and World Bank, *The Drive to Partnership*, 2001.
54   Booth 2003.
55   Development Assistance Committee 2005.
56   Development Assistance Committee 2005, 52.
57   Soederberg 2001, 859.
58   Abed and Gupta 2002, 8.
59   International Monetary Fund/World Bank 2002, 8.
60   Development Assistance Committee 2003. See www.countryanalyticwork.net (accessed 9 September 2005). These include: Country Financial Accountability Assessment (WB), Public Expenditure Reviews (WB), Country Procurement Assessment Reviews (WB), HIPC Expenditure Tracking Assessment (WB and IMF), against 15 public financial management benchmarks; Fiscal Transparency Review (IMF), a module of the Reports on Observance of Standards and Codes, based on the Code of Good Practices on Fiscal Transparency; Diagnostic Study of Accounting and Auditing (ADB); *ex ante* Assessment of Country Financial Management (EU); Country Assessment in Accountability and Transparency (UNDP).
61   World Bank, *Making Services Work for Poor People*, 2004b, 19.
62   World Bank 2004b, Chapter 2.
63   World Bank 2004b, 46.
64   Development Assistance Committee 2005, 59.

5   Vietnam: framing the community, clasping the people

1   These caricatures are explored and contested in a range of sources. See Jamieson 1993; Mus 1949; Taylor 1983; and Craig 2002.
2   See World Bank, *Vietnam Social Indicators*, 2004. See www.worldbank.org.vn/data/s_indicator.htm (accessed 19 September 2005).
3   The PRGF was approved by the International Monetary Fund, 6 April, 2001. The total commitment is around $368 million. The PRSC sports the by now familiar 'legs'; first, 'efforts to strengthen competition in the economy, harden the budget constraint on state-owned enterprises, and level the playing field for the private sector, therefore leading to increased efficiency and job creation; second, it will promote investments in people through better access by the poor to education and health care, more secure asset ownership in the form of land titles, and a greater role for public involvement in environmental management; finally, it will increase the efficiency and accountability in public financial management and improve legal transparency and accessibility, laying the foundations for modern governance', Rohland.
4   Porter and Kerkvliet 1995.
5   Valdelin *et al.* 1992.
6   Craig 2002.
7   ADUKI Pty Ltd 1995.
8   Denny 1996.
9   Kerkvliet 1995, 71.
10   Vietnam Communist Party 1990.
11   Vietnam Communist Party 1993.
12   Craig 2002, 26.
13   Any resemblance to actual characters involved in RIDEF PRA work is completely accidental.

14   Gourou 1936.
15   Nguyen 1993.
16   E.g. PRA's wide provenance is illustrated in the following; Conway 1985; Colletta and Perkins 1995; Cornwall and Fleming 1995; Guijit 1995.
17   During the time period of this story, Robert Chambers published important retrospectives and elaborations on PRA; See Chambers 1994a, b, c.
18   See, for example, Mosse 1994; Christoplos 1995.
19   See Craig and Porter 1997.
20   Narayan *et al.* 2000.
21   Fritzen 2002.
22   World Bank, 'Vietnam: Fiscal Decentralization and the Delivery of Rural Services', 1996; World Bank and others, 'Vietnam: Attacking Poverty', 1999; United Nations Development Program, *Looking Ahead*, 1999.
23   Fritzen 2002.
24   United Nations Capital Development Fund 1991, 5.
25   But as will become evident, this chapter does draw on first-hand and reflective accounts of the RIDEF, and the quotations to follow are drawn from these sources. These include Porter, 'Technical Mission Report', June 1993b; Porter, 'Monitoring and Review Mission Report', September 1993c; Porter, 'Technical Review Mission Report', January 1994; Porter, 'The Planning, Budgeting and Budgeting and Project Management Process', 1994; Porter, 'RIDEF: A Case Study', 1995c; Porter, 'Sustainable Financing and Participatory Planning', 1995e; Gardener *et al.* 1998; Stanley and Fritzen 2001 and Fritzen 2001.
26   Fforde 1993, 73.
27   Indexes (derived for each of 216 communes) include six elements, each of which includes a scale of three categories: (1) Number of school aged children per classroom, ranging from > 600 persons per classroom to <400 per classroom; (2) Status of the commune health clinic (with qualitative judgments on conditions); (3) Status of the market facility and category of market (commune or inter-commune); (4) Percentages of households with electric light; (5) Road density, ranging from < 3m per km^2 to > 10m per km^2; and (6) Percentage of paddy land which is irrigated, ranging from < 10 per cent through to > 50 per cent. Scores for each indicator were combined to create a total infrastructure development index. The index was developed by the CERPAD.
28   These data are fully presented in Porter, 'Economic Liberalization, Marginality and the Local State', in Porter and Kerkvliet 1995a.

6   Uganda: telescoping of reforms, local-global accommodation

1   Reno 2002, 416.
2   Between 300,000 to 800,000 people were killed during the regimes of Idi Amin and then Milton Obote. See Human Rights Watch 1999, 32; Museveni 1997, 201.
3   Uganda as a decentralization pioneer, see Steffensen and Trollegaard 2000; World Bank 2000b; Therkildsen 2002b.
4   Bayart, Ellis and Hibou 1999.
5   This is no more than a 'snapshot' of the period 1995–2000. A recent assessment of political, administrative and fiscal trends (showing continuity of DDP initiatives [the grant system, the approach to demand led capacity building, the performance incentive system], as well as the outstanding issues) is Steffensen, Ssewankambo and Tidemand 2004.
6   Toye 1987.
7   Museveni 1990, 3. See Hooper and Pirouet 1989; Hansen and Twaddle 1988.

8 Museveni 1997, 30. The way these 'people's committees' (Resistance Councils) were made to work had been partially borrowed from Marx's ideas of socialist democracy, as reflected in his writings on the Paris Commune of 1871. Tidemand 1994; Ddungu 1989.

9 Estimates vary, but in 1986, the National Resistance Army may have controlled only one third of territory in the country. Questions of political authority, stability and security were pressing. cf. Museveni 1997, 92.

10 Davidson 1992, 302; Seftel 1994, 285.

11 Museveni 1997, 35.

12 This first plan of the NRM government was inspired by the Lagos Plan of Action, issued about the same time as the Berg Report, and which retained the broad state-led territorial model of accumulation, highlighted the debilitating effects of external factors (terms of trade, etc.) and promoted indigenous solutions, self reliance and mutual African integration. Amaza 1998, 62; Organisation of African Unity 1981; Holmgren *et al.* 1999.

13 Holmgren *et al.* 1999, 16.

14 Amaza 1998, 162; Museveni 1990, 175; Holmgren *et al.* 1999, 36, 24.

15 Museveni 1997, 180–81.

16 The ERP was supported by 25 policy based loans from the World Bank, amounting to more than US$1 billion over 1987 to 1992. ODA increased from $208 million in 1986 to almost $500 million in 1989 (from a report by *The Economist*). Gross aid flows increased from 2.7 per cent of GDP in the FY 1986/1987 to 13.4 per cent in 1993/1994. The World Bank (1993, xiv) declared the country had a solvency ratio of 450 per cent and warned that 'a country with a solvency ratio of higher than 200 is considered to be severely indebted'. In 1980, Uganda's debt stock stood at $689 million. By 1987, it had increased to $1.9 billion and by June 1995 stood at $3.6 billion, over 75 per cent of which was owed to the World Bank and IMF, representing about 64 per cent of GNP and around six times the value of merchandise exports which, coincidentally, were steeply falling. World Bank 1996a.

17 E.g. Mamdani 1994.

18 Amaza 1998, 165; Mamdani 1990.

19 Museveni 1997.

20 Mamdani 1990.

21 Regan 1998, 164.

22 Uganda Constitution Commission 1993; Regan 1998, 166; Oloka-Onyango 1998, 21.

23 Uganda Constitutional Commission 1993, 251.

24 The Danish government was one notable exception, providing crucial support to local government reforms since before 1990.

25 Mamdani 1996, 216.

26 A feature of Uganda's decentralization is repeated efforts by the government to refine its provisions in law. The following list gives some appreciation of this: the Resistance Councils and Committee Statute of 1987 and 1988; the Local Government (Resistance Councils) Statutes of 1993; the Constitution of 1995; the Local Government Act, 1997; the Local Government Act (amendments), 1999. See Villadsen and Lubanga 1996; Nsibambi 1998; Mugaju 2000, 8–23; Steffensen, Ssewankambo and Tidemand 2004.

27 Accountability relationships thus develop over time. Here, then, is a stylized account, but one we contend is quite consistent with how Ugandan officials around DDP saw things. As we will show later, in Pakistan, key elements of this three-dimensional accountability framework would be lost in the 2004 WDR's flat three-cornered accountabilities between policy-maker, service provider and citizen.

28 A wide range of documents, not all published, are referenced in constructing this account. Carino, Flaman and Kulessa 2000; Decentralisation Secretariat 1998; Department for International Development (DFID) 1999, 2002; Ministry of Local Government 1999a; Kullenberg and Porter 1998, 11–15; Ministry of Local Government 1999b; Musissi 1998; Omoding *et al.* 1998; Onyach-Olaa 2003, 41–52; Onyach-Olaa and Porter 1999; Porter 1995d; 2001; Porter and Onyach-Olaa 1999, 56–67; Rugamayo 1999.

29 The pilot districts were Arua, Kotido, Mukono, Jinja and Kabale.

30 The principle of subsidiarity tends most often to be put in this way, that is, in terms of 'responsibility for particular services should be given to that level of government whose jurisdiction best corresponds to the benefits of the decision'. But it is complex in most contexts for it requires decisions on multiple issues about when its best to decentralize – e.g. typically included are judgments about whether the diffusion of power will improve protection of individual liberties; when it is desirable to promote experimentation and diversity; whether duplication or economies of scale are of concern; where multiple policies and fragmentations of decision making will undermine market operations, or movement of people, or security of property.

31 The principle of non-subordination rests alongside that of subsidiarity. In Uganda, District and Sub-County governments are non-subordinate to higher levels. For instance, the Local government Act 1997, mentions no less than 20 ways in which district governments are responsible to 'mentor', 'supervise' and 'regulate' – but not command or control – the activities of lower councils (Onyach-Olaa and Porter 1999, 15).

32 Ugandan Local Government responsibilities include Trade and Industry; Agriculture; Lands, Housing and Urban Development; Education and Sport; Animal Industry and Fisheries; Works, Transport and Communications; Information; Labour and Social Rehabilitation; Internal Affairs, Tourism, Wildlife and Antiquities; Local government; Gender and Development; Energy, Water, Minerals and Environmental Protection.

33 The PEAP preparatory process, acknowledged as one of the better examples of public consultation, lent the document a high degree of legitimacy amongst donors and government development officials. Gariyo 2000; Robb and Scott 2001; Nyamugasira and Rowden 2002; Brock *et al.* 2002.

34 Government of Uganda and UNCDF 1998.

35 International Technical and Agricultural Development (ITAD) 1999.

36 Ministry of Local Government 1999b.

37 In 1998/1999, Uganda was scheduled to repay $175 million. This was reduced to $132 million after a $650 million decline in Uganda's overall $3.5 billion foreign debt. By 2000 total debt service relief of $1.95 billion had been granted.

38 On social funds, see Chapter 4. A useful analysis of social funds is Tendler and Serrano 1999; Narayan and Ebbe 1997.

39 International Monetary Fund/World Bank 2002, 8.

40 Bevan and Palomba 2000.

41 Government of Uganda 2000b, 17.

42 Obwona *et al.* 2001, 149; Williamson 2003, 9.

43 Williamson 2003, viii. The impact was highlighted in Porter and Onyach-Olaa 1999; and Onyach-Olaa and Porter 1999; and later in Steffensen *et al.* 2004.

44 Porter and Onyach-Olaa 1999.

45 Amis 2000, 1.

46 James *et al.* 2002, 17.

47 Space does not permit an analysis of these impacts on local tax efforts. Uganda's local governments have a low degree of self-financing and own sourced revenue as a per cent of total local government revenue declined from

36.4 per cent in FY 1997/1998, to 13.2 per cent in FY 2002/2003. Steffensen *et al.* 2004, Table 4.2.

48   The PMA, key plank in the Presidents' election manifesto, was being supported by 11 donors. The PERC, quipped then as the 'Big Friendly Grant', was about to be renamed the Poverty Reduction Strategy Credit – another Ugandan 'first'.

49   Ministry of Finance, 'Planning and Economic Development', 2001.

50   Ibid.

51   Ibid.

52   Which is why Uganda's decentralization has been referred as resting on the 'patronage mode' of political economy (Francis and James 2003, 325–37).

53   The Fiscal Decentralization Strategy was adopted in June 2002.

54   Judging by increased resources – to health and education especially – service delivery should have improved. And the bulk of funds allocated, to education for instance, do now reach 'front line service providers' (Reinikka 2001). On how Uganda's decentralization reduces corruption see Lubanga 1998, 59–68.

55   Government of Uganda 2001a; Appleton 2000.

56   Ablo and Reinikka's well-quoted study of health and education sector spending in Uganda noted 'the large discrepancy between the official data and the survey is simply stunning' (Ablo and Reinikka1998, 30).

57   Williamson 2003, viii, 19, 23. The generic problem of 'disconnect' is observed in IMF 2003, 8; World Bank 2004c, 16–17.

58   Ugandans voted overwhelmingly in support of the NRM no-party system in a 2000 referendum. The NRM won the Presidential election in April 2001, and parliamentary elections three months later. Uganda's PPA found widespread support for decentralization (Ministry of Finance, Planning and Economic Development 2000).

59   Brett 1991.

60   Golooba-Mutebi 2000, 22; Lenz 2002; Bouckaert 1999; Tukahebwa 1997, 28.

61   We have here pointed only to the fiscal-administrative means of recentralization – the PAF vertical programmes and grant systems. The council electoral process probably increased NRM control over local governments – the system of indirect elections for council positions above the village council facilitated a 'hidden politic' of the Movement, and village 'democracy' may have made it easier to organize the buying or intimidation of small numbers. Therkildsen 2002; Watt *et al.* 1999, 37–64.

62   Hopes for an end to this 18-year war in Uganda's northern districts continue to flicker and fade. At this writing, about 1.6 million people have been displaced.

63   Uganda's involvement in the Congo began in earnest in 1996. United Nations reports accused Army officers of organizing trafficking in stolen vehicles, agricultural and mining products; cf. Reno 2002, 423; Weiss 2000.

64   Reno 2002, 426.

65   Reno 2002, 432.

7   **Pakistan: a fortress of edicts**

1   Iqbal 1932, 1245–255.

2   Jalal 1999, Chapter 5; Khan 2001, 60–64; Wolpert 1998, 254–79; and Wolpert 2001, 218ff, 221–22.

3   Jalal 1999, 178.

4   Events following the day of 'direct action', see Jalal 1999, 215–16, 223. Cf. Feldman 2001, 28–29.

5   Jalal 1999, 244; Jinnah, in McGrath 1996, 1. On the horror of Partition, see Singh 1993, 298–316.

6   Samad 1995; Feldman 2001, 30, 97ff.

7   Cohen 1998; Cloughley 2000.
8   Bennett Jones 2002, 278; Ali, T. 2003, 5–28; Siddiqa-Agha 2003, 124–42.
9   Newberg 1995, 18.
10   Meyer 2004, 88ff, 98.
11   Gilani 2001, 49–64; Ziring 1980, 131–32; and Islam 2004, 311–30.
12   Ziring 1998; Talbot 1999; and Bennet Jones 2002. On ethnic diversity and investing in public services, Goldin and Katz 1999, 37–62; and Alesina *et al.* 1999, 1243–284.
13   Feldman 2001, 26.
14   Feldman 2001, 46.
15   Ali, T. 2003.
16   Dewey 1991, 255–83.
17   Jalal 1999, 144; Ali 1991, 29–52. Cf. Herring 1983, 86.
18   Bennet Jones 2002, 245.
19   Wilder 1999, 4–5, 17.
20   On early party politics, Jalal 1999, 143.
21   On *biraderi*, Khan *et al.* 2002, 2–3; Wilder 1999, 6; Tinkler 1954, 337.
22   Wilder 1999, Chapter 9, summarizes trends in 'horizontal' party identities versus 'vertical' alliances.
23   On these traditions before colonialism, Ali, T. 2003; Michel 1967.
24   Press communiqué by Punjab Government issued 8 December 1914, Ali, I. 2003, 95; see Ali 2002, 24–42.
25   In this vein, see Hussain 1999, 359.
26   Ali, I. 2003, 104; Mason 1985, 40ff.
27   Jalal 1999, 119.
28   On how this occurred Ali 2003, Chapter 5; Ali 2002.
29   Ali, I. 2003, 84.
30   On the roots of this, see Rabbani 1996.
31   Marshall 1959, 255, referring to Ayub Khan's 'revolution of 1958'; cf. Newberg 1995, 4, 19; McGrath 1996, 11, 153.
32   Over half of the British Indian army came from Punjab, which comprised less than 1 per cent of the subcontinent's population (Dewey 1996, 264).
33   The DC's office was rooted in the Mughal period. Strong traces of this can be found in Lord Lugard's *Dual Mandate*. Dewey 1996; Mason 1985, 54ff; McGrath 1996, 7.
34   Thanks to Musharraf Cyan and Raza Ahmad, both once incumbents of these positions, and Shoaib Suddle, a senior police officer, for explaining these points.
35   Abuse of powers by DCs led to the first torture commission in the world in 1856, just ahead of the Indian revolt in 1857.
36   The Code of Criminal Procedure covered persons suspected of being likely to cause a breach of peace, or to disturb public tranquillity, and persons suspected of disseminating, orally or in writing, seditious matter calculated to cause communal tension.
37   Feldman 2001, 237, and Raza Ahmad, personal communication.
38   The utilitarian value of active, self-regulating subjects in times of crisis – fiscal or political – was an early realization of British colonialism; after the 1857 crisis, then again in 1882 (Tinkler 2001, 45–6; David 2002).
39   Feldman 2001, 93.
40   Tinker 1954, 70.
41   Later recounted in Government of Pakistan 2001b, 2. Military speeches had trod this ground before; Ayub Khan, on 8 October 1958; then Yayha Khan's address of 26 March 1969; 5 July 1977, Zia ul Haq, and Pervaiz Musharraf, 13 October 1999. On crisis and legitimacy, see Rodrik 1996, 27; Drazen and Easterly 2001, 129–57.

42 Benazir Bhutto, 1 December 1988; Ghulam Mustafa Jatoi, 6 August 1990; Nawaz Sharif, 6 November 1990; Balakh Shere Mazari, 18 April 1993; Nawaz Sharif, 26 May 1993; Moeen Qureshi, 26 May 1993; Benazir Bhutto, 19 October 1993; Malik Mehraj, 5 November 1996; Nawaz Sharif, 17 February 1997; Pervaiz Musharraf, 12 October 1999.

43 In 2001, debt and defence amounted to two-thirds of public spending. Reflecting consolidated fiscal deficits (including grants) averaging 6.5 per cent of GDP during the 1980–1990s, the ratio of public debt to GDP rose from 66 per cent of GDP in 1980 to 101 per cent by 2000. The interest burden of public debt grew even more sharply, from a little less than 11 per cent of total revenues over 1980–1985 to 46 per cent by 1999–2000 (World Bank 2002c, 3).

44 Easterly 2001.

45 This debate continues. See Social Policy and Development Centre 2003; 2004.

46 Easterly 2001; Asian Development Bank 2002. Overall net enrolment rates declined from 46 per cent in 1991/1992 to 42 per cent by 2001/2002, while net enrolments in rural areas – a SAP focus – fell from 41 per cent in 1991/1992 to 38 per cent in 2001/2002.

47 Government of Pakistan, *Three Year Poverty Reduction Programme, 2001– 2004*, 2001b Interim PRS.

48 Zaman 2002, 166; Easterly 2001.

49 Hussain 1999.

50 Government of Pakistan 2001a.

51 World Bank 2002c, 2. Brackets added.

52 Data drawn from ADB and World Bank records.

53 The Constitution, 1973 mandates by Article 175(3) that 'the judiciary shall be separated progressively from the executive within three years from the commencing day'; Government of Sindh v. Sharaf Faridi, PLD1994 SC.105; The Constitution, 1973, Article 37(i), reads: 'The state shall decentralize government administration so as to facilitate expeditious disposal of its business to meet the convenience and requirements of the public.'

54 Local government elections occurred between December 2000 and September 2001 (Pattan Development Organization 2003). A 'typical' district in Pakistan includes an area of 8–12,000 square kilometres and a population of 700– 1,500,000 (Federal Bureau of Statistics, Statistical Information, Government of Pakistan 2003c). Currently there are 6,458 local self-governments for the population of 146 million: 97 districts and 4 city districts; 306 tehsil municipal administrations and 29 city towns; and 6,022 union administrations.

55 Personal communication, November 2002.

56 Personal communication, June 2003.

57 We draw heavily on *Development in Pakistan* and material prepared during the study, albeit in a much abbreviated way.

58 World Bank 2003a, Chapter 2.

59 World Bank 2004, 32.

60 Government of Pakistan 2003a.

61 On fiscal efficiency; readers may find the following useful, Smoke 2001; Burki *et al.* 1999; Smoke and Schroeder 2003; Shah 1998; Haggard and Webb 2001; Azfar *et al.* 2002; Bahl 1999.

62 One indicator of this political process is that from 2001/2002 to 2004/2005, in three of four provinces, the size of the Annual Development Plan budget (for province politicians' own spending) has increased, in general, at a much faster rate than resources passed on to local governments, through the 'Provincial Allocable'.

63 Other constraints on local council budget decisions include the requirement that councils to reserve 25 per cent of development budgets for spending by non-elected Citizen Community Boards.

64 See India experience (Crook and Manor 1998; Manor 1999).
65 Pattan Development Organization 2003.
66 Local journalist, Bahawalpur, personal communication, 30 July 2003.
67 Service delivery impact of devolution was not the Study's focus (i.e. the focus was on 'incentive' structures). At time of writing, no systematic study has been made of the impact on service delivery performance in those sectors where devolution 'according to the book' has occurred. See National Reconstruction Bureau 2005.
68 *The Nation*, 3 January 2003.
69 *The Dawn*, 4 January 2004, 'Donors to Expedite Financial Support'.
70 World Bank 1997a, 80.
71 Local governments are given constitutional protection under the 17th Constitutional amendment. The MMA is a six-party alliance, with the Jamaat-Ulema-Islam – Party of Islamic Scholars – and the Jamaat-i-Islami, or Islamist Party, its two main pillars.
72 Social Policy and Development Centre 2003.
73 *The News*, 2 January, 2004; *Business Recorder*, 'State Bank's quarterly report', 1 January 2004; *The Nation*. 'Government Fails to Control Poverty', 18 January 2004.
74 Social Policy and Development Centre 2003, 15.
75 This confidence was immediately contested. Official statistics, early 2004, announced that foreign private investment had fallen by about 60 per cent from November 2002 to November 2003, including an outflow of $12.5 million under portfolio investment (*The Nation* 'SAARC Leaders Focus on Terror, poverty', 5 January 2004).
76 The annual meeting for all poor countries at which government and donors showcase their achievements and solidify renewed commitments to harmonization around the MDGs.
77 Asian Development Bank 2001, 18.
78 Established under the Police Order 2002 (Article 37 and Article 49).
79 Personal diaries, Bahawalpur, 2 August 2003.
80 World Bank 1997a, Chapter 6.
81 The 'social regulation' function of the state includes the following: property rights; land use, zoning and municipal management; public health and hygiene; market regulation; natural resource conservation and use; labour relations; NGOs and civic associations; and irrigation and agriculture.
82 On 'politics' and fiscal sharing, Johansson 2003, 883–915; Grossman 1994, 295–303; Khemani 2003.
83 This table reflects on a presentation made by Musharraf Cyan to the NWFP Province Finance Commission during June 2004.
84 Musharraf speaking to UN Conference on Anticorruption. During 2000, the Chairman of the Central Board of Revenue reported to the General that only 105 of his staff were honest and recommending sacking the remaining 90 per cent (Islam 2004, 327; Feldman 2001, 79).
85 Government of Pakistan 1997; Society for the Protection of the Child 1998, 1999; Amnesty International 2002.
86 Newburg 1995.
87 Government of Pakistan 2003b.

8   New Zealand: joining up governance after New Institutionalism

1 Unless otherwise referenced, interview quotes in this chapter are drawn from a wider research project addressing 'Local Partnerships and Governance' (LPG) in Waitakere: see the project website. Available online: www.arts.auckland.ac.NewZealand/lpg/ (accessed 10 September 2005).

2   Elwood 2000.
3   Williamson 1994, see also Kelsey 1997.
4   Kelsey remarks 'New Zealand consultants were estimated to have earned between $70 and $100 million from international consultancy and business during 1992, primarily for international agencies such as the World Bank and Asian Development Bank and overseas governments engaged in similar restructuring programmes' (Kelsey 1993, 140).
5   Schick 1996, 1.
6   For example, Transparency International's annual Corruption Index has placed New Zealand first or second for being corruption free for the past decade: see example www.transparency.org/pressreleases_archive/2004/2004.10.20.cpi.en. html.
7   Skilling and Waldegrave 2004.
8   Esping-Andersen 1990; 1999.
9   In Australia and New Zealand, this primary orientation resulted in a distinctive form of Welfare regime, the 'wage earners' welfare state'. Workers were able to consolidate their position by means of a number of state sanctioned, cross-sectorally benchmarked national wage fixing arrangements, including compulsory arbitration and national awards. But these benefits could be residual rather than universal, because only those with no labour market connection needed them. See Castles 1994; Castles *et al.* 1996.
10 Kelsey 1997, 130.
11 See New Zealand Treasury website, 'The Gorringe Papers'. Available online: www.treasury.govt.NewZealand/gorringe/bibliography.asp (accessed 10 September 2005). On NZ reforms, see Walker 1989; Russell 1996, Easton 1989.
12 New Zealand Treasury 1987.
13 Gorringe 1991.
14 See 'The Gorringe Papers' website, as above.
15 Contractualism had major implications for citizenship rights. See Allars 2001.
16 See e.g. Schick 1996; 1998; 2001. For health reform impact critiques see Easton 1989; Wellington Health Action Group 1993; Ashton 1999.
17 See Ashton 1999; Easton 1989.
18 See Waslander and Thrupp 1995.
19 New Zealand's gini coefficient rose from 2.6 to 3.3, an extraordinary and OECD unparalleled rise over such a short period. Despite employment recovery, it has not dropped since that time, but continued to rise slowly. See further Ministry of Health 2000; Ministry of Social Development 2004.
20 Again, see Schick 1996; 1998, 2001.
21 Early awareness of this eventual outcome is in Taggart 1991, and Allars 2001.
22 Howden-Chapman and Tobias 1999.
23 The quotes here and throughout this chapter are drawn from an extensive series of interviews conducted under the local partnerships and governance research programme. Most are included in Craig, 'From the Wild West to the Waitakere Way', 2003b.
24 Waitakere's income is overwhelmingly property tax (rates) and fee for service based: of recurrent revenues of *c.*NZ $120 million, $90 million is from rates, $20 million from water and sanitation supply, $7 million from waste collection, $5 million from recreation services, $3.5 million from roads, $3 million from building consents, $2 million from parking fines. Of the total, across all categories, central government grants and subsidies amount to just below $8 million (of which nearly half relates to roads).
25 For example, its safety outcome – ensuring the safety of the child or young person and the safety of the community – includes physical, cultural and emotional/psychological safety.

26   See Smith 2002.
27   According to a senior regional manager, the problem is that without strong, locally focused compulsion for e.g. the Waitakere Police to share accountability for outcomes with Waitakere CYFS, Ministry of Social Development and others, youth programmes struggle painfully to deliver outcomes.
28   Waitakere City Council 1996, 1.
29   Craig, 'From the Wild West to the Waitakere Way', 2003.
30   See Craig and Courtney 2004.
31   A list of Calls to Action was debated and mandated by 250 attendees at Waitakere's 2002 Wellbeing Summit. A 'Waitakere wellbeing Collaboration project coordinating committee' was formed with government, council and community representation, and a coordinator's salary was contributed to by a large number of different agencies.
32   For a review of call to action projects and outputs, see the Wellbeing Collaboration Project website: www.waitakere.govt.nz/OurPar/collabproj.asp (accessed 10 September 2005).
33   See Craig, 'Building on Partnership', 2004.
34   Department of the Prime Minister and Cabinet 1999.
35   Unemployment has declined remarkably, but high-income inequalities have not reduced after six years of Third Way governance; see Ministry of Social Development 2004.
36   See Ministry of Social Development, *Mosaics: Whakaahua Papariki*, 2003; State Services Commission 2001.
37   Ministry of Social Development, 'Pathways to Opportunity', 2001a, 16.
38   Ministry of Social Development 2001b.
39   Department of Prime Minister and Cabinet 2003.
40   Between 1989 and 2000, New Zealand's successive health reforms saw governance shift from Area Health Boards, to Regional Funders and marketized local provision, back to National interim scale, and then returning to District scale. See Gauld 2001.
41   See Schick 2001.
42   Italics are our emphasis. Available online: www.ssc.govt.nz/display/document. asp?docid=24292pageno=2.
43   See the website www.ssc.govt.nz/display/document.asp?NavID=105 (accessed 10 September 2005). See especially Alan Schick's commentary, Schick 2001; practical reform constraints notwithstanding, a limited Public Service culture change is arguably underway. By 2003, the State Services Commission's Statement of Intent would recognize 'an increased appetite among government agencies for collaborative approaches to address major public policy challenges; a greater expectation that central government will work more in partnership with others such as communities, the voluntary sector and local government'.
44   A web search will find 'Statements of Intent' with all their nested outcomes and outputs in each and every New Zealand government department.
45   Available online: www.msd.govt.nz/documents/publications/msd/statementof-intent-2004-part-a.pdf
46   Again, see www.msd.govt.nz/documents/publications/msd/statement-ofintent-2004-part-a.pdf
47   See Schick 2001.
48   See www.ssc.govt.nz/upload/downloadable_files/Statement_of_Intent_2003.pdf
49   Schick 2001.
50   Some more devolved funding for pilot collaborative projects in violence prevention and 'sustainable development' (environmentally progressive building standards, community development again) has been allocated.

51 New Zealand's post-structural adjustment growth depends on how you look at it: 9 per cent GDP growth versus an OECD average of 29 per cent over 1986–1999, or an impressive annual average rate of well over 3 per cent, 1990–2004. Recent growth tracks closely to record high (largely agricultural) commodity prices, consumption and debt.
52 They do feature in a minor way in DHB's population based funding.
53 Schick 2001, op. cit.

9 Conclusions: accountability and Development beyond neoliberalism?

1 Pakistan Development Forum was opened with these remarks by the Prime Minister, 25 April, 2005.
2 Recent discussion suggests a range of possibilities for second-generation PRSP. See McKinley 2005.
3 Bakan 2004.
4 See Hardt and Negri 2000.
5 Scoring systems against standard indicators will ensure 'rigorous' but 'fair' decisions to include or exclude poor countries. The Millennium Challenge Account will be managed by a Corporation comprised of US Cabinet officials. See the Millennium Challenge Account website: www.whitehouse.gov/infocus/developingnations/millennium.html (accessed 10 September 2005).
6 Department for International Development 2005, 5.
7 Department for International Development 2005, 8. A country at $250 GDP per capita has an average 15 per cent risk of experiencing civil war in the next five years, compared to less than 1 per cent at $5000 GDP per capita.
8 Wade, 'Is Globalization Reducing Poverty and Inequality?', 2004.
9 Bhalla 2002.
10 At the moment, poor countries pay the cost of these concessions themselves; the arrangement is called Export Processing Zones.
11 Indeed the World Bank's and IMF's own recent evaluations of the PRSP experience recommend eliminating uniform requirements, better customization to country requirements, better domestic country policy formulation and that a wider range of productive/economic policy options be considered. E.g. World Bank, 'The Poverty Reduction Strategy Initiative', 2004c; International Monetary Fund, *Report on the Evaluation of Poverty Reduction Strategy Paper*, 2004.
12 International Monetary Fund, *Evaluation Report*, 2003.
13 As McKinley 2005, 12, points out, gross capital formation rates (including public and private investment) slowed precipitously through the late 1990s and early 2000s. Whereas there is not much that poor country governments can do to stimulate private sector investment, public investment in productive capital assets is declining in many poor countries, and remains well below the GDP averages needed (that is, around 5 per cent of GDP).
14 Cline 2004.
15 See Sands 2005.
16 Rodrik 2004.
17 Hayter 2000.
18 Rodrik 2004.
19 Investments required to deal with the inter-generational, population- and economy-wide implications of HIV/AIDS have proven to be beyond the political reach of the typical 'community based' approaches to HIV mitigation strategies and programmes. See Linge and Porter 1997.
20 See Sachs 2005.
21 See, for example, A Social Health Atlas of Australia, www.publichealth.gov.au/atlas.htm (accessed 10 September 2005).

22   See, for example, Safe Waitakere's 'Protecting our Tamariki' summary document: www.waitakere.govt.nz/ourpar/pdf/tamariki-summary03.pdf (accessed 10 September 2005). See also Bennett 2003; National Public Health Partnership 2000; Hanson *et al.* 2003; Marmot, and Wilkinson 1999.

23   Radaelli 2003.

24   See United Kingdom Department of Health 2003.

25   Craig and Courtney 2004; Craig 2003a.

26   See United Kingdom HM Treasury 2002; Thurley 2002.

27   Information drawn from the following sources: Kenya: Kenyan Urban Bribery Survey 2002, cited in Transparency International's *Global Corruption Report*, London: Transparency International, 2003, p. 242; Pakistan: Corruption Survey South Asia, cited in Transparency International *Global Corruption Report*, London: Transparency International, 2004, p. 301, and Government of Pakistan, Pakistan Economic Survey, 2003–04, Islamabad: Finance Division, 2004b, p. 43; Indonesia: Indonesian prosecutor's figure ($2.3bn) for cases investigated between January 2002 and April 2004. Quoted in Agence France Presse, June 2003; Nigeria: Phil Mason, DFID, personal communication, 2004.

28   See Giddens 1984, 16; Held 2000; Falk 1999.

29   See, for example, the following Oxfam on WTO www.oxfam.org; the Centre for International Environmental Law, www.ciel.org; the International Centre for Trade and Sustainable Development www.ictsd.org/; the IFI/Bank Information Centre http:www.bicusa.org/; the Bretton Woods Project, www.brettonwoodsproject.org/; WTO WATCH www.ngos.net, www.wtowatch.org/; Trade Information Project linked to www.iatp.org/; Our World is Not for Sale www.ourworldisnotforsale.org/ (all accessed 10 September 2005).

30   Deleuze and Guatarri 1977.

Bibliography

Abed, G. and Gupta, S., *Governance, Corruption and Economic Performance*, Washington DC: International Monetary Fund, 23 September 2003.

Ablo, E. and Reinikka, R., 'Do Budgets Really Matter? Evidence from Public Spending on Education and Health in Uganda', *Policy Research Working Paper, 1926*, Washington DC: World Bank, June 1998.

Acheson, D., *Present at the Creation: My Years in the State Department*, New York: Norton, 1969.

ADUKI Pty Ltd, 'Poverty In Vietnam', *Report for Swedish International Development Agency (SIDA)*, Stockholm: SIDA, 1995.

Ahmed, A., *Jinnah, Pakistan and Islamic Identity: The Search for Saladin,* London: Routledge, 1997.

Ahrens, J., 'Governance, Conditionality and Transformation in Post-Socialist Countries', in Hoen, H. (ed.) *Good Governance in Central and Eastern Europe*, Cheltenham: Edward Elgar, 2001, 54–90.

Alavi, H., 'Pakistan and Islam: ethnicity and ideology', in Halliday, F. and Alavi, H. (eds) *State and Ideology in the Middle East and Pakistan*, London: Macmillan, also published in New York: Monthly Review Press, 1987.

Alesina, A., Baqir, R. and Easterly, W., 'Public Goods and Ethnic Divisions', *Quarterly Journal of Economics* CXIV, November 1999, 1243–284.

Alesina, A., Devleeschauwer, A., Easterly, W., Kurlat, S. and Wacziarg, R., 'Fractionalization', *Journal of Economic Growth* 8(2), 2003, 155–94.

Ali, I., 'The Punjab and the Retardation of Nationalism', in Low, D. (ed.) *The Political Inheritance of Pakistan*, London: Macmillan, 1991, 29–52.

—— 'Past and Present: the making of the State in Pakistan', in Mumtaz, S., Racine, J-L. and Ali, I. (eds) *Pakistan: The Contours of State and Society*, Oxford: Oxford University Press, 2002, 24–42.

—— *The Punjab Under Imperialism: 1885–1947*, Oxford: Oxford University Press, 2003.

Ali, T., 'The Colour Khaki', *New Left Review* 19, January–February 2003, 5–28. Available online: www.newleftreview.net/NLR25301.shtml (accessed 19 September 2005).

Allars, M., 'Citizenship Rights, Review Rights and Contractualism', *Law in Context* 18(2), 2001, 79–110.

Amaza, O., *Museveni's Long March: From Guerrilla to Statesman,* Kampala: Fountain Publishers, 1998.

Amin, A., 'Regulating Economic Globalization', *Transactions of the Institute of British Geographers* 29, 2004, 217–33.

Amis, P., 'Decentralization and Central Local Funding Mechanisms in Uganda', report for Danida, Kampala, October 2000.

Amjad, R. and Kemal, A. R., 'Macro-Economic Policies and their Impact on Poverty Alleviation in Pakistan', *Pakistan Development Review* 36, Spring 1997, 39–68.

Amnesty International, 'Pakistan: Insufficient Protection of Women', 2002. Available online: www.web.amnesty.org/library/print/ENGASA330062002 (accessed 19 September 2005).

Appleton, S., 'Poverty In Uganda: preliminary estimates from Uganda national household survey', Kampala: Ministry of Finance, Planning and Economic Development, mimeo, 2000.

Apter, A., 'The Subvention of Tradition: a genealogy of the Nigerian Durbar', in Steinmetz, G. (ed.) *State/Culture: State Formation after the Cultural Turn*, London: Cornell University Press, 1999, 213–52.

Arrighi, G., *The Long Twentieth Century,* London: Verso, 1994.

Arrighi, G. and Silver, B. J., *Chaos and Governance in the Modern World System,* London: University of Minnesota Press, 1999.

Ashton, T., 'The Health Reforms: to market and back?', in Boston, J., Dalziel, P. and St John, S. (eds) *Redesigning the Welfare State in New Zealand: Problems, Policies, Prospects,* Auckland: Oxford University Press, 1999, 134–53.

Asian Development Bank, 'Governance: Sound Development Management', *Board Paper*, Manila: Asian Development Bank, 1995.

—— 'Governance: Promoting Sound Development Management', *30th Annual Meeting of Board of Governors*, Fukuoka, 10 May 1997.

—— 'Pakistan Legal and Judicial Reform Project', *Integrated Report*, Manila: Asian Development Bank, 1999.

—— 'Performance-based Allocation of ADF Resources', *Board Working Paper 9–00*, Manila: Asian Development Bank, 2 November 2000.

—— *Supporting Access to Justice Under the Local Government Plan*, SSTA 3640 Pak., Manila: Asian Development Bank, 2001.

—— 'Evaluation of Social Action Program', Operations Evaluation Department, Manila: Asian Development Bank, 2002.

—— *Intergovernmental Fiscal Transfers in Asia: Current Practice and Challenges for the Future*, Manila: Asian Development Bank, 2003.

Asian Development Bank (ADB), World Bank, Department for International Development, *Devolution in Pakistan* (3 vols), Islamabad, 2004. Available online: www.adb.org/Documents/Studies/Devolution-in-Pakistan/default.asp also at www.decentralization.org.pk/pakdevolution.asp (accessed 19 September 2005).

Ayres, R. L., *Banking on the Poor,* London: MIT Press, 1983.

Azfar, O., Kahkonen, S. and Meagher, P., 'Conditions for Effective Decentralized Governance: a synthesis of research findings', College Park MD: Center for Institutional Reform and the Informal Sector, 2002. Available online: www.iris.umd.edu/Reader.aspx?TYPE=FORMAL_PUBLICATION&ID=760f4bad-62f0-4659-b631-b6a849552a6d.

Bahl, R., 'Implementation Rules for Fiscal Decentralization', *Working Paper 99–1,* Atlanta, GA: Georgia State University, 1999. Available online: www.isp-aysps.gsu.edu/papers/ispwp9901.pdf (accessed 19 September 2005).

Bakan, J., *The Corporation: The Pathological Pursuit of Profit and Power,* New York: The Free Press, 2004.

Bates, R., *Markets and States in Tropical Africa: The Political Basis of Agricultural Policy,* Berkeley: University of California Press, 1981.

Bauman, Z., *Society Under Siege,* Cambridge: Polity Press, 2002.

Bayart, J-F., Ellis, S. and Hibou, B., *The Criminalization of the State in Africa,* Bloomington: James Currey, Oxford and Indiana University Press, 1999.

Beck, U., *Democracy Without Enemies,* Oxford: Blackwell, 1998.

Bennet Jones, O., *Pakistan: Eye of the Storm,* New Delhi: Penguin, 2002.

Bennett, J., *Investment in Population Health in Five OECD Countries,* OECD Health Working Papers, No. 2. Paris: OECD, 2003. Available online: www.oecd.org/dataoecd/30/39/2510907.pdf (accessed 19 September 2005).

Berger, M., *The Battle for Asia: From Decolonisation to Globalization,* London: Routledge, 2004.

Berlin, I., *Four Essays on Liberty,* London: Oxford University Press, 1969.

Berman, P., *Terror and Liberalism,* New York: Norton, 2003.

Bevan, D. and Palomba, G., 'Uganda: the budget and medium term expenditure framework set in a wider context', background paper for World Bank's Poverty Reduction Support Credit, mimeo, London: Department for International Development, October 2000.

Bhabha, H., *The Location of Culture,* London: Routledge, 1994.

Bhalla, S., *Imagine There's No Country: Poverty Inequality and Growth in the Era of Globalization,* Washington DC: Institute for International Economics Press, 2002.

Blair, T., *The Third Way: New Politics for the New Century,* London: The Fabian Society, 1998.

—— 'The Blair Doctrine', speech to the Chicago Club, 22 April 1999. Available online: www.pbs.org/newshour/bb/international/jan-june99/blair_doctrine4-23.html

—— 'Foreword by the Prime Minister', in *DFID Eliminating World Poverty: A Challenge for the 21st Century,* Government White Paper on International Development, Cm 3789, London, 2000, 6.

Blanchard, O., Froot, K. and Sachs, J., *The Transition in Eastern Europe Vol. 1,* Chicago: University of Chicago Press, 1994.

Blustein, P., *The Chastening,* New York: Public Affairs, 2001.

Bolt, R. and Fujimura, M., 'Policy Based Lending and Poverty Reduction: an overview of processes, assessment and options', *Economic Research Division (ERD) Working Paper,* No. 2, Manila: Asian Development Bank, January 2002.

Booth, D., 'Introduction and Overview, Special Issue on Poverty Reduction Strategy Papers (PRSPs)', *Development Policy Review* 21(2), 2003, 131–59.

Bouckaert, P., *Hostile to Democracy: The Movement System and Political Repression in Uganda,* New York: Human Rights Watch, 1999.

Branson, W. and Hanna, H., 'Ownership and Conditionality', Office of Equal Opportunity (OEO) *Working Paper,* No. 8, Washington DC: The World Bank, 2000.

Bratton, M., 'Beyond the State: Civil Society and Associational Life in Africa', *World Politics* 41(3), 1989, 407–30.

Brett, E., 'Rebuilding Survival Structures for the Poor: organizational options for the 1990s', in Hansen, H. and Twaddle, M. (eds) *Developing Uganda,* London: James Currey, 1991.

Brock, K., McGee, R. and Ssewakiryanga, R., *Poverty, Knowledge and Policy Processes: A Case Study of Ugandan National Poverty Reduction Policy,* Brighton:

Institute for Development Studies, University of Sussex, 2002. Available online: www.ids.ac.uk/ids/bookshop/rr/rr53.pdf (accessed 19 September 2005).

Brodhead, T., 'NGOs: in one year, not the other?', *World Development* 15, 1987, 1–6.

Brommelhorster, J. and Paes, W-C. (eds) *The Military as an Economic Actor: Soldiers in Business*, Basingstoke: Palgrave Press, 2003.

Brunner, E. and Marmot, M., 'Social Organisation, Stress and Health', in Marmot, M., and Wilkinson, R. (eds) *Social Determinants of Health*, Oxford: Oxford University Press, 1999, 17–43.

Brzezinski, Z., *Out of Control: Global Turmoil, on the Eve of the 21st Century*, New York: Collier Books, 1993.

Burchell, G., Gordon, C. and Miller, P. (eds) *The Foucault Effect: Studies in Governmentality*, Chicago: University of Chicago Press, 1991.

Burki, J. and Perry, G., *Beyond the Washington Consensus: Institutions Matter*, Washington DC: World Bank, 1998.

——, and Dillinger, W., *Beyond the Center: Decentralizing the State*, Washington DC: World Bank, 1999.

Burnside, C. and Dollar, D., 'Aid, Policies, and Growth', *American Economic Review* 90(4), 2002, 847–68.

Bush, G., President's Priorities for Fall: Education, Economy, Opportunity, Security, 2001. Available online: www.secure.pcmac.org/cgi-bin/nph-prcv-employment. cgi/000000A/ www.whitehouse.gov/news/releases/2001/08/20010831-3.html

—— 2002 State of the Union Address. Available online: www.secure.pcmac.org/ cgi-bin/nph-prcv-employment.cgi/000000A/ www.whitehouse.gov/news/releases/ 2002/01/20020129-11.html

Caiden, G., 'Administrative Reform', in Baker, R. (ed.) *Comparative Public Management: Putting US Public Policy and Implementation in Context,* Westport: Praeger, 1994, 107–17.

Cain, P. and Hopkins, A., *British Imperialism 1688–2000*, London: Longman, 2002.

Callinicos, A., *Against the Third Way*, Cambridge: Polity, 2000.

Camdessus, M., *International Monetary Fund Survey*, 14 December 1992.

Camilleri, J. and Falk, J., *The End of Sovereignty? The Politics of a Shrinking and Fragmenting World*, Aldershot: Edward Elgar, 1992.

Cammack, P., 'What the World Bank Means by Poverty Reduction, and Why It Matters', *New Political Economy* 9(2): 23, June 2004, 189–211.

Carino, L., Flaman, R. and Kulessa, M., 'Evaluation of Decentralization and Local Governance', Uganda Report, New York: German Ministry of Economic Cooperation and UNDP, 2000.

Carlsson, I. and Ramphal, S. (eds) *Our Global Neighbourhood. The Report on the Commission on Global Governance*, Oxford: Oxford University Press, 1995.

Cassels, A., *A Guide to Sector-wide Approaches for Health Development: Concepts, Issues and Working Arrangements*, Geneva: World Health Organization, 1997.

Castles, F. G., 'The Wage Earners' Welfare State Revisited: refurbishing the established model of Australian social protection, 1883–1993', *Australian Journal of Social Issues* 29(2), 1994, 120–45.

——, Gerritsen, R. and Vowles, J. (eds) *The Great Experiment: Labour Parties and Public Policy Reform in Australia and New Zealand*, Sydney: Allen & Unwin, 1996.

Chabal, P., *Political Domination in Africa: Reflections on the Limits of Power*, Cambridge: Cambridge University Press, 1986.

Chambers, R., 'The Origins and Practice of Participatory Rural Appraisal', *World Development* 22(7), 1994a, 953–69.

—— 'Participatory Rural Appraisal (PRA): analysis of experience', *World Development* 22(9), 1994b, 1253–68.

—— 'Participatory Rural Appraisal (PRA): challenges, potentials and paradigm', *World Development* 22(10), 1994c, 1437–54.

—— *Whose Reality Counts? Putting the First Last,* London: Intermediate Technology Publications, 1997.

Chandhoke, N., *The Conceits of Civil Society*, Delhi: Oxford University Press, 2002.

Chang, H-J., *Kicking Away the Ladder? Policies and Institutions for Economic Development in Historical Perspective,* London: Anthem Press, 2002.

Chanock, M., 'A Peculiar Sharpness: an essay on property in the history of customary law in colonial Africa', *Journal of African History* 32, 1991, 65–88.

Chien, N., 'Flowers from Hell: forty-four quatrains', *Vietnam Forum* vol. 2, 1983, 141–53.

Christoplos, I., 'Representation, Poverty and PRA in the Mekong Delta', *Research Programme on Environmental Policy and Society*, Linkoeping, Sweden: Linkoeping University, 1995.

Clark, H., Prime Minister's Address to the London School of Economics, 2002. Available online: www.executive.govt.nz/speech.cfm?speechralph=37394& SR=1 (accessed 19 September 2005).

Cline, W., *Trade Policy and Global Poverty,* Washington DC: Institute of International Economics Press, 2004.

Cloughley, B., *A History of the Pakistan Army: Wars and Insurrections,* Karachi: Oxford University Press, 2000.

CNN, New US administrator in Baghdad CNN online. Available online: www.cnn. com/2003/WORLD/meast/05/12/sprj.irq.intl.main/ 12 May 2003, (accessed 19 September 2005).

Cohen, S., *The Pakistan Army,* Karachi: Oxford University Press, 1998.

Colletta, N. and Perkins, G., 'Participation in Education', Environment Department Papers, *Participation Series No. 1*, Washington DC: World Bank, 1995.

Collier, P. and Gunning, J., 'The IMF's Role in Structural Adjustment', *The Economic Journal* 109, November 1999, 634–51.

Collier, P. and Reinikka, R. (eds) *Uganda's Recovery: the role of farms, firms and government,* Washington DC: World Bank.

Commins, S. (ed.) *Africa's Development Challenges in the World Bank,* Boulder: Lynne Reiner, 1988.

Conway, G., 'Agro-ecosystem Analysis', *Agricultural Administration* 20, 1985, 31–55.

Cornia, G., 'Convergence on governance issues, dissent on economic policies', in Evans, A. and Moore, M. (eds) 'The Bank, The State and Development: dissecting the 1997 World Development Report', *IDS Bulletin* 29(2), April 1998, 32–38.

Cornwall, A. and Fleming, S., 'Context and Complexity: anthropological reflections on PRA', *Participatory Learning and Action (PLA) Notes* 24, 1995, 8–12.

Cowen, M. and Shenton, R., *Doctrines of Development,* London: Routledge, 1996.

Craig, D., *Familiar Medicine: Everyday Health Knowledge and Practice in Today's Vietnam*, Honolulu: University of Hawaii Press, 2002.

—— 'Re-territorializing Health: inclusive partnerships, joined up governance, and common accountability platforms in Third way New Zealand', *Policy and Politics* 31(3), 2003a.

—— 'From the Wild West to the Waitakere Way', *Local Partnerships and Governance (LPG) Working Paper*, no. 8, 2003b.

—— 'Building on Partnership: sustained local collaboration and devolved coordination', *Local Partnerships and Governance (LPG) Working Paper*, no. 15, 2004, Available online: www.arts.auckland.ac.nz/lpg/researchpaper15.pdf

Craig, D. and Courtney, M., *The Potential of Partnerships*, Auckland: Local Partnerships and Governance Research Project, 2004. Available online: www.arts. auckland.ac.nz/lpg/plainenglishguide.cfm (accessed 19 September 2005).

Craig, D. and Porter, D., 'Framing Participation: development projects, professionals and organizations', in Tegegn, M. (ed.) *Development and Patronage. A Development in Practice Reader,* Oxford: Oxfam, 1997, 50–57.

—— 'Poverty Reduction Strategy Papers: a new convergence', *World Development* 31(1), 2003, 53–70.

Crampton, P., Salmond, C. and Kirkpatrick, R., *Degrees of Deprivation in New Zealand: An Atlas of Socioeconomic Difference,* Auckland: David Bateman, 2000.

Crook, R. and Manor, J., *Democratization and Decentralization in South Asia and West Africa: Participation, Accountability and Performance*, Cambridge: Cambridge University Press, 1998.

David, S., *The Indian Mutiny*, London: Penguin Books, 2002.

Davidson, B., *The Black Man's Burden: Africa and the Curse of the Nation State,* New York: Times Books, 1992.

Davis, M., *Late Victorian Holocausts: El Nino Famines and the Making of the Third World,* London: Verso, 2001.

—— 'Planet of Slums', *New Left Review* 26, March–April 2004, 5–34.

Ddungu, E., *Popular Forms and the Question of Democracy: The Case of Resistance Councils in Uganda,* Kampala: Center for Business Research Publications, no 4, 1989.

—— 'The Crisis of Democracy in Africa: a case of resistance councils in Uganda', Unpublished M.A. thesis, Department of Sociology, University of Dar-es-Salaam, January 1989.

Decentralisation Secretariat, 'Report on the National Forum on the Implementation of Decentralisation', Kampala: International Conference Centre, 18–19 June 1998.

Deleuze, G. and Guattari, F., *Anti-Oedipus Capitalism and Schizophrenia,* Minneapolis: University of Minnesota Press, 1977.

Denny, S., 'Rich and Poor in Vietnam', mimeo, 1996.

Department for International Development (DFID) 'Examples of Good (and Bad) Practice in Uganda', School of Public Policy, University of Birmingham, Government of Uganda, May 1999.

—— 'Local Government Decision Making: citizen participation and local accountability', mimeo, London: DFID, 2002.

—— *Partnerships for Poverty Reduction: Rethinking Conditionality*, London: DFID, 2004.

—— *Fighting Poverty to Build a Safer World: A Strategy for Security and Development*, London: DFID, March 2005.

Department of the Prime Minister and Cabinet, 'Key Government Goals to Guide Public Sector Policy and Performance', 1999. Originally available: www. dpmc.govt.nz/publications/key_goals.html, now no longer available.

——, 'Sustainable Development for New Zealand: Programme of Action', Wellington, January 2003.

Development Assistance Committee, Development Committee Communique, September, CD/99–29, Paris: Organisation for Economic Cooperation and Development, 1999.

—— 'Harmonising Donor Practices for Effective Aid Delivery: Good Practice Papers', *A DAC Reference Document*, Paris: OECD, 2003.

—— *Harmonising Donor Practices for Effective Aid Delivery*, OECD Development Assistance Committee, Working Party on Aid Effectiveness and Donor Practices, High Level Forum Report, 2 March, 2005, Paris.

Devarajan, S., Miller, M. and Swanson, E., 'Goals for Development: History, Prospects and Costs', *Discussion Paper 2819*, Washington DC: World Bank, April 2002.

Dewey, C. 'The Rural Roots of Pakistani Militarism', in Low D. (ed.) *The Political Inheritance of Pakistan*, London: Macmillan, 1991, 255–83.

—— *The Mind of the Indian Civil Service*, Delhi: Oxford University Press, 1996.

Dezalay, Y. and Garth, B. G., *The Internationalization of the Palace Wars: Lawyers, Economists, and the Contest to Transform Latin American States*, Chicago: University of Chicago Press, 2002.

Dollar, D. and Levin, V., 'The Increasing Selectivity of Foreign Aid, 1984–2002', *World Bank Policy Research Working Paper* 3299, Washington DC: World Bank, May 2004.

Dollar, D. and Svensson, J., 'What Explains the Success or Failure of Structural Adjustment Programs?', *World Bank Working Paper 1938*, Washington DC: World Bank, 1998. Available online: www.worldbank.org/html/dec/Publications/Workpapers/WPS1900series/wps1938/wps1938-abstract.html (accessed 19 September 2005).

Douglas, I., 'Globalization as Governance: towards and archaeology of contemporary political reason', in Prakash, A. and Hart, J. (eds) *Globalization and Governance*, London: Routledge, 2000, 134–60.

Drazen A. and Easterly, W., 'Do Crises Induce Reform? Simple Empirical Tests of Conventional Wisdom', *Economics and Politics* 13(2), 2001, 129–57.

Easterly, W., 'Pakistan's Critical Constraint: not the financing gap but the social gap', mimeo, Development Research Group, Washington DC: World Bank, February 2001.

—— *The Illusive Quest for Growth: Economist's Adventures and Misadventures in the Tropics*, Cambridge: MIT Press, 2001.

—— 'The Cartel of Good Intentions: The Problem of Bureaucracy in Foreign Aid'. *Journal of Policy Reform*, 2003, 1–28.

Easton, B. (ed.) *The Making of Rogernomics*, Auckland: Auckland University Press, 1989.

Edwards, M., and Hulme D. (eds) *NGOs, States and Donors: Too Close for Comfort?*, Basingstoke: Macmillan in association with Save the Children, 1997.

Eichengreen B., 'The Bretton Woods System: Paradise lost?', in Eichengreen, B. and Flandreau, M. (eds) *The Gold Standard in Theory and History* (2nd ed.), London: Routledge, 1997, 313–28.

Eliot, C., *The East Africa Protectorate*, London: Frank Cass, (1905) 1966.

Elwood, B., 'Government by Contract: Its Impact Upon Good Administrative Conduct – An Ombudsman's Perspective', in Finn, C. (ed.) *Sunrise or Sunset?*

Administrative Law in the New Millennium, Papers from the 2000 Administrative Law Forum, 2000, 37. Available online: www.lawlink.nsw.gov.au/lawlink/supreme_court/ll_sc.nsf/pages/SCO_speech_mason_150600,

Escobar, A., *Encountering Development: The making and unmaking of the Third World*, Princeton, New Jersey: Princeton University Press, 1995.

Esping-Andersen, G., *The Three Worlds of Welfare Capitalism*, Cambridge: Polity Press, 1990.

—— *Social Foundations of Post Industrial Economies*, Oxford: Oxford University Press, 1999.

Evans, A. and Moore, M., 'Editorial Introduction in 'The Bank, The State and Development: Dissecting the 1997 World Development Report', *IDS Bulletin* 29(2), April 1998, 4–13.

Falk, R. *Predatory Globalization: a critique,* Cambridge: Polity Press, 1999.

Federal Bureau of Statistics, Yearbook, Government of Pakistan, 2003. Available online: www.statpak.gov.pk/depts./fbs/publications/yearbook2003.html.

Feldman, H., *The Herbert Feldman Omnibus,* Oxford: Oxford University Press, 2001.

Fergusson, J., *The Anti-politics Machine: "Development", Depoliticization, and Bureaucratic Power in Lesotho*, Cambridge: Cambridge University Press, 1990.

Fforde, A., 'The Political Economy of "Reform" in Vietnam: some reflections', in Ljunggren, B. (ed.) *The Challenge of Reform in Indochina*, Cambridge: Harvard Institute for International Development, Harvard University Press, 1993.

Fisher, S., 'Closing Remarks', IMF Conference on Macroeconomic Policies and Economic Growth, Washington DC: International Monetary Fund, 13 April 2001.

Foster, M. and Mackintosh-Walker, S., 'Sector Wide Programmes and Poverty Reduction', *Working Paper 157*, London: Centre for Aid and Public Expenditure, Overseas Development Institute, 2001. Available online: www.odi.org.uk/pppg/publications/working_papers/157.pdf (accessed 19 September 2005).

Foster, M., Brown, A. and Conway, T., *Sector-wide Approaches for Health Development: A Review of Experience*, Geneva: World Health Organization, 2000. Available online: www.eldis.org/static/DOC10735.htm

Fowler, A., 'Distant Obligations: speculations on NGO funding and the global market', *Review of African Political Economy* 55, 1992, 9–29.

Francis, P. and James, R., 'Balancing Rural Poverty Reduction and Citizen Participation: the contradictions of Uganda's decentralization program', *World Development* 31(2), 2003, 325–37.

Frank, T., *One Market Under God: Extreme Capitalism, Market Populism and the End of Economic Democracy*, London: Vintage Books, 2000.

Friedman, T., *The Lexus and the Olive Tree*, London: Harper Collins, 2000.

Fritzen, S., 'Donors, Local Development Groups and Institutional Reform over Vietnam's Development Decade', paper to Vietnam Update Conference, Canberra: Australian National University, November 2001.

—— 'The "foundation of Public Administration"? Decentralization and its Discontents in Transitional Vietnam', paper presented at the International Conference on Governance in Asia: Culture, Ethics, Institutional Reform and Policy Change, City University of Hong Kong, 5–7 December 2002.

Galbraith, J. K., 'Don't Turn the World Over to the Bankers', *Le Monde Diplomatique,* May 2003.

Gardener, J., Guild, R. and Fforde, A., 'Mid-term Evaluation of the Rural Infrastructure Development Fund', mimeo, New York: United Nations Capital Development Fund, 1998.

Gariyo, Z. 'The Poverty Reduction Strategy Paper Process in Uganda', 2000, Available online: www.eurodad.org (accessed 19 September 2005).

Gauld, R., *Revolving Doors: New Zealand's Health Reforms*, Wellington: Institute of Policy Studies and Health Services Research Centre, Victoria University, 2001.

Geddes, M. and Bennington, J., *Local Partnerships and Social Exclusion in the European Union: New Forms of Local Social Governance?*, London: Routledge, 2001.

Ghai, D. and de Alcantara, C., 'The Crisis of the 1980s in Africa, Latin America and the Carribean: an overview', in Ghai, D. (ed.) *The IMF and the South*, Geneva, UN Research for Social Development, 1991, 13–42.

Giddens, A. *The Constitution of Society: Outline of the Theory of Structuration*, Cambridge: Polity Press, 1984.

Gilani, S. Z., 'Personal and Social Power in Pakistan', in Weiss, A. and Gilani, S. (eds) *Power and Civil Society in Pakistan,* London: Oxford University Press, 2001, 49–64.

Gill, S., 'The Emerging World Order and European Change: the political economy of the European Union', in Pantich, L. and Miliband, R. (eds) *The New World Order in International Finance*, London: Merlin Press, 1999, 190.

Girishanker, N., *Evaluating Public Sector Reforms: Guidelines for Assessing Country-Level impact of Structural Reform and Capacity Building in the Public Sector*, Washington DC: World Bank, 2001, quoted in Santiso, C., *Governance Conditionality and The Reform of Multilateral Development Finance: The Role of the Group of Eight*, 2002, Available online: www.g8.utoronto.ca/governance/santis.2002-gov7.pdf

Goldin, C. and Katz, L., 'The Shaping of Higher Education: the formative years in the United States 1890 to 1940', *Journal of Economic Perspectives* 13(1), winter 1999, 37–62.

Goldstein, M., 'International Monetary Fund, Structural conditionality: how much is too much?', *Institute for International Economics Working Paper* 01-4, Washington DC: Institute for International Economics, 2001.

Golooba-Mutebi, F., 'Reassessing Participation in Uganda', draft paper prepared for Development Studies Institute 10th Anniversary conference, 'New Institutional Theory, Institution Reform and Poverty Reduction', 7–8 September 2000.

Gorringe, P., 'Reviewing the Methodology of Treasury's Policy Advice', *Working Paper, The Treasury*, Wellington: New Zealand, 1991. Available online: www.treasury.govt.New Zealand/gorringe/papers/gp-1991.pdf

Gourou, P., *Les Paysans du Delta Tonkinoise*, Paris: Les Editions d'Art et d'Histoire, 1936.

Government of Cambodia, Poverty Reduction Strategy Paper. Available online: www.siteresources.worldbank.org/INTCAMBODIA/Data%20and%20Reference/20182374/Cambodia_PRSP.pdf (accessed 19 September 2005).

Government of New Zealand, 'Sustainable Development for New Zealand: Programme of Action', Department of Prime Minister and Cabinet, New Zealand, January 2003. Available online: www.mfe.govt.New Zealand/publications/susdev/sus-dev-programme-of-action-jan03.html

—— *The Social Report*, 2004. Available online: www.socialreport.msd.govt.New Zealand/

—— *Review of the Centre*, State Services Commission, Available online: www.ssc.govt.NewZealand/display/document.asp?NavID=105

—— Available online: www.ssc.govt.nz/display/document.asp?docid=2429&
pageno=2

Government of Pakistan, *Report of the Law Reform Commission 1958–59*, Karachi:
Government of Pakistan, 1959.

—— *Report of the Inquiry Commission on Women*, Islamabad: Government of
Pakistan, 1997.

—— *Three Year Poverty Reduction Programme, 2001–2004*, Islamabad: Planning
Commission, February 2001a.

—— *Interim Poverty Reduction Strategy*, Islamabad: Finance Division and Planning
Commission, Government of Pakistan, November 2001b.

—— *Accelerating Economic Growth and Reducing Poverty: The Road Ahead,
Poverty Reduction Strategy Paper*, Islamabad: Finance Division and Planning
Commission, Government of Pakistan, 2003a.

—— *Between Hope and Despair: Pakistan Participatory Poverty Assessment*,
Islamabad: Planning Commission, 2003b.

—— *Statistical Information*, Islamabad: Federal Bureau of Statistics, Government
of Pakistan, 2003c.

—— *Poverty Reduction Strategy*, Finance Division and Planning Commission,
Government of Pakistan, Islamabad: December 2004a.

—— Pakistan Economic Survey, 2003–04, Islamabad: Finance Division, 2004b.

Government of Uganda*, The Report of Commission on Local Government*, Kampala:
Government of Uganda, 1997.

—— *Learning from the Poor: Uganda Participatory Poverty Assessment Report*,
Kampala, Ministry of Finance, Planning and Economic Development, 2000a.

—— *Poverty Reduction Strategy Paper: Uganda's Poverty Eradication Action
Plan: Summary and Main Objectives*, Kampala: Ministry of Finance, Planning
and Economic Development, 2000b.

—— *Poverty Reduction Strategy Paper: Progress Report*, Kampala: Ministry of
Finance, Planning and Economic Development, March 2001a.

—— *Fiscal Decentralization in Uganda: the Way Forward*, Kampala: Ministry of
Finance, Planning and Economic Development/Donor Sub-group on Decentralization, January 2001b.

—— and United Nations Capital Development Fund (UNCDF), *District
Development Project: Evaluation Review*, mimeo, 1998.

Graham, H. (ed.) *Understanding Health Inequalities*, Buckingham: Open University,
2000.

Gray, J., *False Dawn: the Delusions of Global Capitalism*, New York: The New
Press, 1998.

Grossman, P., 'A political theory of intergovernmental grants', *Public Choice* 78,
1994, 295–303.

G7 Finance Ministers, 'Strengthening the International Financial Architecture', report
from G7 finance Ministers to the Heads of State and the Government for their
Meeting at Fukuoka, Japan, 8 July 2000.

Guijit, I., *Moving Slowly and Reaching Far: Institutionalizing Participatory Planning
for Child-centred Community Development*, Kampala: International Institute for
Environment and Development (IIED) and Redd Barna, August 1995.

Haggard, S., *Pathways from the Periphery: the Politics of Growth in Newly
Industrialising Countries*, Ithaca, New York: Cornell University Press, 1990.

—— and Kaufman, R. (eds) *The Politics of Economic Adjustment*, Princeton, NJ:
Princeton University Press, 1992.

Haggard, S. and Webb, S., *Political Incentives and Intergovernmental Fiscal Relations: Argentina, Brazil and Mexico Compared,* Washington DC: World Bank, 2001.

Hanmer, L., Pyatt, G. and White, H., 'What Do the World Bank's Poverty Assessments Teach us About Poverty in Sub-Saharan Africa?', *Development and Change* 30, 1999, 795–823.

Hansen, H. and Twaddle, M., *Uganda Now: Between Decay and Development,* London: James Currey, 1988.

Hanson, D., Vardon, P. and Lloyd, J., *Safe Communities: An Ecological Approach to Safety Promotion,* 2003. Available online: www.jcu.edu.au/school/sphtm/documents/rimnq/Paper2.pdf (accessed 19 September 2005).

Hardt, M. and Negri A., *Empire*, Cambridge MA: Harvard University Press, 2000.

Harriss, J., *Depoliticizing Development: the World Bank and Social Capital,* London: Anthem Press, 2001.

Harrold, P., 'The Broad Sector Approach to Investment Lending: Sector Investment Programmes', *World Bank Discussion Paper* 302, Washington DC: World Bank, August 1995.

Havinden, M. and Meredith, D., *Colonialism and Development: Britain and its Tropical Colonies, 1850–1960,* London: Routledge, 1993.

Hayter, T., *Open Borders: the Case Against Immigration Controls,* London: Pluto Press, 2000.

Hearn, J., 'The Uses and Abuses of Civil Society in Africa', *Review of African Political Economy* 87, 2001, 45–53.

Held, D., *Models of Democracy*, Cambridge: Polity Press, 2000.

Helleiner, G., 'The IMF and Africa in the 1980s', *Essays in International Finance* 152, Princeton NJ: Princeton University Press, 1983.

Hellinger, D., Hansen-Kuhn, K. and Fehling, A., 'Stripping Adjustment Policies of their Poverty Reduction Clothing: A New Convergence in the Challenge to Current Global Economic Management', 2001. Available online: www.developmentgap.org/UN%20paper.pdf (accessed 19 September 2005).

Hellinger, S., Hellinger D., and O'Regan, F., *Aid for Just Development*, Boulder CO: Lynne Reinner, 1988.

Herring, R., *Land to the Tiller: The Political Economy of Agrarian Reform in South Asia*, New Haven: Yale University Press, 1983.

Hill, P., 'The Rhetoric of Sector-wide Approaches for Health Development', mimeo, Brisbane: University of Queensland, 2002. (Available from author: peter.hill@mailbox.uq.edu.au)

Hirschman, A., *Development Projects Observed*, Washington DC: The Brookings Institute, 1967.

Hoff, K. and Stiglitz, J., 'Modern Economic Theory and Development', in Meier, G. and Stiglitz, J. (eds) *Frontiers of Development Economics: The Future in Perspective,* Washington DC: World Bank and Oxford University Press, 2001 419–20.

Holmgren, T., Kasekende, L., Atingi-Ego, M. and Ddamulira, D., 'Aid and Reform in Uganda', *Country Case Study, Research Series No. 23*, Kampala: Economic Policy Research Centre, 1999.

Holzmann, R. and Jorgensen, S., *Social Risk Management: A New Conceptual Framework for Social Protection and Beyond*, Washington DC: World Bank, 2002.

Hoogvelt, A., *Globalisation and the Post-colonial World: The New Political Economy of Development*, London: Macmillan, 1997.

Hooper, E. and Pirouet, L., 'Uganda', The Minority Rights Group Report 66, *Africa Confidential,* London, 1999.

Hooper, E. and Hamid, A., 'Scoping Study on Social Exclusion in Pakistan', for Department for International Development, UK Government, mimeo, August 2003. Available online: www.Auckland, New Zealand. www.developmentgap.org/UN%20paper.pdf

Hopkins, T. K. and Wallerstein, I., *The Age of Transition: Trajectory of the World-system 1945–2025,* Atlantic Highlands, NJ: Zed Books, 1996.

Hoselitz, B., 'Non-economic Barriers to Economic Development', *Economic Development and Cultural Change* 1, 1952, 8–21.

Howden-Chapman P. and Tobias, M. (eds) *Social Inequalities in Health: New Zealand,* Wellington: Ministry of Health, 1999.

Human Rights Watch, *Hostile to Democracy: The Movement System and Political Repression in Uganda,* New York: Human Rights Watch, 1999.

Hussain, I., *The Economy of an Elitist State,* Karachi: Oxford University Press, 1999.

—— 'Institutions of Restraint: The Missing Element in Pakistan's Governance', *The Pakistan Development Review* 38(4), 1999, Part I.

Hutchful, E., 'Adjustment in Africa and Fifty Years of the Bretton Woods Institutions: change or consolidation?', *Canadian Journal of Development Studies* 16(3), 1995, 391–417.

Hyden, G., 'Why Africa Finds It So Hard to Develop? Reflections on Political Representation and Accountability'. Available online: www.ethno.unizh.ch/csf conference/files/abstracts/Hyden_Abstract.pdf (accessed 19 September 2005).

Ingham, K., 'Some Aspects of the History of Buganda', *The Uganda Journal* 20(1), 1956, 1–12.

Integrated Regional Information Networks (IRIN), 'Poverty Reduction Strategy Papers', UN Office for the Coordination of Humanitarian Affairs, 2001. Available online: www.irinnews.org/webspecials/drought/pprs.asp (accessed 19 September 2005).

International Monetary Fund, 'IMF builds on Initiatives to Meet Challenges of Globalisation', *IMF Survey*, September 1997.

—— 'IMF Concessional Financing Through the ESAF'. Available online: www.imf.org/external/np/exr/facts/esaf.htm (accessed April 2004).

—— 'The IMF's Approach to Promoting Good Governance and Combating Corruption – A Guide', 2002. Available online: www.imf.org/external/np/gov/guide/eng/index.htm (accessed 19 September 2005).

—— *Aligning the Poverty Reduction Growth Facility (PRGF) and the Poverty Reduction Strategy Paper (PRSP) Approach: Issues and Options,* Washington DC: International Monetary Fund, 25 April 2003.

—— 'Operational Guidance on the New Conditionality Guidelines', 2003. Available online: www.imf.org/External/np/pdr/cond/2003/eng/050803.htm. (accessed 19 September 2005).

—— *Evaluation Report: Fiscal Adjustment in IMF-Supported Programs,* Washington DC: International Monetary Fund, 2003.

—— *Report on the Evaluation of Poverty Reduction Strategy Papers (PRSPs) and the Poverty Reduction Growth Facility (PRGF),* Independent Evaluation Office, Washington DC: IMF, July 2004.

—— International Development Association, 'Heavily Indebted Poor Countries (HIPC) Initiative – strengthening the link between debt relief and poverty reduction', Washington DC: World Bank and IMF, 1999.

—— and World Bank, *Poverty Reduction Strategy Papers – Operational Issues*, Washington DC: IMF 10 December 2001.

—— *Review of the Poverty Reduction Strategy Paper (PRSP) Approach: Early Experience with Interim PRSPs and Full PRSPs*, Washington DC: IMF, 26 March 2002.

—— *Poverty Reduction Strategy Papers – Detailed Analysis of Progress in Implementation*, Washington DC: IMF, September 15 2003.

International Technical and Agricultural Development (ITAD), 'District Development Project', *Evaluation Review Report*, New York: UNCDF Evaluation, November 1999.

Iqbal, M., 'Divine Government', The Sphere of Mercury, *Javid-Nama*, 1932. A. Arberry (trans.), lines 1245–255. Available online: www.allamaiqbal.com/works/poetry/persian/javidnama/translation/ (accessed 19 September 2005).

Isaacson, W. and Thomas E, *The Wise Men: Six friends and the World They Made*, London: Faber & Faber, 1986.

Islam, N., '*Sifarish*, Sycophants, Power and Collectivism: administrative culture in Pakistan', *International Review of Administrative Sciences* 70(2), 2004, 311–30.

Jalal, A., *The Sole Spokesman: Jinnah, the Muslim League and the Demand for Pakistan*, Lahore: Sang-e-Meel Publications, also published by Cambridge University Press, 1999.

James, R., Francis, and Pereza, G., 'The Institutional Context of Rural Poverty Reduction in Uganda: decentralisation's dual nature', mimeo, 2002.

Jamieson, N., *Understanding Vietnam*, Berkeley: University of California Press, 1993.

Jessop, B., 'The Regulation Approach and Governance Theory: alternative perspectives on economic and political change', *Economy and Society* 24(3), 1995, 307–33.

—— 'Capitalism and its Future: remarks on regulation, government and governance', *Review of International Political Economy* 4(3), 1997, 567–91.

—— 'Liberalism, Neo-liberalism and Urban Governance: A State-Theoretical perspective', *Antipode* 34(3), 2002a, 455.

—— *The Future of the Capitalist State*, Cambridge: Polity, 2002b.

Johansson, E., 'Intergovernmental Grants as a Tactical Instrument: empirical evidence from Swedish municipalities', *Journal of Public Economics* 87, 2003, 883–915.

Johnson, J. and Wasty, S., 'Borrower Ownership of Adjustment Programs and the Political Economy of Reform', *Discussion Paper*, No. 199, World Bank, 1993.

Kakwani, N. and Pernia, E., 'What is Pro-poor growth?', *Asian Development Review* 18(1), 2000, 1–16.

Kanbur, R., 'The Strange Case of The Washington Consensus: a brief note on John Williamson's "What Should the Bank Think About the Washington Consensus?"', 1999. Available online: www.people.cornell.edu/pages/sk145/papers/Washington%20Consensus.pdf (accessed 19 September 2005).

Kaplan, R., 'The Coming Anarchy', *The Atlantic Monthly* 273, 2, 1994, 44–76. Available online: www.theatlantic.com/politics/foreign/anarchy.htm

Kapur, D. and Webb, R., 'Governance Related Conditionalities of the International Financial Institutions', *UNCTAD G-24 Discussion Paper 6*, 2000.

Kapur, D., Lewis, J. and Webb, R., *The World Bank: Its First Half Century, Vol. 1: History*, Washington DC: The Brookings Institution, 1997.

Kaufmann, D., Kraay, A. and Zoido-Lobaton, P., 'Governance Matters', *World Bank Policy Research Working Paper*, no. 2196, Washington DC: World Bank, 1999.

Kelsey, J., *Rolling Back the State: Privatisation of Power in Aotearoa/New Zealand*, Wellington: Bridget Williams Books, 1993.

—— *The New Zealand Experiment: A World Model for Structural Adjustment?*, Auckland: Bridget Williams Books and Auckland University Press, 1997.

Kerkvliet, B., 'Rural Society and State Relations', in Kerkvliet, B. and Porter, D., (eds) *Vietnam's Rural Transformation*, Boulder CO and Singapore: Westview Press and Institute of South East Asian Studies, 1995, 65–96.

—— and Porter, D. (eds) *Vietnam's Rural Transformation*, Boulder and Singapore: Westview Press and Institute of South East Asian Studies, 1995.

Khan, H., *Eight Amendment: Constitutional and Political Crises in Pakistan* (2nd edn), Lahore: Maktaba Jadeed Press, 1995.

—— *Constitutional and Political History of Pakistan*, Oxford: Oxford University Press, 2001.

Khan, S., Akhtar, F. and Khan, S., 'Investigating the Importance of Landed Power and Other Determinants of Local Body Election Outcomes', Sustainable Development Policy Institute, Monograph No. 17, 2002.

Khemani, S., 'Partisan Politics and Intergovernmental Transfers in India', *Working Paper*, No. 3016, Development Research Group, Washington DC: World Bank, 2003.

Killick, T., *IMF Programmes in Developing Countries: Design and impact*, London: Overseas Development Institute, 1995.

—— *Aid and the Political Economy of Policy Change*, London: Routledge for Overseas Development Institute, 1998.

Krippner, G., 'The Elusive Market: embeddedness and the paradigm of economic sociology', *Theory and Society* 30(99), 2001, 775–810.

Kuczynski, P. and Williamson, J., *After the Washington Consensus: Restarting Growth and Reform in Latin America*, Washington DC: Institute for International Economics, 2003.

Kullenberg, L. and Porter, D., 'Decentralization and Accountability: recent experience from Uganda', *Agriculture and Rural Development* 3, 1998, 11–15.

Land, T. and Hauck, V., 'Building Coherence Between Sector Reforms and Decentralization: do SWAps provide the missing link?', *Discussion Paper* no. 49, Maastricht: European Centre for Development Policy Management, September 2003.

Lefebvre, H., *The Production of Space,* Oxford: Blackwell, 1991.

Lenz, M., 'Assessing the impact of Uganda's Poverty Action Fund – a participatory rural appraisal of Kamuli District', mimeo, 2002.

Levitas, R., *The Inclusive Society? Social Exclusion and New Labour*, London: MacMillan, 1998.

Lewis, W., *The Theory of Economic Growth*, London: Allen & Unwin, 1955.

Leys, C., *The Rise and Fall of Development Theory*, Bloomington: Indiana University Press, 1996.

Linge, G. and Porter, D. (eds) *No Place for Borders: HIV and AIDS and Development in Asia and the Pacific*, Sydney: Allen & Unwin, 1997.

Lubanga, F., 'On-going Reforms Relevant to Curbing Corruption: the contribution of decentralization', in Ruzinda, A., Langseth, and Gakwandi, A. (eds) *Fighting Corruption in Uganda: the Process of Building a National Integrity System,* Kampala: Fountain Publishers, 1998, 59–68.

Lugard, F., *The Dual Mandate in British Tropical Africa*, Edinburgh: Frank Cass and Co. Ltd, 1922.

Lutz, W., Sanderson, W. and Scherbov, S., 'Doubling of World Population Likely', *Nature* 387, 19 June 1997, pages.

McGrath, A., *The Destruction of Pakistan's Democracy*, Oxford: Oxford University Press, 1996.

McKinley, T., 'MGD-Based PRSPs Need More Ambitious Economic Policies', *Policy Discussion Paper*, New York: UNDP, 2005. Online available: www.undp.org/poverty/docs/MDG-based%20PRSPs%201-05%20Background%20Paper%20(New%20York).doc (accessed 19 September 2005).

McMaster, H., *Dereliction of Duty: Johnson, McNamara, The Joint Chiefs of Staff, and the Lies that Led to Vietnam*, New York: Harper Collins, 1997.

McNamara, R., *The Essence of Security: Reflections in Office*, New York: Harper & Row, 1968.

—— *In Retrospect: the Tragedy and Lessons of Vietnam*, New York: Vintage. 1995.

Malaluan, J. and Guttal, S., 'Structural Adjustment in the Name of the Poor: the PRSP experience in the Lao PDR, Cambodia and Vietnam', Focus on the Global South Working Paper, Bangkok, January 2002. Available online: www.focusweb.org.

Mamdani, M., 'Uganda: contradictions of the IMF Programmes and perspective', *Development and Change* 21(3), 1990.

—— *Critical Reflections on the NRM*, Kampala: Monitor Publications Ltd., 1994.

—— *Citizen and Subject: Contemporary Africa and the Legacy of Late Colonialism*, Princeton: Princeton University Press, 1996.

Manor, J., *The Political Economy of Decentralization*, Washington DC: World Bank, 1999.

Marmot, M. and Wilkinson, R., (eds) *Social Determinants of Health*, Oxford: Oxford University Press, 1999.

Marshall, C., 'Reflections on a Revolution in Pakistan', *Foreign Affairs* 37(2), 1959, 255.

Marshall, G., 'Against Hunger, Poverty, Desperation and Chaos', The Harvard Address, 5 June 1947. Available online: www.willamette.edu/~jlgreen/ids150/george%20c.%20marshall

Marshall, O., *A Many Coated Man*, Dunedin: Longacre Press, 1995.

Martinez-Vazquez, J. and Jameson B., *Russia's Transition to a New Federalism*, Washington DC: World Bank Institute, 2001.

Mason, P., *The Men Who Ruled India*, New Delhi: Rupa & Co. Publishers, 1985.

Mawhood, P. (ed.) *Local Government in the Third World: the Experience of Tropical Africa*, Chichester: Wiley, 1983.

Meyer, K., *The Dust of Empire: The Race for Supremacy in the Asian Heartland*, London: Abacas, 2004.

Michel, A., *The Indus Rivers: A Study of the Effects of Partition*, New Haven CT: Yale University Press, 1967.

Ministry of Finance, Planning and Economic Development, *Uganda Participatory Poverty Assessment Project, 2000; 'Perceptions of Poverty: Key Findings'*, Kampala: Ministry of Finance, Planning and Economic Development, 2000.

Ministry of Finance, Planning and Economic Development/Ministry of Local Government and Donor Sub-group on Decentralization, 'Fiscal Decentralization in Uganda: the Way Forward', Kampala: Government of Uganda, January 2001.

Ministry of Local Government, 'District Development Project: Evaluation Review', 17 November to 20 December 1998, *Findings and Recommendations*, Kampala: Ministry of Local Government, Government of Uganda, 1999a.

—— 'District Development Project: Evaluation Review, 11 November to 11 December', *Findings and Recommendations,* Kampala: Ministry of Local Government, Government of Uganda, 1999b.

Ministry of Social Development, 'The Social Development Approach', Wellington: Ministry of Social Development, 2001a.

—— 'Pathways to Opportunity: from social welfare to social development', Wellington: Ministry of Social Development, 2001b.

—— *Mosaics: Whakaahua Papariki. Key Findings and Good Practice Guide for Regional Coordination and Integrated Service Delivery,* Wellington: Ministry of Social Development, 2003.

—— *The Social Report,* New Zealand, 2005. Available online: socialreport.msd. govt.nz/

Momani, B. 'American Politicization of the International Monetary Fund', *Review of International Political Economy* 11, 2004, 880–904.

Moore, M., 'Towards a Useful Consensus? The 1997 World Development Report and the Role of Government in Development', in Evans, A. and Moore, M. (eds) 'The Bank, The State and Development: Dissecting the 1997 World Development Report', *IDS Bulletin* 29(2), April 1998, 40.

Morawetz, D., 'Twenty-five Years of Economic Development', *Finance and Development* 14(3), 1977, 10–13.

Mosley, P., Harrington J. and Toye, J., *Aid and Power: the World Bank and Policy-Based Lending,* London: Routledge, 1991.

Moss, L., 'Implementing Site and Services: the institutional environment of comprehensive development projects', unpublished PhD dissertation, Berkeley CA: University of California, 1984.

Mosse, D., 'Authority, Gender and Knowledge: theoretical reflections on the practice of Participatory Rural Appraisal', *Development and Change* (25)3, 1994, 497–526.

Mouffe, C., *The Democratic Paradox,* London: Verso Books, 2000.

Mugaju, J., 'An Historical Background to Uganda's No-Party Democracy', in Mugaju, J. and Oloka-Onyango, J. (eds) *No-Party Democracy in Uganda: Myths and Realities,* Kampala: Fountain Publishers, 2000, 40–59.

Mukand, S. and Rodrik, D., 'In Search of the Holy Grail: Policy Convergence, Experimentation and Economic Performance', *John F. Kennedy School of Government Faculty Research Working Papers Series* RWP02–027, Boston, July 2002, published in *American Economic Review*, 95(1) March 2005, 374–83. Available online: www.ideas.repec.org/p/nbr/nberwo/9134.html (accessed 19 September 2005).

Mumtaz, S., Racine, J. and Ali I., *Pakistan: The Contours of State and Society,* Oxford: Oxford University Press, 2000.

Mus, P., 'The Role of the Village in Vietnamese Politics', *Pacific Affairs* 22, 1949, 265–72.

Museveni Y., 'Theoretical Justification of NRM Struggle', in *Mission to Freedom: Uganda Resistance News 1981–85,* Directorate of Information and Mass Mobilization, NRM Secretariat, Kampala, 1990.

—— *Sowing the Mustard Seed: The Struggle for Freedom and Democracy in Uganda,* London: Macmillan, 1997.

Musissi, K., 'Decentralization in Uganda', *Report of the National Forum on the Implementation of Decentralization,* Kampala, September 1998.

Naim, M., 'Latin America: The Second Stage of Reform', *Journal of Democracy* 5(4), October 1994, 32–48.

—— 'Fads and Fashion in Economic Reforms: Washington Consensus or Washington Confusion?' Working Draft of a paper prepared for the IMF Conference on Second Generation Reforms, Washington DC: International Monetary Fund, 26 October 1999. Available online: www.imf.org/external/pubs/ft/seminar/1999/reforms/Naim.HTM (accessed 19 September 2005).

—— 'Washington Consensus or Washington Confusion', *Foreign Policy* 118, Spring 2000, 87–103.

Narayan, D., Chambers, R., Shah, M. and Petesch, P., *Voices of the Poor: Crying Out for Change,* Oxford: Oxford University Press, 2000.

—— and Ebbe, K. 'Design of Social Funds: Participation, Demand Orientation, and Local Organizational Capacity', *World Bank Discussion Paper No. 375,* Washington DC: World Bank, 1997.

National Public Health Partnership, *Integrated Public Health Practice: Supporting and Strengthening Local Action,* Melbourne, Victoria, 2000. Available online: www.dhs.vic.gov.au/nphp/publications/strategies/integph-bground.pdf (accessed 19 September 2005).

National Reconstruction Bureau, 'Local Government Plan', Islamabad: Government of Pakistan, 2000.

—— presentation to *Pakistan Development Forum*, 26 April, Islamabad: Prime Minister's Secretariat, 2005.

Newberg, P., *Judging the State: Courts and Constitutional Politics in Pakistan*, New Delhi: Cambridge University Press, Foundation Books, 1995.

Newman, J., *Modernising Governance: New Labour, Policy and Society*, London: Sage, 2001.

New Zealand Treasury, *Government Management: Brief to the incoming Government*, 1987, 2 vols, Wellington: New Zealand Treasury, 1987.

—— *Towards an Inclusive Economy*, 2001, 15. Available online: www.treasury. govt.nz/gorringe/bibliograhy.asp

—— 'The Gorringe Papers'. Available online: www.treasury.govt.nz/gorringe/bibli-ography.asp

Nguyen Thi Chien, 'Flowers from Hell: forty-four quatrains', *Vietnam Forum Vol. 2*, 1983, 141–53 quoted in Jamieson, N., *Understanding Vietnam*, Berkeley: University of California Press, 1993.

North, D., *Structure and Change in Economic History*, New York: W. W. Norton, 1981.

—— *Institutional Change and Economic Performance,* Cambridge: Cambridge University Press, 1990.

North, D., Calvert, R. and Eggertsson, T., 'Institutions, Institutional Change and Economic Performance', *Political Economy of Institutions and Decisions Series,* Cambridge: Cambridge University Press, 1990.

Norton, A., with others, *A Rough Guide to PPAs: An Introduction to the Theory and Practice*, London: Overseas Development Institute, Centre for Aid and Public Expenditure, 2001.

Nsibambi A., 'Financing Decentralization', in Nsibambi, A. (ed.) *Decentralisation and Civil Society in Uganda: The quest for good governance*, Kampala: Fountain Publishers, 1998, 47–68.

Nyamugasira, M. and Rowden, R. 'New Strategies, Old Loan Conditions: Do the New IMF and World Bank loans support countries' Poverty Reduction Strategies? The case of Uganda'. 2002. Available online: www.brettonwoodsproject.org/topic/adjustment/a28ugandaprsp.pdf (accessed 19 September 2005).

O'Brien, R., Goetz, A., Scholte, J. and Williams M., *Contesting Global Governance: Multilateral Institutions and Global Social Movements*, Cambridge: Cambridge University Press, 2000.

Obwona, M., Steffensen, J., Trollegaard, S. *et al.*, 'Fiscal Decentralization and Local Government Finance in Relation to Infrastructure and Service Provision in Uganda', prepared for World Bank and Danida, Final Report, March 2000.

Obwona, M., *et al.*, *Fiscal Decentralization and Sub-national Government Finance in relation to Service Provision in Uganda*, Washington DC: NALAD and World Bank, 2001.

OECD, Voluntary Aid for Development, *The Role of Non-government Organizations*, Paris: OECD, 1988.

—— *Recommendation of the Council of the OECD on Improving the Quality of Government Regulation*, OCDE/GD (95) Note 1, 1995.

—— 'Harmonising Donor Practices for Effective Aid Delivery: Good Practice Papers', *A DAC Reference Document*, Paris: OECD, 2003. Available online: www.countryanalyticwork.net (accessed 19 September 2005).

Ofcansky, T., *Uganda: Tarnished Pearl of Africa*, Boulder CO: Westview Press, 1996.

Ohno, K., 'Development with Alternative Strategic Options: A Japanese View on the Poverty Reduction Drive and Beyond', mimeo, Tokyo: National Graduate Institute for Policy Studies, 3 May 2004.

Oloka-Onyango, J., *Governance, State Structures and Constitutionalism in Contemporary Uganda,* Kampala: Centre for Basic Research, May 1998.

Olowu, D., 'African Decentralization Policies and Practices from the 1980s and beyond', *Working Paper*, no. 334, The Hague: Institute of Social Studies, March 2001.

Olson, M., *Logic of Collective Action: Public Goods and the Theory of Groups,* Cambridge MA: Harvard University Press, 1965.

Omoding, J., Mpabulungi, A. and Johnson, D., 'Commitment, Participation and Trust: lessons from the District Development Project', *District Development Project Working Briefs Series*, Kampala: Ministry of Local Government, 1998.

Onyach-Olaa, M., 'The Challenges of Implementing Decentralization: recent experiences from Uganda', *Public Administration and Development* 23, 2003, 41–52.

—— and Porter, D., 'Local Government Performance: implications for the centre and donors. Some lessons from the District Development Project', Address to the African Local Government Association Conference, Kampala, 22 July 1999.

Organisation of African Unity, *The Lagos Plan of Action for the Economic Development of Africa: 1980–2000*, Geneva: OAG, 1981.

Owens, J. M. R., 'New Zealand Before Annexation', in Oliver, W.H. (ed.) *The Oxford History of New Zealand*, Wellington: Oxford University Press, 1981.

Oxfam International, 'From "Donorship" to "Ownership"?' *Oxfam Briefing Paper*, Oxford: OXFAM, January 2003.

Oxford Policy Management, *Sub-Saharan Africa's Poverty Reduction Strategy Papers from Social Policy and Sustainable Livelihoods Perspectives*, for Department Fund for International Development, London, March 2001.

Palat, R., 'Eyes Wide Shut: reconceptualizing the Asian crisis', *Review of International Political Economy* 10(2), May 2003, 169–195.

Parker, A. and Serrano, R., *Promoting Good Local Governance Through Social Funds and Decentralization*, Washington and New York: World Bank and United Nations Capital Development Fund, 2000.

Pascal, B., 1670, as quoted by C. Santiso, *Governance Conditionality and The Reform of Multilateral Development Finance: The Role of the Group of Eight*, 2002. Available online: wwwg8.utoronto.ca/governance/santiso2002-gov7.pdf

Pattan Development Organization, *Local Government Elections 2001–Phase III, IV and V*, Islamabad: Pattan Development Organization, 2003.

Peck, J. and Tickell, A., 'Neo-liberalizing Space', *Antipode* 34(3), June 2002, 380–404.

Pender, J., 'From "Structural Adjustment" to "Comprehensive Development Framework": conditionality transformed?', *Third World Quarterly* 22(3), 2001, 397–411.

Perham, R., *Lugard, The Years of Authority, 1898–1945*, London: Harper Collins, 1960.

Petesch, P., 'Managing Aid for National Development – Moving Towards Ownership, Participation and Results', Bonn: Bread for the World and Friedrich Ebert Foundation, 1996.

Pieper, U. and Taylor, L., 'The Revival of the Liberal Creed: The IMF, the World Bank, and inequality in a globalized economy', Center for Economic Policy Analysis, New School for Social Research, *Working Paper Series,* New York, 1 January 1998.

Pieterse, N., 'After Post-development', *Third World Quarterly* 21(2), 2000, 175–91.

Polanyi, K., *The Great Transformation: the political and economic origins of our time*, Boston: Beacon Press, 1944.

Porter, D., 'Cutting Stones for Development', in Zivetz L. *et al., Doing Good: The NGO Community*, Sydney: Allen & Unwin, 1991, 54–75.

—— 'Limits and Beyond: Global Collapse or a Sustainable Future', *Sustainable Development: People, Economy and Environment* 1(2), 1993a, 53–67.

—— 'Technical Mission Report', VIE/90/CO9, New York: United Nations Capital Development Fund, mimeo, June 1993b.

—— 'Monitoring and Review Mission Report', VIE/90/CO9, New York: United Nations Capital Development Fund, mimeo, September 1993c.

—— 'Technical Review Mission Report', VIE/90/CO9, New York: United Nations Capital Development Fund, mimeo, January 1994.

—— 'The Planning, Budgeting and Project Management Process', Discussion Paper 2, Small Scale Rural Infrastructure Project, Quang Nam Da Nang, New York: United Nations Capital Development Fund, 1994.

—— 'Economic Liberalization, Marginality and the Local State', in Porter, D. and Kerkvliet, B. (eds) *Vietnam's Rural Transformation,* Boulder CO and Singapore: Westview Press and Institute of South East Asian Studies, 1995a, 215–46.

—— 'Scenes from Childhood: The Homesickness of Development Discourses', in J. Crush (ed.) *The Power of Development,* London: Routledge, 1995b, 63–86.

—— 'RIDEF: A Case Study', New York: United Nations Capital Development Fund, *mimeo*, 1995c.

—— 'District Development Program – Pilot, Briefing Note', New York: United Nations Capital Development Fund, mimeo, August 1995d.

—— 'Sustainable Financing and Participatory Planning: the small-scale infrastructure project, Quang Nam Da Nang, Vietnam', *Development Bulletin* 29, December 1995e.

—— 'On Summits and Unfortunate Attitudes', in Farrar, A. and Inglis, J. (eds) *Keeping it Together: State and Civil Society in Australia,* Melbourne: Pluto Press, 1996, 18–41.

—— 'District Development in Uganda: The Formulation Process for a Pilot Project', 2001. Available online: www.uncdf.org/english/countries/uganda/local_governance/other_project_related_reports/dporter_pilot.php

—— and Craig, D., 'Devolving the Hydra, Respecting Local Domains: decentralized approaches to Governance in Indigenous Australia', *mimeo*, 2002.

—— 'The Third Way and the Third World: Poverty Reduction and Social Inclusion Strategies in the rise of 'inclusive' liberalism', *Review of International Political Economy* 11(2) 37, May 2004, 387–423.

Porter, D. and Kerkvliet, B., 'Rural Vietnam in Rural Asia', in Kerkvliet, B. and Porter, D. (eds) *Vietnam's Rural Transformation*, Boulder CO and Singapore: Westview Press and Institute of South East Asian Studies, 1995, 1–38.

Porter, D. and Onyach-Olaa, M., 'Inclusive Planning and Allocation for Rural Services', *Development in Practice* February 1999, 56–67; also published in Eade, D. (ed.) *Development and Management*, Oxford: Oxfam and Open University, 2000, 104–20.

Porter, D., Allen, B. and Thompson, G., *Development in Practice: Paved with Good Intentions*, London: Routledge, 1991.

Prakash, A. and Hart, J. (eds) *Globalization and Governance,* London: Routledge, 2000.

Pritchett, L. and Woolcock, M., 'Solutions When the Solution is the Problem: arraying the disarray in Development', *World Development* 32(2), February 2004, 191–212.

Rabbani, M., *I Was the Quaid's ADC,* Karachi: Oxford University Press, 1996.

Radaelli, C. M., *The Open Method of Coordination: A New Governance Architecture for the European Union?,* Stockholm: Swedish Institute for European Policy Studies, 2003.

Ranger, T., 'The Invention of Tradition in Colonial Africa', in Hobsbawn, E. and Ranger, T. (eds) *The Invention of Tradition*, Cambridge: Cambridge University Press, 1983, 211–62.

Regan, A., 'Decentralization Policy: reshaping state and society', in Hansen, H. and Twaddle, M. (eds) *Developing Uganda*, Oxford: James Currey, 1998.

Reich, R. *The Work of Nations,* New York: Vintage, 1992.

Reifer, T. and Sudler, J., 'The Interstate System', in Hopkins, T. J. and Wallerstein, I. et al., *The Age of Transition: Trajectory of the World System 1945–2025*, London: Zed Books, 1996, 13–37.

Reinikka, R., 'Recovery in service delivery: evidence from schools and health centers', in Collier, and Reinikka, R. (eds) *Uganda's Recovery: the role of farms, firms and government,* Washington DC: World Bank, 2001, Chapter 11.

Reno, W., 'Uganda's Politics of War and Debt Relief', *Review of International Political Economy* 9(3), 2002, 415–35.

Republic of Bolivia, *Revised Bolivian Poverty Reduction Strategy*, La Paz: Republic of Bolivia, 2004.

Robb, C. and Scott, A., 'Reviewing Some Early Poverty Reduction Strategy Papers for Africa', *IMF Policy Discussion Paper PDP/01/5*, Washington DC: IMF, 2001. Available online: www.imf.org (accessed 19 September 2005).

Robinson, W., 'Remapping Development in Light of Globalisation: from a Territorial to a Social Cartography', *Third World Quarterly* 23(6), 2002, 1047–071.

Rodrik, D., 'Understanding Economic Policy reform', *Journal of Economic Literature* 34, 1996, 9–41.

—— 'Trading in Illusions', *Foreign Policy* March-April 2001. Available online: www.foreignpolicy.com/issue_marapr_2001/rodrick.html

—— 'How to Make the Trade Regime Work for Development', February 2004. Available online: www.ksghome.harvard.edu/~drodrik/How%20to%20Make%20 Trade%20Work.pdf (accessed 19 September 2005).

Roelvink, G. and Craig, D., 'Partnering the Man in the State: re-gendering the Social through partnership', Local Partnerships and Governance, Research Paper 13, University of Auckland. Available online: www.arts.auckland.ac.nz/lpg/research papers.cfm (accessed 19 September 2005). Forthcoming in *Studies in Political Economy*, 2005.

Rohland, K., World Bank Country Director for Vietnam. Available online: www. www.worldbank.org.vn/news/press27_01.htm 25 June 2003.

Romeo, L., 'Local Development Funds: promoting decentralized, participatory financing and planning', *UN Capital Development Fund Policy Series*, New York: UNCDF, 1996. Online available: www.uncdf.org/english/local_development/ documents_and_reports/thematic_papers/ldf_romeo.php (accessed 19 September 2005).

Rose, N., 'Inventiveness in Politics', *Economy and Society* 28(3), 1999a, 467–93.

—— *Powers of Freedom: Reframing Political Thought,* Cambridge: Cambridge University Press, 1999b.

Rostow, W. W., *The Stages of Economic Growth: a Non-Communist Manifesto*, Cambridge: Cambridge University Press, 1960.

Royal Government of Cambodia, 'Interim Poverty Reduction Strategy Paper', Phnom Penh: Royal Government of Cambodia, 2000.

Rugamayo, C., 'A Local Government Country Profile Study of Uganda', *Report for NORAD*, Pretoria: Norwegian Institute for Urban and Regional Research, 1999.

Rugendyke, B., 'Unity in Diversity: the changing face of the Australian NGO community', in Zivetz, L. *et al.*, *Doing Good: The NGO Community*, Sydney: Allen & Unwin, 1991.

Ruggie, G., 'International Regimes, Transactions and Change: embedded liberalism in the post-war economic order', *International Organization* 36, Spring 1982, 379–415.

Russell, M., *Revolution: New Zealand from Fortress to Free Market,* Auckland: Hodder Moa Beckett, 1996.

Sachs, J., 'The Strategic Significance of Global Inequality', *Washington Quarterly* 24(3), 2001, 187–98.

——, *The End of Poverty*, London: Penguin, 2005.

Sachs, W. (ed.) *The Development Dictionary*, London: Zed Books, 1992.

—— 'Development: the rise and decline of an ideal', article for the Encyclopaedia of Global Environmental Change, Wuppertal Papers, Wuppertal Institute for Climate, Environment and Energy, Wuppertal, Germany, August 2000.

Samad, Y., *A Nation in Turmoil: Nationalism and Ethnicity in Pakistan, 1937–1958*, Delhi: Sage Publications, 1995.

Sands, P., *Lawless World: America and the Making and Breaking of Global Rules*, London: Allen Lane, 2005.

SAPRI, *Structural Adjustment, The SAPRI Report: The Policy Roots of Economic Crisis, Poverty and Inequality*, 2001.

Schacter, M., *Sector Wide Approaches, Accountability and CIDA: Issues and Recommendations*, Ottawa: Institute on Governance, 2001.

Schick, A., 'The Spirit of Reform: managing the New Zealand State Sector in a time of change', report prepared for the SSC and Treasury, August 1996. Available online: www.ssc.govt.nz/display/document.asp?navid=82&docid=2845&pageno=9

—— 'Why Most Developing Countries Should Not Try New Zealand Reforms', *The World Bank Research Observer* 13(1), February 1998, 123–31.

—— *Reflections on the New Zealand Model*, based on a lecture at the New Zealand Treasury, August 2001. Available online: www.treasury.govt.nz/academiclink ages/schick/schick-rnzm01,pdf

Schurmann, F., *The Logic of World Power: An Inquiry into the Origins, Currents and Contradictions of World Politics*, New York: Pantheon, 1974.

Seftel, A. (ed.) *The Rise and Fall of Idi Amin, from the Pages of Drum*, Bailey's African Photo Archives Production, Uganda: Pretoria, 1994.

Sen, A. *Inequality Re-examined*, Oxford: Oxford University Press, 1992.

—— *Development as Freedom*, Oxford: Oxford University Press, 1999.

Shah, A., 'The Reform of Intergovernmental Fiscal Relations in Developing and Emerging Market Economies', Washington DC: World Bank, 1994.

—— 'Balance, Accountability, and Responsiveness: lessons about decentralization', *Working Paper*, no. 2021, Washington DC: World Bank, 1998.

—— 'Fiscal Decentralization in Developing and Transition Economies', in Blindenbacher, R. and Koller, A. (eds) *Federalism in a Changing World – Learning from Each Other*, Montreal and Kingston, Canada: McGill-Queen's University Press, 2000.

Short, C., 'Foreword', in DFID *Eliminating World Poverty: A Challenge for the 21st Century*, Government White Paper on International Development, Cm 3789, London, 1997.

Siddiqa-Agha, A., 'Power, Perks, Prestige and Privileges: the military's economic activities in Pakistan', in Brommelhorster, J. and Paes, W. (eds) *The Military as an Economic Actor*, Basingstoke: Palgrave and Bonn International Center for Conversion, 2003, 124–42.

Singh, A., 'Asian Capitalism and the Financial Crisis', in Michie, J. and Grieve Smith, J., *Global instability: The Political Economy of World Economic Governance*, London: Routledge, 1999, 9–36.

Singh, K., *Not a Nice Man to Know: The Best of Khushwant Singh*, New Delhi: Penguin, 1993.

Sinha, S. and Lipton, M., 'Damaging Fluctuations, Risk and Poverty: a review', *Background Paper for the World Development Report 2000–01*, Brighton: University of Sussex, 2000.

Skidelsky, R., *John Maynard Keynes: Fighting for Britain 1937–1946*, London: Papermac, 2000.

Skilling, D. and Waldegrave, A. M., *The Wealth of a Nation*, Auckland: The New Zealand Institute Ownership Society Papers, 2004. Available online: www.

nzinstitute.org/index.php/ownershipsociety/paper/the_wealth_of_a_nation (accessed 19 September 2005).

Slater, D. and Bell, M., 'Aid and the Geopolitics of the Post-colonial: critical reflections on the New Labour's overseas development strategy', *Development and Change* 33(2), 2002, 3353–360.

Smith, A., *The Theory of Moral Sentiments,* Oxford: Oxford University Press, 1976 (1759).

——— *The Wealth of Nations,* Oxford: Oxford University Press, 1976 (1759).

Smith, C., The 'Programme of Change', Conference, Department of Child, Youth and Family Services (CYFS), 2002. Available online: www.conferenz.co.nz/2004/library/s/smith_craig.htm

Smoke, P., *Fiscal Decentralization in Developing Countries: A Review of Current Concepts and Practice*, New York: United National Research Institute for Social Development, 2001.

——— and Schroeder, L. 'Intergovernmental Fiscal Transfers: concepts, international practice and policy issues', in Kim, Y-H. and Smoke, P. (eds) *Intergovernmental Fiscal Transfers in Asia: Current Practice and Challenges for the Future,* Manila: Asian Development Bank, 2003.

Smuts, J. C., *Africa and Some World Problems*, including the Rhodes Memorial Lectures delivered in Michaelmus Term, 1929, Oxford: Oxford Univesity Press, 1930.

Social Policy and Development Centre (SPDC), *Social Development in Pakistan: Annual Review 2002–03*, Karachi: Social Policy and Development Centre, 2003.

——— *State of the Economy: A Shift Towards Growth, Annual Review 2004*, Karachi: Social Policy and Development Centre, 2004.

Society for the Protection of the Child (SPARC), *The State of Pakistan's Children*, Islamabad: SPARC, 1998.

——— *The State of Pakistan's Children*, Islamabad: SPARC, 1999.

Soederberg, S., 'Grafting Stability onto Globalisation? Deconstructing the IMF's Recent Bid for Transparency', *Third World Quarterly* 22(5), 2001, 849–64.

Soja, E., *Postmodern geographies: the reassertion of space in critical social theory,* London: Verso, 1989.

de Soto, H., The *Mystery of Capital*, London: Bantam Press, 2000.

Spiro, 'New Global Potentates: nongovernmental organizations and the "unregulated" marketplace', *Cardozo Law Review* 18, 1996, 957–69.

Standing, G., 'Brave New Words? A Critique of Stiglitz's World Bank Rethink', *Development and Change* 31, 2000, 737–63.

Stanley, J. and Fritzen, S., 'Vietnam Technical Review Mission Report: Rural Infrastructure Development Fund', United Nations Capital Development Fund, mimeo, 2001.

State Services Commission, 'Report of the Advisory Group on the Review of the Centre', Wellington: State Services Commission, November 2001.

Steffensen, J., Ssewankambo, E. and Tidemand, P., 'A Comparative Analysis of Decentralization in Kenya, Tanzania and Uganda', draft report prepared for the World Bank, Danida and the Government of Uganda, May 2004.

Steffensen, J. and Trollegaard, S., 'Fiscal Decentralization and Sub-national Government Finance in Relation to Infrastructure and Services Provision: synthesis report of six sub-Saharan African country studies', prepared for World Bank, Danida and USAID, The National Association of Local Authorities, Denmark, 2000.

Sterling, R. G., *Dollar Diplomacy in Current Perspective: The Origins and Prospects of our International Economic Order,* New York: Columbia University Press, 1980.

Stewart, F. and Wang, M., 'Do PRSPs Empower Poor Countries and Disempower The World Bank, Or Is It The Other Way Round?', *QEH Working Paper*, No. 108, Oxford: Oxford University Press, 2003.

Stiglitz, J., 'More Instruments and Broader Goals: moving toward the post-Washington Consensus', The 1998 WIDER lecture, Helsinki, Finland, 7 January 1998. Available online: www.global policy.org/socecon/bwirlwto/stig.htm

—— 'Whither Reform? Ten Years of the Transition': Washington DC: Paper prepared for the Annual Bank Conference on Development Economics, Washington DC, April 28–30 1999.

—— 'Unraveling the Washington Consensus: an interview with Joseph Stiglitz', *Multinational Monitor* 21(4), April 2000. Available online: www.essential.org/monitor/mm2000/00april/interview.html (accessed 19 September 2005).

—— Foreword in *The Great Transformation: The Political and Economic Origins of Our Time*, Polanyi, K. (ed.) Boston: Beacon Press, 2001, vii–xvii.

—— *Globalization and its Discontents,* New York: Norton, 2002.

—— 'Karl Polanyi and the writing of The Great Transformation', *Theory and Society* 32(4), June 2003, 275–306.

—— 'Democratising the International Monetary Fund and the World Bank: Governance and Accountability', *Governance: An International Journal of Policy, Administration and Institutions* 16(1), 2003, 111–39.

Taggart, M., 'Corporatisation, Privatisation and Public Law', *Public Law Review* 2, 1991, 77–108.

Talbot, I., *Pakistan: a Modern History,* Lahore: Vanguard Books, 1999.

Taylor, K., *The Birth of Vietnam*, Berkeley CA: University of California Press, 1983.

Tendler, J. *Turning Private Voluntary Organizations into Development Agencies: Questions for Evaluation,* Washington DC: US Agency for International Cooperation and Development, April 1982.

—— and Serrano, R., 'The Rise of Social Funds: what are they a model of?' prepared for United Nations Development Program, 1999. Available online: www.un.org/esa/socdev/poverty/tendler2.pdf (accessed 19 September 2005).

Thacker, S., 'The High Politics of IMF Lending', *World Politics*, 52(1), 1999, 38–37.

The Commission for Africa, *Our Common Interest: Report of the Commission for Africa*, London: Commission for Africa, 2005.

The Republic of Uganda, 'Decentralisation in Uganda – The Policy and its Implications', Decentralisation Secretariat, Kampala, 1994.

—— 'The Constitution of the Republic of Uganda', Entebbe: Government Printer, 1995.

—— *Fiscal Decentralisation: The Way Forward*, Kampala: Ministry of Local Government, 2002.

Therien, J.P. and Lloyd, C., 'Development Assistance on the Brink', *Third World Quarterly* 21(1), 2000, 2–38.

Therkildsen, O., 'Uganda's Radical Decentralization: an example for others?', unpublished working paper, Copenhagen: Centre for Development Research, 2002.

Thompson, E., *Whigs and Hunters,* Harmondsworth: Penguin, 1975.

Thurley, D., 'Tax Credits and Income Changes', *Welfare Rights Bulletin*, no. 169, Child Poverty Action Group, August 2002.

Tidemand, P., 'The Resistance Councils in Uganda – A Study of Rural Politics and Popular Democracy in Africa', unpublished PhD dissertation, Copenhagen: Roskilde University, 1994.

Tinkler, H. R., *The Foundations of Local Self-Government in India, Pakistan and Burma*, London: The Athlone Press, University of London, 1954.

Toye, J., *Dilemmas of Development: Reflections on the Counter-Revolution in Development Theory and Policy* (1st edn), Oxford: Blackwell, 1987.

Transparency International, *Global Corruption Report*, London: Transparency International, 2003.

—— *Global Corruption Report*, London: Transparency International, 2004.

Truman H., Inaugural Speech, 1949. Available online: www.juntosociety.com/inaugural/truman.html (accessed 19 September 2005).

Tukahebwa, G., 'The Role of District Councils in Decentralisation', in Nsibambi, A., (ed.) *Uganda Government The Report of Commission on Local Government*, Kampala: Government Printer, 1997, pp. 12–30.

Uganda Constitution Commission, 'The Report of the Uganda Constitution Commission: Analysis and Recommendations', Kampala: United Republic of Uganda, 1993.

United Kingdom Department of Health, *Tackling Health Inequalities, 'A Programme for Action'*, 2003. Available online: www.dh.gov.uk/PolicyAndGuidance/Health AndSocialCareTopics/HealthInequalities/ProgramForAction/ProgramForAction GeneralArticle/fs/en?CONTENT_ID=4072948&chk=%2B0wc2o (accessed 19 September 2005).

United Kingdom HM Treasury, *The Child and Working Tax Credits, The Modernisation of Britain's Tax and Benefit System*, no. 10, London: H.M. Treasury, 2002. Available online: www.hm-treasury.gov.uk/mediastore/otherfiles/new_tax_credits.pdf (accessed 19 September 2005).

United Nations Capital Development Fund, 'PPC of QNDN, Project Agreement: CDF Programme': Implementation Unit, VIE/90/C09, mimeo, February 1991.

—— 'Province People's Committee of Quang Nam Da Nang, "Project Agreement"', Implementation Unit VIE/90/C09, New York: UNCDF, mimeo, February 1991.

—— *Taking Risks*, Background Papers, New York: UNCDF, 1999.

United Nations Conference on Trade and Development, *The Least Developed Countries Report*, Geneva: United Nations Conference on Trade and Development, 2002.

United Nations Development Program (UNDP), *Looking Ahead: A United Nations Common Country Assessment of Viet Nam*, Hanoi: United Nations Development Program, December 1999.

—— *Human Development Report*, New York: The United Nations, 1999.

United Nations Office for Project Services, Contract for Professional Services: United Nations Office for Project Services and CERPAD, Terms of Reference, mimeo, 1991.

UN Millennium Project, *Investing in Development: A Practical Plan to Achieve the Millennium Development Goals*, New York: United Nations, 2005.

Valdelin, J. *et al.*, *Doi Moi and Health: Evaluation of Health Sector Cooperation Between Viet Nam and Sweden*, Stockholm: Swedish International Development Cooperation Agency (SIDA), 1992.

Van der Heijden, H., *Efforts and Programmes for Non-Governmental Organizations (NGOs) in the Least Developed Countries*, Report prepared for UNCTAD,

UNCLDC II/NGO 2, Second United Nations Conference on the Least Developed Countries, 1989.

Van Nieuwenhuijze, C., *Development: A Challenge to Whom?*, The Hague: Mouton, 1969.

Vietnam Communist Party, 'Bao Cao Tinh Tranh Chap Ruong Dat o Nong Thon Hien Nay', Report about the Current Agricultural Land Conflict Situation in the Countryside, Agricultural Committee of the Central Committee, 27 September 1990.

—— 'Tiep Tuc Doi Moi va Phat Trien Kinh Te Xa Hoi Nong Thon', Continued Socio-economic Renovation and Development in the Countryside, Central Committee, Fifth Plenum Resolution, 1 July 1993. (BBC Service, 14 July 1993).

Vietnam Government, 'Comprehensive Poverty Reduction and Growth Strategy, (CPRGS)', Fourth Draft, mimeo, 2002.

Villadsen, S. and Lubanga, F. (eds) *Democratic Decentralization in Uganda: A New Approach to Local Governance*, Kampala: Fountain Publishers, 1996.

Wade, R., *Governing the Market: Economic Theory and the Role of Government in East Asian Industrialization*, Princeton NJ: Princeton University Press, 1990.

—— 'From "Miracle" to "Cronyism": explaining the great Asian slump', *Cambridge Journal of Economics* 22(6), November 1998, 693–706.

—— 'The US Role in the Long Asian Crisis of 1990–2000', in Batista-Riveria, F. and Lukauskis, A. (eds) *The East Asian Crisis and its Aftermath*, Cheltenham: Edward Elgar, 2001, 195–226.

—— 'Economic Growth and the Role of Government: or how to stop New Zealand from falling out of the OECD', Paper presented to the Catching the Knowledge Wave Conference, Auckland, New Zealand, 1–3 August 2001a.

—— 'Winners and Losers', *Economist* (Print edition), 26 April 2001b. Available online: www.economist.com/opinion/displayStory.cfm?Story_ID=587251 (accessed 19 September 2005).

—— 'The Rising Inequality of World Income Distribution', *Finance and Development* 38(4), December 2001c.

—— 'Showdown at the World Bank', *New Left Review* 7, January-February 2001d, 124–37.

—— 'Reply', in 'Symposium on Infant Industries', *Oxford Development Studies* 31(1), 2003a, 8–14.

—— 'What Strategies are Viable for Developing Countries Today? The World Trade Organization and the Shrinking of "Development Space", Crisis States Programme, *Working Papers Series* 1(4), 3 June 2003b.

—— 'Is Globalization Reducing Poverty and Inequality?', *World Development* 32(4), 2004, 567–89.

Waitakere City Council, *Towards Wellbeing in Waitakere,* Waitakere: Waitakere City Council, 1996.

Walker, S. (ed.) *Rogernomics: Reshaping New Zealand's Economy*, Wellington: Government Printing Books, 1989.

Walton, J., 'Urban Poverty in Latin America', *Latin America Program Working Papers*, No. 202, Washington DC: Woodrow Wilson International Center for Scholars, 1993.

Wapenhans, W., 'Efficiency and Effectiveness: is the World Bank Group prepared for the Task Ahead?', Internal Working Party, Washington DC: World Bank, 1993, Chapter 8.

Waslander, S. and Thrupp, M., 'Choice, Competition and Segregation: an empirical analysis of a New Zealand secondary school market, 1990–93', *Journal of Education Policy* 10, 1995, 1–26.

Watt, D., Flanary, R. and Theobald, R., 'Democracy or the Democratization of Corruption? The Case of Uganda', *Commonwealth and Comparative Politics* 37(3), 1999, 37–64.

Weber, H., 'The "new economy" and social risk: banking on the poor?', *Review of International Political Economy* 12(2), 2004, 356–86.

Weiss, H., 'War and Peace in the Democratic Republic of Congo', *Current African Issues* 22, Uppsala: Nordiska Afrikainstitutet, 2000.

Wellbeing Collaboration Project, New Zealand. Available online: www.waitakere.govt.nz/OurPar/collabproj.asp (accessed 19 September 2005).

Wellington Health Action Group, *The Health Reforms: a Second Opinion*, Wellington: WHAG, 1993.

Went, R. 'Economic Globalization Plus Cosmopolitanism?', *Review of International Political Economy* 11(2), 2004, 337–55.

Whitehead, A. and Lockwood, M., 'Gender in the World Bank's Poverty Assessments: six cases from Sub-Saharan Africa', *Development and Change* 30(3), 1999, 409–33.

Wilder, A., *The Pakistani Voter: Electoral Politics and Voting Behaviour in the Punjab*, Karachi: Oxford University Press, 1999.

Wilkinson, R., *Unhealthy Societies: The Afflictions of Inequality,* London: Routledge, 1996.

Wilks, A. and Lefrancois, F., *Blinding with Science or Encouraging Debate? How World Bank Analysis Determines PRSP Policies,* London: Bretton Woods Project, and World Vision, 2002.

Williamson, J., 'What Washington Means by Policy Reform', in Williamson, J. (ed.) *Latin American Adjustment: How Much Has Happened?* Washington DC: Institute for International Economics, 1990, 18–23. Available online: www.worldbank.org/research/journals/wbro/obsang00/pdf/(6)Williamson.pdf

—— 'In Search of a Manual for Technopols', in Williamson, J. (ed.) *The Political Economy of Policy Reform,* Washington, Institute for International Economics, 1994, 11–28.

—— 'What should the World Bank think about the Washington Consensus?', paper prepared as a background to the World Bank's World Development Report 2000, *Institute for International Economics Papers*, 1999, 2. Available online: www.207.238.152.36/papers/williamson0799.htm

—— 'Did the Washington Consensus Fail?', outline of remarks at the Center for Strategic and International Studies, 6 November 2002. Available online: www.iie.com/publications/papers/williamson1102.htm (accessed 19 September 2005).

Williamson, O. E., 'The New Institutional Economics: taking stock, looking ahead', *Journals of Economic Literature*, 38, 595–613, 2000.

Williamson, T., 'Targets and Results in Public Sector Management: Uganda Case Study', *Working Paper*, no. 205, London: Overseas Development Institute, 2003.

Wohl, R., 'Editorial', *Economic Development and Cultural Change* 1, 1952, 3–7.

Wolfensohn, J., 'The Other Crisis', Address to the Board of Governors, Washington DC: World Bank, October 6 1998. Available online: www.worldbank.org/html/extdr/am98/jdw-sp/am98-en.htm (accessed 19 September 2005).

324   Bibliography

—— Speech to 9th International Anti-Corruption Conference, Durban, October 1999.

—— 'Empowerment, Security and Opportunity Through Law and Justice', St. Petersburg, 9 July 2001.

—— and Fischer, S., *The Comprehensive Development Framework and Poverty Reduction Strategy Papers*, Washington DC: World Bank, 5 April 2000.

Wolpert, S., *Jinnah of Pakistan,* Karachi: Oxford University Press, 1998.

—— *Gandhi's Passion: The Life and Legacy of Mahatma Gandhi*, Oxford: Oxford University Press, 2001.

World Bank, *Sub-Saharan Africa: From Crisis to Sustainable Growth*, Washington DC: World Bank, 1981.

—— *Accelerated Development in Sub-Saharan Africa: An Agenda for Action*, Washington DC: World Bank, 1981.

—— 'Toward Sustained Development in Sub-Saharan Africa: a joint programme of action', Washington DC: World Bank, 1984.

—— *Financing Adjustment with Growth in Sub-Saharan Africa: 1986–1990*, Washington DC: World Bank, 1986.

—— 'Analysis of New and Completed Cases of NGO Collaboration in Bank Projects – FY 1989', *Office Memorandum*, Washington DC: World Bank, 7 September 1989.

—— *Long Term Perspective Study for Sub-Saharan Africa*, Washington DC: World Bank, 1990.

—— *The Challenge of Development: World Development Report 1991*, Washington DC: World Bank and Oxford University Press, 1991.

—— 'Managing Development: The Governance Dimension', *Discussion Paper*, Washington DC: World Bank, 25 June 1991.

—— *Governance and Development*, Washington DC: World Bank, 1992.

—— 'Operational Directive 4.15', Washington DC: World Bank, 1992.

—— *States and Markets*. 1992. Available online: www.worldbank.org/data/wdi2002/statesmkts.htm (accessed 19 September 2005).

—— *Uganda: Growing Out of Poverty*, Washington DC: World Bank, 1993.

—— *Governance: the World Bank's experience*, Washington DC: World Bank, 1994.

—— *Uganda: The Challenge of Growth and Poverty Reduction*, Washington DC: World Bank, 1996a.

—— 'Vietnam: Fiscal Decentralization and the Delivery of Rural Services', *Country Economic Memorandum*, Washington DC: World Bank, 1996b.

—— *The State in a Changing World, World Bank Development Report 1997*, Washington DC and New York: Oxford University Press for the World Bank, 1997a.

—— *Helping Countries Combat Corruption – Progress at the World Bank since 1991*, Washington DC: World Bank, September 1997b. Available online: www.worldbank.org/publicsector/anticorrupt/corruptn/coridx.htm (accessed 19 September 2005).

—— *Assessing Aid, What Works, What Doesn't, and Why?*, New York: Oxford University Press, 1998.

—— *Attacking Poverty, World Development Report 2000*, Washington DC: Oxford University Press, 2000a.

—— *Can Africa Claim the 21st Century*, Washington DC: World Bank, 2000b.

—— 'Participation in Poverty Reduction Strategy Formulation', Uganda Participation Processes, Full Poverty Reduction Strategy Paper, April 2000c. Available online: www.worldbank.org/participation/prspcase2.htm (accessed 19 September 2005).

—— *Country Assessments and IDA Allocations,* Washington DC: World Bank, November 2000d.

—— *Reforming Punjab's Public Finances and Institutions*, Washington DC: World Bank, 2001.

—— *Community Driven Development in Africa: a Vision of Poverty Reduction Through Empowerment*, Washington DC: World Bank Group, Africa Region, 2001.

—— *Reviewing Poverty Reduction Strategy Program,* Washington DC: World Bank, 2001.

—— 'Adjustment Lending Retrospective: Final Report', Operations Policy and Country Services, Washington DC: World Bank, 15 June 2001.

—— 'Report and Recommendation of the President of the International Development Association to the Executive Directors on a Proposed Poverty Reduction Strategy Credit of SDR 197.2 million ($US 250 million equivalent) to the Socialist Republic of Vietnam', Report No: P-7446-VN., Washington DC: World Bank, 2001.

—— 'Reviewing Poverty Reduction Strategy Program', 2001. www.worldbank.org/developmentnews/stories/html/080601a.htm (accessed 19 September 2005).

—— *The Drive to Partnership: Aid Coordination and the World Bank*, Washington DC: Operations Evaluation Department, World Bank, 2001.

—— *Building Institutions for Markets, World Development Report 2002,* Washington DC: World Bank, 2002a.

—— 'Brazil: Issues in Fiscal Federalism', *Report No. 22525-Br.*, Washington DC, World Bank, 2002b.

—— *Poverty in Pakistan: Vulnerabilities, Social Gaps, and Rural Dynamics*, Washington DC: World Bank, 2002c.

—— *Poverty Reduction and the World Bank: Progress in Operationalizing the World Development Report 2000/2001*, Washington DC: The World Bank, 2002d.

—— 'Poverty Assessments', World Bank Poverty net website, 2002e. Available online: www.worldbank.org/poverty/wbactivities/pa/index.htm (accessed 19 September 2005).

—— *Sustainable Development in a Dynamic World*, World Development Report 2003, Washington DC: World Bank, 2002f.

—— *Globalization, Growth and Poverty: Building an Inclusive World Economy,* Washington DC: World Bank and Oxford University Press, 2002g.

—— States and Markets, 2002h. Available online: www.worldbank.org/data/wdi2002/statesmkts.htm (accessed 19 September 2005).

—— Note from the President of the World Bank, Prague: Development Committee of the World Bank, 18 September 2002i.

—— 'Pakistan Development Policy Review: A New Dawn?', *Report No. 23916-Pak*, Washington DC: World Bank, 2002j.

—— *Pakistan Public Expenditure Management: Strategic Issues and Reform Agenda*, Washington DC: World Bank, 2003a.

—— *Toward Country-led Development: A Multi-Partner Evaluation of the Comprehensive Development Framework*, World Bank, 4 June 2003b.

—— 'The World Bank Regional Public Expenditure Review for Indonesia: Decentralizing Indonesia', Jakarta: World Bank, 2003c. Available online: www.lnweb18.worldbank.org/eap/eap.nsf/0/7a59aff0cdda926047256dc800138fa7?OpenDocument (accessed 19 September 2005).

—— *Getting Serious About Meeting the Millennium Development Goals: A Comprehensive Development Framework Progress Report*, Washington DC World Bank, 2004a.

—— *Making Services Work for Poor People*, World Development Report 2004, Washington DC: World Bank, 2004b.

—— 'The Poverty Reduction Strategy Initiative: an independent evaluation of the World Bank's support through 2003', *Operations Evaluation Department VII*, Washington DC: World Bank, 2004c.

—— *PRSPs – Operational Issues*, Washington DC: World Bank, 2004d.

—— *Vietnam Social Indicators*. Available online: www.worldbank.org.vn/data/s_indicator.htm (accessed 19 September 2005).

—— and International Monetary Fund, *Strengthening the Link between Debt Relief and Poverty Reduction*, Washington DC: World Bank and IMF, August 1999.

—— *Building Poverty Reduction Strategies in Developing Countries*, Washington DC: World Bank and IMF, September 1999.

World Bank and others, 'Vietnam: attacking poverty', *Joint Report of the Government of Vietnam – Donor – NGO Poverty Working Group*, 14–15 December 1999.

World Economic Forum, 'Poverty Reduction: what works, what doesn't, and why', 27 January 2003. Available online: www.weforum.org/site/knowledgenavigator.nsf/Content/_57646?open

Wunch, J., 'Refounding the African State and Local Self-governance: the neglected foundation', *Journal of Modern African Studies* 38 (3), 2000, 487–509.

—— 'Decentralization, Local Governance and "Decentralization" in Africa', *Public Administration and Development* 21, 2001, 277–88.

—— and Olowu, D. (eds) *The Failure of the Centralized State in Africa: Institutions and Self-governance in Africa,* Boulder CO: Westview Press, 1990.

Zaman, A., 'Economic Strategies and Policies in Pakistan, 1947–1997' in Mumtaz, S., Racine, J-L. and Ali, I. (eds) *Pakistan: The Contours of State and Society*, Oxford: Oxford University Press, 2002, 155–86.

Ziring, L., *Pakistan: The Enigma of Political Development,* Boulder CO: Westview Press, 1980.

—— *Pakistan in the Twentieth Century*, Karachi: Oxford University Press, 1998.

Zizek, S., *The Sublime Object of Ideology,* London: Verso, 1989.

Index

Please note that page references to non-textual material such as tables are in *italic* print. References to notes have the letter 'n' following the page number.

Printed and bound by CPI Group (UK) Ltd, Croydon, CR0 4YY

01/11/2024

01782635-0012